**Protection of Global Biodiversity**

# PROTECTION OF GLOBAL BIODIVERSITY

## Converging Strategies

Lakshman D. Guruswamy and Jeffrey A. McNeely, editors

Duke University Press    Durham and London 1998

© 1998 Duke University Press

All rights reserved

Printed in the United States of America on acid-free paper ∞

Typeset in Minion by Running Feet Books, Inc.

Library of Congress Cataloging-in-Publication Data

appear on the last printed page of this book.

*This book is dedicated to Michael Cusanovich,*
*who sponsored an international conference*
*on this subject at the University of Arizona*
*and inspired us with his vision.*

# Contents

## V   Moral Responses

## VI   Legal Implementation

# Acknowledgments

We are most grateful to our colleagues at the University of Arizona who were members of the organizing committee of the conference, Biological Diversity: Exploring the Complexities, the chrysalis from which this book emerged. Hans Bohnert, Codirector, displayed cheerful and perceptive leadership despite his frenetic schedule. The others, including Kathleen Maloney-Dunn (Coordinator), Elizabeth Baker, David Galbraith, Marl Jensen, Rita Manak, Juanita Simpson, Robert Robichaux, Edella Schlager, and Barbara Timmermann, made significant contributions in their different ways. A special debt of gratitude is owed to Mike Cusanovich, Vice President for Research at the University of Arizona, to whom this volume is dedicated. The support of the Government of Switzerland (Swiss Development Corporation) for Jeffrey A. McNeely during the time he was working on this book is gratefully acknowledged.

Brent Hendricks and Jason Aamodt, Research Fellows at the National Energy-Environment Law and Policy Institute (NELPI) served as creative, skillful, and dedicated in-house editors. This book would not have been possible without their expert help, or the assistance of Sue Lorenz, NELPI Administrative Assistant, and Terri Lynn Carr and Catina Drywater, NELPI Research Associates. We owe a special debt to Carol Arnold, Research Librarian, College of Law Library, University of Tulsa, for her willing and creative research help.

Cindy Milstein did an excellent copyediting job, and we are very grateful to Reynolds Smith and Paula Dragosh at Duke University Press for giving this endeavor such invaluable support.

**Protection of Global Biodiversity**

# Introduction

by Lakshman D. Guruswamy and Jeffrey A. McNeely

## Organizational Framework

E.O. Wilson (1993, 243) vividly describes a small ridge called Centinela hidden among the western Andean foothills of Ecuador: "When the forest on the ridge was cut a decade ago, a large number of rare species were extinguished. They went just like that, from full healthy populations to nothing, in a few moments." For them there were no epitaphs. *Biodiversity* in our title stands for a portfolio of diverse life-forms, including all species whose survival is currently threatened.

Thus a familiar, nonetheless frightening, scenario depicts the present as an age of mass extinction rivaling the Cretaceous period 66 million years ago, when 12 percent of the species resident on earth disappeared in spasmodic catastrophe (Wilson 1993, 31). This apocalyptic vision emerges within the context of exponential population growth, in which a global population of 2.515 billion in 1950 exploded to 5.291 billion in 1990 and is estimated to rise to 6.251 by the year 2000. Yet human beings must be fed, clothed, and housed. When considering this increase along with the further stress placed on forests, wetlands, and coral reefs by an improved standard of living and continued economic growth, commentators commonly arrive at one conclusion: extinctions are increasing inexorably.

In reaction to this conclusion, we have tried to construct a single volume of intellectual coherence and thematic integrity that contributes to the larger human enterprise of designing environmental policies and actions. Because the problem looms as a comprehensive and complex one, we have attempted to create a holistic response with two compelling features. First, the book displays a broad conceptual canopy that brings together and facilitates interdisciplinary discourse ranging from the physical sciences to the social sciences, and from philosophy to law. Second, and more important, our authors have moved past a mere understanding of this interdisciplinary relationship to reach conclusions beyond the static boundaries of traditional fields of inquiry. We believe that only an interdisciplinary discourse can advance the case for practicable policies and effective laws.

A review of contemporary writings on the loss of biodiversity quickly reveals the extent to which different disciplines remain fragmented in their outlook.

Whether emanating from the physical, natural, social, or political sciences, the norm is for scholarly works to address the cognoscenti within their own specialties. Disciplines by their very nature are devoted to discrete areas of learning, and authors are usually acclaimed or decried by those within, not outside, their fields. We consider it vitally important, therefore, to introduce a wide range of disciplines and mind-sets examining the loss of biodiversity, and to calibrate them within a conceptual framework that makes sense to disparate perspectives. This may prove to be a Sisyphean task, but it is something that must be attempted.

While many offer answers to the problems they perceive, disciplinary narrowness often prevents scholars from seeing the whole, leading to a neglect of issues falling outside their own intellectual pale. In the "Conclusion" to this volume, we expand on how this fractured outlook is further distorted by remarkable differences between the mind-sets of physical and natural scientists on the one hand, and political and social scientists on the other. This dissonance is illustrated by the expectations of physical and natural scientists that policymakers should act on the scientific theories pertaining, for example, to climate change. When this does not happen as envisioned, many physical and natural scientists are troubled by the "politics" of such decisionmaking, but show little interest in the explanations of political and social scientists about how and why such policymakers fail to do so, and instead, behave in the way they do.

We worry that the disciplinary training and instincts of some readers will direct them toward chapters containing materials with which they are familiar, where they might look for advances in their own science, discipline, or particular specialty. If they see nothing pathbreaking in their specific field of inquiry, such readers might be inclined to dismiss the rest of the book as insignificant or irrelevant. Yet the purpose of this book is not to advance the frontiers of biology, ecology, or economics, *per se*, but rather to weave a tapestry out of the collective wisdom of these different strands of thinking. Consequently, this book has little to offer those who fail to see the forest for the trees or only recognize a particular species but not an intricate ecosystem.

The book begins with the natural and physical sciences that provide a bedrock biophysical understanding of the problems besetting the biodiversity of the planet. We would have no understanding of the nature of the crisis we face without the light shed by biologists, ecologists, paleontologists, and other natural scientists. The chapters written by Peter Raven and Jeffrey McNeely and Ariel Lugo address whether the loss of biodiversity constitutes a planetary peril rather than just another episode in the evolutionary process or a chapter in the history of life. Raven and McNeely provide a road map for the issues to be examined in the following sections and traverse subjects such as the importance of intraspecies diversity, the interfaces between marine and terrestrial environments, and the need for ecological management.

Emerging from this introduction, we move to the empirical world of "Scientific Responses," in which Robert Horsch and Robert Fraley, Laura Jackson, and Gregory Benford deal with a variety of issues ranging from the cataloging of species, and more productive agriculture based on biotechnological innovations, to *in situ* preservation and better land use.

Scientific responses by themselves, however, do not address the difficulties confronting the implementation of their prescriptive remedial measures and strategies. Nor do the strategies, policies, and laws adopted by decisionmakers always honor, or display fidelity to, scientific findings or conclusions. Concrete steps to arrest the depletion of global biodiversity can only be taken within a political matrix, and are often based on socioeconomic analysis that frequently, though not always, falls outside the expertise and predilection of biologists, ecologists, and natural and physical scientists. Having established a horizontal axis of scientific analysis and prescription, we have attempted to construct a vertical sociopolitical axis, conscious that we are dealing with unpredictable and factional political issues. Moreover, as risk analysts keep reminding us, even individuals—quite apart from political entities—do not always behave rationally.

Economists analyze and prescribe ways of dealing with scarce resources, and therefore form an important part of this vertical axis. In "Economic Responses," a distinguished array of economists—Graciela Chichilnisky, Geoffrey Heal, and David Simpson, Roger Sedjo, and John Reid—examine the extent to which markets are capable of capturing the social values residing in biodiversity, and whether genetic resources located in less-developed countries have the potential to be commercially exploited.

Markets cannot function without institutions, be they national or international, and these institutions are the second component of our vertical axis. In view of the well-established misfit between institutions and goals, and the malfunctioning of institutions and bureaucracies, pressing questions emerge about the nature and functioning of such human systems of cooperation and governance. In "Institutional Responses," Elinor Ostrom, Walter Reid, Anil Gupta, and Gary Toenniessen each deal with how international, national, and local institutions should reflect the complexities that lie at the core of biodiversity. They are followed by two case studies, one by James Anaya and Todd Crider concerning the Awas Tingni in Nicaragua, and another by Ana Sittenfeld and Annie Lovejoy on the National Biodiversity Institute (INBio) in Costa Rica.

While economic utility and institutions are important factors that motivate human behavior, these are not the only mechanisms that affect it. Our perceptions and responses to the destruction of biodiversity are also based on philosophical or moral foundations, and it is these foundations that Bryan Norton and Mark Sagoff seek to uncover in the section dealing with "Moral Responses."

Yet whatever the ideology underpinning remedial actions, it is not possible to implement or enforce such actions without laws. In the global context, we are dealing with at least two systems of law: international and national. International law, such as the Convention on Biological Diversity (CBD; see appendix), is the consensual law made between states. States assuming duties under instruments of international law are obliged to fulfill their duties. Such implementation is usually carried out through national legislation, or executive or administrative action. As social instruments that affect human behavior, laws dealing with the protection of biodiversity must necessarily interface with the relevant social and political thinking. To ensure acceptance at the international level, and permit effective and efficient implementation, laws ought to be based on, incorporate, or institutionalize the most important scientific assessments, philosophical premises, and socioeconomic policies or strategies.

Thus, in the final section on "Legal Implementation," Christopher Stone, Yvonne Cripps, Mark Sagoff, Lakshman Guruswamy, and Brent Hendricks span a wide range of legal analysis and conclusions, ranging from the extent to which the earth's biological assets lie beyond the ambit of markets, to the patenting of novel organisms, and ultimately, to the CBD.

In the remainder of this introduction, we would like to highlight three important themes that emerge in this book: the reasons why there is a need for immediate concern about the extinction of species, the importance of biodiversity to biotechnology, and the relevance of biodiversity to global change research.

## Rationales

Raven and McNeely, echoing the belief of the majority of our authors, are deeply troubled by the rate of extinctions. They estimate that we could be losing an annual average of 50,000 species—including mammals, birds, insects, and plants—over the next several decades. If this rate continued, we could lose more than two-thirds of all living species over the course of the twenty-first century. Lugo, however, contends that such estimates are difficult to support with hard data and suggests instead that attempts to preserve biodiversity should be based on the need to preserve genetic codes—the seeds of all life and information. Significant disagreements about the extent to which biodiversity is being lost arise for two reasons. First, the number of species known to science is only a small fraction—perhaps just 1 to 15 percent—of the total species thought by experts to exist. And second, scientists are leery about claiming that a species has actually gone extinct, as the lack of observation of a species can be explained by many factors; for example, while most scientists consider that the thylacine, or Tasmanian wolf, has been extinct since 1934, sightings of this elusive creature are still reported from time to time. Thus, the estimates of extinction rates tend to be based on models and assumptions that are open to some challenge. Yet the reality of increasing human impact on the environment is impossible to ignore and, inevitably, has influences on other creatures.

Apart from asymmetrical views about rates of extinction, a question that has been touched on by a number of our authors concerns the rationale or rationales for protecting biodiversity. Raven and McNeely graphically and movingly describe extinctions as "a ripping apart—desecration, in the literal sense—of the fabric of the living world," vividly reflecting the outrage of biologists and natural scientists who have arrived at an almost spiritual or ethical basis for protecting biodiversity. At one level, it is an "I-Thou relationship with nature" (Ashby 1978, 82) inspired by their awe of evolution over millions of years and the interdependence of ecosystems. They condemn the destruction of species with the revulsion felt over the sacrilegious desecration of a temple, church, or icon.

Raven and McNeely's awe arises as much from their wonder at contemplating nature as from fear of the unknown consequences of ignorantly, and perhaps wantonly, interfering with the unreplicable outcome of a multimillion-year drama. The current environment is the grand product of millions of years of evolution, natural extinction, and networking. Species have adapted to changes in environmental conditions or they have vanished. In addition to genetic adaptation, species have also formed a complex network among themselves. The various species have an interactive order in the long-term evolutionary sense as well as in their daily existence. When humans disrupt that order, they may forever change not only the present, but the future of the world. Raven and McNeely's ethical assertions about biodiversity and nature are premised on what ought to be done and are value laden, not value free. Their rationale resonates with the views of Sagoff, who advocates the superiority of ethical, as distinct from economic or instrumental, reasons for protecting nature.

But Raven and McNeely, unlike Sagoff, place equal or greater emphasis on the material consequences accompanying the extinction of species. At the simplest level, humans directly use organisms and living things as material for food, clothing, and shelter. Indirectly, humans utilize the ecosystem services provided by nature, such as the provision of oxygen and the natural cleansing properties of the environment. These services constitute the difference between life and death, and their importance does not require a lengthy discussion.

At a different, perhaps less-understood level, nature is a storehouse of knowledge, and the loss of species permanently expunges genetic, physiological, and behavioral information about them. A number of the papers, including that of Benford, discuss how the loss of genetic codes can preclude the discovery of valuable applications just as we are becoming aware of the crucial importance of genetic information, and of how it could be used. Currently publicized techniques that enable us to transfer genes and complexes of genes are only the beginnings of the information highway that is immutably being destroyed.

Examples abound. *Artemisia annua*, annual wormwood, led to the development of Artemisin, a drug that is effective against all strains of the malaria parasite *Plasmodium*. Another example involves transferring the disease-resistant genes of the maize species *Zea diploperennis* to domestic corn to increase production. Furthermore, assorted varieties of corn are one of the primary food sources worldwide, supporting a multibillion-dollar industry.

Raven and McNeely offer other utilitarian reasons for protecting biodiversity. Many of today's pharmaceuticals trace their heritage to natural sources, either directly as a source or indirectly through structure modeling; 25 percent of prescriptions are filled by drugs whose active ingredients are derived or extracted from plants. The pharmaceutical industry, which thus owes its existence to the environment, is an immensely profitable enterprise for which forests are enormous storehouses of novel chemical compounds that may be prospected and developed. This supports the need to protect tropical forests in order to facilitate prospecting for useful, perhaps life-saving, materials.

Lugo, Sagoff, and Stone provide a different perspective. Pointing to the resiliency of the tropical forest, Lugo arrives at the conclusion that human activity (such as cutting down tropical forests) is not incompatible with the maintenance of biodiversity. Stone shows how people have regularly converted biological assets into other forms and assets without regret. Sagoff drives his point home by asserting that in 1600, the carrying capacity of the what is now the U.S. continent for urban people was close to zero: "If 250 million of us were suddenly blessed with pristine and healthy ecological systems, such as existed in the United States many centuries ago, very few of us would survive more than a few weeks."

In dealing with how to slow down the rate of consumption, conversion, or destruction of biological diversity, a number of our authors—including Stone, Chichilnisky, Heal, and Simpson et al.—have in different ways assessed the loss of biological diversity from an economic perspective, and inquired whether biodiversity could be treated as a commodity. Stone asks the question, what can we do to ensure that the rate of consumption/conversion is efficient? After a searching examination of the options, he deduces that the full social benefits of biological resources lie beyond appropriation by markets, so new methods are needed to ensure that society is able to evaluate the full value of these resources. Chichilnisky states that habitats are destroyed because of economic reasons and, along with Stone, examines ways in which their true social costs might be reflected in the products consumed. Simpson et al. conclude that the economic value of unknown genetic resources located in developing countries is negligible, while Heal and Stone concur that we cannot rely on markets to preserve biodiversity.

What is striking about our authors is that they are not "fundamentalist" microeconomists. They do not appear to subscribe to the worldview that all humans seek only to maximize the satisfaction of their own wants in a world where others are engaged in similar pursuits. Neither do they consider that the paramount objective of public policy is the maximization of human satisfaction or utility expressed through choices made in a free market economy. They do not treat economic value as a "metavalue" that comprehends all others; on the contrary, they all appear to agree with the conclusion of Simpson et al. that apart from economic values there are "important ecological, ethical, and aesthetic reasons" for preserving biodiversity.

## Biotechnology and Biodiversity

The potential economic value of biodiversity for biotechnology has sparked an interesting discussion. It has been suggested that "biodiversity prospecting" is an essential partner to biotechnology, the techniques by which changes made to DNA or genetic materials in plants, animals, and microbial systems leading to useful products and technologies. Horsch and Fraley assert that biotechnology has opened the door to even greater pharmaceutical and agricultural uses of genetic resources. Commercial applications of genetic engineering will soon include plants, animals, fungi, and invertebrates, as well as the genes of bacteria and viruses. With biotechnology, according to Horsch and Fraley, "the genes from the entire world's biota are within reach."

Sagoff offers three reasons for suggesting that advances in biotechnology—absent policy direction—will hinder rather than support efforts to protect biodiversity. First, genetic engineers create a lot of their own materials, for example, by the computer-assisted design of molecules. Second, biotechnology may encourage the domestication of nature, as in the replacement of wild habitats with bioindustrial systems of aquaculture, silviculture, and agriculture. Third, it seems impossible for biotechnology firms to prospect—nearly at random—among the trillions and trillions of bits of unknown genetic information in millions and millions of unidentified kinds of organisms. Accordingly, biotechnology can preserve only a very small part of the biodiversity we ought to and want to save.

Thus, we may need to attach a "dignity" as distinct from a "price" for biodiversity. Norton, concurring with Sagoff, underlines the importance of values outside those based on utilitarian or market forces by pointing to the "tranformative" character of biodiversity and nature. He contends that the existence of biodiversity can transform irrational and indefensible felt preferences into more considered ones.

Biotechnology is also crucial in preserving biodiversity through more productive agriculture. Horsch and Fraley, supported in this conclusion by Chichilnisky, argue that the loss of habitat to nonsustainable agriculture is the single largest threat to biodiversity. The pressure to clear forests and other biologically rich habitats is driven by the need to feed human populations that are increasing at the rate of nearly 100 million per year. Biotechnology can produce more food on the same land by developing insect-resistant crops, thereby eliminating the need to convert petroleum resources into pesticides and saving energy otherwise used in the packaging, transport, and application of pesticides. The development of herbicide-resistant crops will accelerate the adoption of nontill practices and reduce the need for expending energy to manufacture and apply herbicides. Releases of carbon dioxide would thereby be limited, and agriculture would contribute to averting global warming that would affect farming adversely. Toenniessen accepts and endorses the Horsch and Fraley thesis.

Jackson assails these arguments on several fronts. First, she points out that any reduction in energy use in industrial agriculture (the kind practiced in the developed world) scarcely touches the fringes of the massive consumption of energy in agriculture. North American feed grains, for example, must pass through livestock before they become food for people, and this involves a 50–90 percent reduction in their original caloric value depending on whether the animal is a chicken, a hog, or a steer. Therefore, maize fields apparently yielding 100 bushels to the acre are actually yielding 10 to 50 bushels of human food per acre. Second, the agricultural industry is heavily dependent on fossil fuel and nonrenewable energy. She adopts Amory Lovin's estimate of 9.8 kcal of fossil energy per kcal of food energy. Third, an accurate assessment of agricultural productivity needs to internalize the costs of soil erosion, ground and surface water degradation, and agricultural subsidies. Fourth, the type of industrial agriculture most likely to earn the greatest benefits from biotechnology eliminates the wild biodiversity located within agricultural landscapes—in hedgerows, farm lanes, ponds, and amidst cultivated crops. Such a pushing away of nature from agricultural landscapes into forests and protected reserves destroys the traditional way in which agriculture is a complex mutualism between people and nature. It denies rural people a necessary bond to nature, and reduces the quality of life for farmers on a grand scale.

Is biotechnology the silver bullet that could save biodiversity? What emerges is that the answer to this question is locked in contention. Whether the argument is premised on greater yields and productivity removing the pressure to open up forests, or on biotechnology's reliance on biodiversity, our authors make the case for and against with unusual cogency and sophistication.

### Relevance to Global Change Research

Physical, chemical, and biological scientists, in a remarkable display of interdisciplinary cooperation, have confronted the interrelated character of global problems by successfully establishing huge billion-dollar initiatives on global change, such as the International Geosphere Biosphere Program (IGBP) and the World Climate Research Program (WCRP), which intend to produce a comprehensive and integrated model of the physical, chemical, and biological processes that regulate the world. These studies involve the biosphere, the geosphere, and all the interactions within and between them, and include "biogeochemistry" models that track the cycles of chemical elements flowing through ecosystems, "biogeophysics" models that determine the way energy flows through them, and additional models that predict the effects of land use (Problem 1994; Schlesinger 1993).

The fact remains, however, as we discuss more fully in the "Conclusion," that these efforts simply underline the case for incremental progress, and provide another substantial cluster of reasons why the analyses and strategies advanced in this book should not be postponed until the full results and conclusions of these computer models are available. It is doubtful if any computer models could satisfactorily simulate all the physical, chemical, biological, and social forces at work in the real world. Even if they could, it is now recognized that the measurements of the initial conditions can never, in practice, be fully accurate. Chaos theory has shown how small initial differences can lead to hugely different results.

While producing vast quantities of data, these models offer no answers on how to solve the big global questions. The task of finding answers will, of course, be greatly facilitated by all this new information, but the hard fact is that extinctions inexorably increase even as these studies labor on, and prolonged postponement of decisionmaking is not a practicable option. It is precisely because the international community wanted to avoid paralysis by analysis that treaties dealing with climate change, biodiversity, and ozone depletion require programs and measures designed to address these global perils.

But global strategies, such as those embodied in the CBD, make sense only if they are capable of supporting the more detailed strategies and actions that implement them regionally, nationally, and locally. The authors of this book have attempted *inter alia* to transform vague generalities and objectives, such as those found in the CBD, into workable strategies.

## References

A problem as big as a planet. 1994. *Economist,* 4 November.
Ashby, E. 1978. *Reconciling Man with the Environment.* Stanford: Stanford University Press.
Schlesinger, W. 1993. Biogeochemistry. *Geotimes* 38:19.
Wilson, E.O. 1992. *The Diversity of Life.* Cambridge, Mass.: Belknap Press of Harvard University Press.

# I  Identifying the Problem: An Overview

# Biological Extinction: Its Scope and Meaning for Us

by Peter H. Raven and Jeffrey A. McNeely

The earth is home to an estimated 10 million species, of which about 1.5 million have been formally described. Due to population growth and increasing rates of consumption, however, the natural wealth of our planet is being lost at an estimated rate of 5 percent per decade. This is a tragic loss of the biological wealth of our planet, for ethical, aesthetic, and economic reasons, as well as for reasons of ecological functioning. A major international initiative is required to make the business community, decisionmakers, and the general public aware of this unparalleled loss of the productive capacity of our planet.

\*

We are confronting an episode of species extinction greater than anything that the world has experienced for the past 65 million years. To understand species extinction in the broadest possible terms, consider this fundamental fact: the earth, our planetary home, is finite. Since the earth, and everything on or in it, is limited, the economic formulas developed over the past few hundred years to keep track of the values involved in human transactions cannot make it any larger, nor give us any more of the productive systems and commodities on which we depend. It does not matter what conversions may be possible: no matter how clever we may be, the earth remains the same. Contrary to the wishful thinking embodied in some cornucopian scenarios (see, for example, Simon and Kahn 1984), the earth and its systems can either be used in such a way as to provide a sustainable context for our operations, or we shall destroy them. We are currently losing the biological diversity on which we depend at a rate that will greatly limit our future options.

## Background

Our species first appeared perhaps 500,000 years ago, at the last instant of the 4.5-billion-year history of the planet earth. As our hunter-gatherer ancestors moved over the face of the earth, they began to exterminate many of the large mammals and birds that they hunted for food—on numerous islands, in the New World 8,000 to 12,000 years ago, and elsewhere (Martin and Klein 1984). After the development of agriculture at scattered localities 11,000 to 6,000 years ago, a population estimated at 5 million people just before the era of recorded history began to increase rapidly. At the same time, with the extensive land clearing and grazing that characterized early agriculture, species became extinct at rapidly increasing rates. For example, in Hawaii, the activities of the Polynesians led to the documented extinction of half the original 100 or so land birds in the 1,200 years before Europeans arrived. Over the subsequent two centuries, 18 additional species have been eliminated, leaving 18 more at such low numbers that they are in immediate danger of extinction and only 9 with populations that appear large enough to be viable over time (Olson 1989, 50–53).

What is occurring now, during the second half of the twentieth century? The human population is in the process of growing from 1.8 billion people at the turn of the century (WRI 1995), a third of them living in industrialized countries, to more than 6 billion by the year 2000, with only a fifth in industrialized countries (United Nations 1992). The population of developing countries during this same period is increasing from about 66 percent of the world's population to roughly 80 percent, or from around 1.2 billion people to perhaps 5 billion (WRI 1995). About 40 percent of those people are living in absolute poverty, dependent on firewood, with no reliable access to clean freshwater, and half of them receive less than the minimum recommended daily amount of food. Many of the women and children who live in these societies are essentially enslaved by these conditions (Sadik 1992). Collectively, the people of developing countries control about 15 percent of the world's cash economy, use some 20 percent of the industrial energy and less of most other materials that contribute to their standard of living, and include among their numbers only about 6 percent of the world's scientists and engineers (United Nations 1992; World Bank 1992).

Over the past 40 years, while the human population increased at unprecedented rates, we have wasted about a fifth of the world's topsoil (Pimental 1993); lost around an eighth of the world's cultivated lands to desertification and salinization; increased greenhouse gases in the atmosphere by over a third, thus setting the world on a course that is leading inexorably to warmer climates (Houghton, Jenkins, and Pephramus 1990); destroyed more than 5 percent of the stratospheric ozone layer, thus increasing the incidence of malignant skin cancer in the latitude of the United States by an estimated 20 percent (EPA 1987; Jones and Wigley 1989); and cut or converted to simplified biological deserts about a third of the forests that existed in 1950 (Brown, Flavin, and Kane 1992). Grain production per capita began to decline in 1986, bringing Thomas Malthus's 200-year-old prediction closer to reality (Pinstrup-Andersen 1993). These grim facts demonstrate that we are not managing the world sustainably now, even as we consider how it might be made sustainable when more people consume more resources in the future.

Since the human populations of developing countries contain such high proportions of young people, even the devotion to family planning that has been exhibited by most of these countries for more than 20 years will not result in population stability for another two or three generations. The most optimistic scenarios for a stable world population range from 8.5 billion to 14 billion people, somewhere toward the end of the next century. If we fail to maintain and strengthen family planning programs worldwide, the situation could be even more serious; if fertility were to remain at the 1990 level of 4.3 children per woman until 2025, the world population would be 109.4 billion by the year 2100, but if fertility were to have dropped in 1990 to the replacement level of 2.1 children per woman, the population would be 8.1 billion in the year 2100 (United Nations 1992).

Perhaps the simplest way to summarize the conditions we face now, and distinguish them from those prevailing at any earlier time, is to point out that human beings, one of an estimated 10 million species on earth (WCMC 1992), are currently estimated to be consuming, wasting, or diverting 40 percent of the total net photosynthetic production on land (Vitousek et al. 1986). Since our population is likely to double or triple over the next century, many of the other species on earth will simply have "no place to hide." There is no rational scenario by which most of them could find a corner somewhere, leading to the inevitable conclusion that species are becoming extinct at levels unprecedented for the past 65 million years. We turn next to a consideration of those rates.

## Future Extinction Rates

How do we project rates of species loss reasonably into the future? First, the 1.4 million species that have been named and classified probably represent no more than 15 percent of the world's total species, or about 10 million species, approximately 85 percent of them terrestrial; some scientists consider far higher numbers to be credible (Erwin 1982; May 1990). Probably at least 7 million of the species exist in the tropics and subtropics, with 1.5 to 2 million in the better-known temperate regions. These estimates refer to species of eukaryotic organisms only (organisms with cells that have a contained nucleus); current methods are unable to estimate the numbers of distinct kinds of bacteria or viruses. Insects have an especially interesting history. Labanderia and Sepkoski (1993) have found that an extensive fossil record demonstrates that the diversity of insects at the family level exceeds that of preserved vertebrate tetrapods through 91 percent of their evolutionary history. The great diversity of insects was achieved not by high origination rates but, rather, by low extinction rates comparable to the low rates of slowly evolving marine invertebrate groups. The great radiation of modern insects began 245 million years ago and was not accelerated by the expansion of angiosperms—flowering plants—during the Cretaceous period. The basic trophic machinery of insects was in place nearly 100 million years before angiosperms appeared in the fossil record, so the loss of these ancient life-forms seems especially tragic.

The geological record suggests that despite recent depredations, we now share the earth with the largest number of species that have ever existed at one time, with the communities and ecosystems of the tropics containing the richest and most complex number of species (WCMC 1992).

Judging from the fossil record, the average life span of a species is approximately 4 million years, and of a mammal is about 2 million years (Raup 1986). If the total number of species in the world is around 10 million, then the background rate of extinction can be calculated at about four species per year. This appears to be a reasonable order-of-magnitude estimate, even though extinction rates (and the longevity of species) vary widely between groups. At such a rate, some 40,000 species of organisms would have become extinct during the first 10,000 years following the appearance of agriculture. For the approximately 250,000 species of plants, it would amount to the disappearance of one species every 10 years; for the 4,500 mammals, one every 275 years; and for the 9,000 bird species, one every 275 years (assuming an avian species life span of 4 million years).

In actuality, at least 115 species of birds and 58 species of mammals have become extinct during the past 400 years (WCMC 1992), during which time the number of human beings has increased 11-fold to its current level of 5.5 billion people. This documents a rate of extinction for these groups more than 50 times the base level. Since 1930, when the global human population reached 2 billion people, at least 19 species of birds and 14 species of mammals have become extinct, with more extinctions probable but not yet finally documented. The rates for the past 64 years, therefore, can be documented at roughly 100 times the base level for these groups, and are probably considerably higher—despite intensive efforts to conserve species of birds and mammals.

In temperate regions, for groups that are well known—such as plants, butterflies, and vertebrate animals—it is generally the case that about 10 percent of the species in a given area are currently regarded by specialists as threatened or endangered (IUCN 1993). The figures are much higher for freshwater organisms than for terrestrial ones; thus, more than 70 percent of the 297 native freshwater mussels in the United States are considered threatened, endangered, or of special concern. Although some industrialized countries, such as the United States, are in a position to put substantial resources into the preservation of biodiversity, other nations are not so fortunate. Therefore, the losses in temperate countries such as Chile and South Africa, with large numbers of species found nowhere else, are likely to be great.

Predictions of extinction rates in the future have been based largely on the demonstrated relationship between species number in a given group of organisms and habitat area. The relationship between species number and area is expressed by the formula $S = CA^z$, where $S$ is the number of species, $A$ is the area of the place where the species live, and $C$ and $z$ are constants. For the purpose of calculating rates of species extinction, $C$ can be ignored, and $z$ is what counts. Generally, the value of $z$ varies between 0.15 and 0.35, with the exact value depending on the kind of organism being considered and the habitats in which those organisms are found. As Wilson (1985) has pointed out, when species are able to disperse easily from one place to another, $z$ is low; when they do not have this ability, $z$ is high. Thus, birds have a low $z$ value, while orchids a high one.

The rule of thumb, which corresponds to the commonly observed $z$ value of 0.3, is that when an area is reduced to one-tenth of its original size, the number of species eventually drops to half (MacArthur and Wilson 1967). The word "eventually" is used because some species may disappear immediately when the forest is cleared, while others decline slowly and constitute what Janzen (1987) called "the living dead." Also, for as many as half of the original species to persist in a habitat that has been reduced to a tenth of its original size, an undivided block of the original habitat in optimal condition must remain. Even now, about two-thirds of the surviving tropical rain forests are highly fragmented. Large blocks of individual forests rarely persist, so that the rate of survival at equilibrium is likely to be much lower, with as few as 30 percent, 20 percent, or even 10 percent of the original species surviving. When the last forest patches are cleared in a given area, the rate of survival suddenly may drop to near zero (Wilson 1988, 3–18). It is worth turning here to the misconceptions that have been presented in this context about extinction rates in Puerto Rico and the eastern United States.

Taking a perspective based on forestry, Lugo (1988a) has asserted that Mac-Arthur and Wilson ignored the concept of equilibrium in their development of the theory of island biogeography, when in fact, their theory is based explicitly on survival and equilibrium. Lugo and his colleagues properly emphasize the role of plantations in maintaining some species, which could then repopulate reestablished forests if the pressures are relaxed. They have also asserted that there has been little extinction in Puerto Rico, despite a great reduction in its forest cover around the turn of the century.

These claims, which in any case, cannot be applied to a world in which the population will be doubling or tripling over the course of the next century, ignore the fact that the Greater Antilles were inhabited by dense populations of native people at the time that Christopher Columbus reached them, and that these islands had already been altered significantly 500 years ago—as documented clearly in the early chronicles of the area (Denevan 1992). By comparison with Hawaii, where the fossil record of birds is better known, one might reasonably expect more than a third of the biota to have gone extinct earlier as a result of human activities. The assertions also ignore the fact that the dozens of Puerto Rican plant species represented by one or a few individuals must be members of a class from which not every species survived—many have clearly been lost. The fossils of birds, mammals, and other vertebrates taken from the caves of Puerto Rico amply document widespread extinction since people first reached the island, so that the encouraging discovery that some species do indeed survive in plantations (Lugo 1988b)—a very good thing for the world of the future—cannot logically be taken to dispute the extremely general relationship between species number at equilibrium and area, once the facts are properly analyzed. What is of key importance, however, is the general point that Lugo has presented so well over the years—that human beings are, in fact, managing the entire biosphere and must make serious choices about how to do it well, or suffer the consequences. The survival of species beyond the twenty-first century will not depend, for the most part, on the preservation of whatever approximations of pristine wilderness remain, but rather, on our collective ability to manage and restore to sustainability the various kinds of degraded lands that our activities have created.

The survival of forest bird species in the eastern United States based on the reduction of forests—extinction rates are said to be lower than expected—exemplifies a second specific argument in which the facts have been misunderstood. Here, the outstanding analysis of Pimm (1991) has put the situation into perspective. Of the slightly more than 200 species of birds that use the forests of this region, only about 40 have the greater part of their ranges in the East, and of these, 12 have ranges that extend across either the forests of the northern United States or often Canada. (The ranges of birds and other organisms in tropical forests are characteristically much smaller, making them considerably more susceptible to extinction when those forests are reduced in extent.) Despite assertions to the contrary, the eastern forests of the United States were never reduced at any one time to 1 to 2 percent of their original area. There was certainly a major wave of deforestation, which started in the northeast and moved south and west. The forests cleared the earliest recovered first, and no more than 40 percent of the eastern forests were ever cleared at any one time (Boyden 1992). Taking this fact into account, and considering the species-area curve, with $z$ at about 0.25, one would expect about 12 percent of the total species to have been lost or seriously endangered. Applying this figure to the 40 species of birds that actually occur mainly in these forests yields an expected total of 5 species. In fact, three—the Carolina parakeet, ivory-billed woodpecker, and Backman's warbler—are extinct, and two more—Kirtland's warbler and the red-cockaded woodpecker—are globally endangered (WCMC 1992; IUCN 1993). These five species represent precisely what theory would have predicted about extinction in the region.

The relationship between species number and the area the species inhabit has been clearly demonstrated for hundreds of different habitats and situations throughout the world. Obviously, exceptional situations exist that prove the rule, but it is time to stop picking at the general pattern in an attempt to argue to a conclusion that defies the substantial body of information that exists in this area. We therefore apply these principles to an analysis of what extinction rates are likely to be in the coming decades.

The rate of clearing of all tropical forests in the 1980s was estimated by the United Nations's main agency dealing with forests, the Food and Agriculture Organization, at 0.8 percent per year (FAO 1995). For tropical rain forests and other moist tropical forests, they calculated the rate for the 1980s at 1.5 percent annually, reaching 1.8 percent by the end of the decade. In other words, an area of the biologically richest and most poorly known ecosystem on earth equal in size to the state of Florida is clear-cut each year. Although forest-loss data are certainly important, they are not able to pick up other types of negative impacts on the habitat, such as the plowing of grasslands, overgrazing of rangelands, draining of wetlands, damming of streams, pollution of soils and water, introduction of exotic organisms, hunting pressures, and so forth. In all forests, at least an equivalent amount is grossly disrupted each year through such practices as selective logging and slash-and-burn agriculture, but let us accept only clear-cutting as a basis for our calculations here.

Following his careful review of the data, Wilson (1988) has calculated that for the area occupied by tropical rain forests alone, and using the most conservative projection of the relationship between species number and area, 2.7 percent of the species in tropical rain forests are being lost per decade, a rate well over 10,000 times the background extinction rate for this area—and doubtless a minimum estimate; while alarming, Smith et al. (1993) have arrived at a similar extinction rate, based on their global analysis of several lines of evidence. Furthermore, rates of forest destruction are accelerating throughout the world, and roughly doubled for tropical rain forests during the 1980s; consequently, rates of extinction can logically be expected to increase rapidly in the future. As mentioned earlier, the global population is not expected to stabilize until it has doubled or tripled, and 90 percent of that growth is projected to take place in the tropics.

On the basis of similar projections, Simberloff (1992, 75–89) has calculated that by the end of the next century, about two-thirds of all plant and bird species in South America will have become extinct. His calculations are conservative, because they unrealistically assume a constant rate of forest clearing similar to that experienced in the 1980s. Although the actions that we may take can still affect the situation profoundly, it is difficult to imagine any area of the tropics persisting with more than a tenth of its original vegetation undisturbed by the end of the twenty-first century, so that conservatively a quarter of global biodiversity will have disappeared by that time due to clearing of tropical rain forests alone.

We can summarize the current situation as follows. Assuming a total of 10 million species in the world and a current extinction rate of 5 percent per decade—a moderate estimate of the loss by people competent to deal with the field—

we would be losing an average of 50,000 species per year over the next several decades—of which only 7,000, on average, would have been recognized and named. These losses would include about 20 species of mammals, 40 species of birds, and 1,250 species of plants per year—staggering figures, but ones consistent with the rates of loss of habitat that are currently observed, and likely to occur in the future if we simply ignore them and get on with "business as usual." Even at the most conservative level ever calculated for extinction rates, we are in the process of losing a fifth that many species per year—no less than 250 species of plants, for example.

Throughout the world, rates of extinction will largely depend on the way we treat individual areas. For example, areas of high biological diversity and threat, which Myers (1988) has termed "hot spots," contain an estimated 20 percent of the world's plant species on about 0.5 percent of the world's total land surface. Taking these and other factors into account, those scientists who have attempted to estimate species extinction—scientists who are competent and active specialists in the field, and understand the factual and theoretical basis of dealing with it—have arrived at rates centering around 5 percent of the world's total species per decade, with some estimates as low as 1 percent (2,500 times the background rate of extinction for the past 65 million years) and others as high as 11 percent per decade (nearly 30,000 times the background rate) (Smith et al. 1993; Nitecki 1984; Soulé 1987; Ehrlich and Ehrlich 1981; Reid 1992, 55–73). As Wilson (1988) has pointed out succinctly, "Clearly we are in the midst of one of the great extinction spasms of geological history."

If two-thirds of living species are indeed lost over the course of the next century, the proportion will be more or less equivalent to that which disappeared at the end of the Cretaceous period—one of the several great extinction events in earth's history (Raup 1986). It took more than 5 million years for the world to regain its ecological equilibrium after that event, a sobering period of time to contemplate since it is more than 10 times the length of history of our own species.

## Why Does It Matter?

Why is the loss of biodiversity important? The answers to this question are not generally well understood by the public, or even by policymakers. For example, a study commissioned in 1993 by Defenders of Wildlife showed clearly that most Americans are unaware of the scope or importance of biodiversity, regarding many local and less-threatening problems, such as pollution, as more serious. Viewed appropriately, the loss of species and their genetic diversity is not a matter of the decimation of rhinos or elephants, the northern spotted owl, the Tennessee snail darter, or the Furbish lousewort—icons that convey very different messages to people who approach them from different standpoints. Instead, the loss of biodiversity should be seen as a ripping apart—desecration, in the literal sense—of the fabric of our living world and the destruction of the machinery that makes our unique planetary home function. Only someone unaware, or unwilling to become aware, of the ways in which living systems function could conceivably view this destruction with indifference, because it will profoundly affect each of us, our children and their children, and our planet for as long as our species exists.

The reasons for being concerned with the loss of biodiversity fall into three basic classes: ethical and aesthetic, economic, and ecological.

## Ethical and Aesthetic Values

As Paul Ehrlich and Edward O. Wilson put it in their acceptance of the 1990 Craford Prize, the Royal Swedish Academy's equivalent of the Nobel Prize for ecology, the first reason is ethical and aesthetic:

> Because *Homo sapiens* is the dominant species on Earth, we and many others think that people have an absolute moral responsibility to protect what are our only known living companions in the universe. Human responsibility in this respect is deep, beyond measure, beyond conventional science for the moment, but urgent nonetheless. The popularity of ecotourism, bird-watching, wildlife films, pet-keeping, and gardening attest that human beings gain great rewards from those companions [and generate substantial economic activity in the process].

A related view, strongly supportive of a reverence for life, was presented by Charles Darwin (1936) in the concluding paragraph of his classic of 1859, *On the Origin of Species by Means of Natural Selection*:

> It is interesting to contemplate a tangled bank, clothed with many plants of many kinds, with birds singing on the bushes, with various insects flitting about, and with worms crawling through the damp Earth, and to reflect that these elaborately constructed forms, so different from each other, and dependent upon each other in so complex a manner, have all been produced by laws acting around us. These laws, taken in the largest sense, being Growth with reproduction; Inheritance which is almost implied by reproduction; Variability from the indirect and direct action of the conditions of life, and from use and disuse; a Ratio of Increase so high as to lead to a Struggle for Life, and as a consequence to Natural Selection, entailing Divergence of Character and the Extinction of less improved forms. Thus, from the war of nature, from famine and death, the most exalted object which we are capable of conceiving, namely, the production of the higher animals, directly follows. There is grandeur in this view of life with its several powers, having been originally breathed by the Creator into a few forms or into one; and that, while this

planet has gone circling on according to the fixed law of gravity, from so simple a beginning endless forms most beautiful and most wonderful have been, and are being evolved.

Aldo Leopold's (1949) words put the matter poetically:

> It is a century now since Darwin gave us the first glimpse of the origin of species. We know now what was unknown to all the preceding caravan of generations: that men are only fellow-voyagers with other creatures in the odyssey of evolution. This new knowledge should have given us, by this time, a sense of kinship with fellow-creatures; a wish to live and let live; a sense of wonder over the magnitude and duration of the biotic enterprise.
>
> Above all we should, in the century since Darwin, have come to know that man, while now captain of the adventuring ship, is hardly the sole object of its quest, and that his prior assumptions to this effect arose from the simple necessity of whistling in the dark.

Many sensitive human beings find the ethical argument for the preservation of life implied by these authors compelling.

## Economic Values

The second class of reasons for being concerned with the loss of biodiversity are economic, in relation to the properties of particular kinds of organisms (Pearce and Moran 1994). Individually, we use organisms as sources of food, medicines, chemicals, fiber, clothing, structural materials, energy (biomass), and for many other purposes. For example, about 100 kinds of plants provide the great majority of the world's food, but there are tens of thousands of kinds of other plants, especially in the tropics, that have edible parts and might be used more extensively for food, and perhaps brought into cultivation (Myers 1983, 1984). Incidentally, our collective investment in the development of appropriate agricultural systems or new crops in the tropics is disgracefully small, despite the extreme importance of such advances for the welfare of a large proportion of our fellow human beings, and their consequences for global stability.

The fundamental relationship involved here is that sustainable productivity is essentially biological productivity. Plants and other organisms are natural biochemical factories, and can provide many products of importance for human welfare. More than 60 percent of the world's people depend directly on plants for their medicines; for example, over 5,000 species are used in China and 1,300 in northwestern Amazonia. In the global drug markets that supply the needs of the remaining people, some 119 drugs in international commerce are derived directly from plants, and the great majority of Western medicines owe their existence to research on the natural products that organisms produce (Principe 1988).

For instance, of the top 20 pharmaceutical products sold in the United States in 1988, a $6 billion market, 2 are taken directly from natural sources, 3 are semisynthetics, 8 are synthetics with their chemical structure modeled on previously used natural compounds, and 7 had their pharmacological activity defined by natural products research. Natural products research, therefore, played a role in the derivation of each one of them. Since there are some 250,000 kinds of plants in the world, and relatively few have been examined for their secondary compounds, it stands to reason that the remaining species contain many unknown compounds of probable therapeutic importance (Akerele, Heywood, and Synge 1989). Against such a background, the projected loss of something like 12,500 plant species per decade seems particularly threatening (Farnsworth and Soejarto 1985): Why have we no intervention scheme to preserve all plants, something that is both feasible and highly desirable?

It is worth considering why these natural products are so important in the development of drugs. Gordon Cragg, chief of the Natural Products Branch of the National Cancer Institute, put it this way: "I still maintain that no chemists can 'dream up' the complex bioactive molecules produced by nature, but once the natural lead compounds have been discovered then the chemists can proceed with synthetic modifications to improve on the natural lead."

A few examples will illustrate the point. Artemisin, the only drug effective against all strains of the *Plasmodium* organisms that cause malaria, has a chemical structure totally different from that of quinine and the other chemicals used against this disease, which afflicts 250 million people annually. It was discovered because the Chinese people have for 2,000 years used an extract of annual wormwood, *Artemisia annua*, to treat fevers.

Taxol, derived initially from the western yew, *Taxus brevifolia*, was found in a random collection by a joint USDA-NIH program screening plants for anticancer activities. The taxol molecule is structurally unique, and it is highly unlikely that it could have been visualized if it had not been discovered in nature. To date, it is the only drug that shows promise against breast cancer and ovarian cancer, two of the major diseases facing women in the United States and throughout the world. From a scientific point of view, taxol's novel mechanism of action is especially interesting. It promotes the polymerization of tubulin to form microtubules, which it then stabilizes, thereby preventing cell division and the spread of cancer; taxol is unique in this respect. Other antimitotic agents interfere with mitosis by inhibiting the polymerization of tubulin. Biochemists are now trying to determine the site of action of taxol on tubulin and microtubules, which may then permit the rational design of other drugs through molecular modeling. Significantly, the natural product led to the discovery of this new mechanism of action; until one determines such mechanisms and sites of action, one cannot proceed with molecular modeling.

A final example concerns Michellamine B, a novel compound from the African vine *Ancistrocladus korupensis*, which was discovered by Missouri Botanical Garden scientists working under contract with the National Cancer Institute to collect random samples of plants for screening as anticancer and anti-HIV activity. Subsequently, the garden has assisted the government of Cameroon in establishing a plantation for this valuable plant, which is rather rare in nature, since it is now to be tested widely in laboratory animals due to its remarkable range of anti-HIV activity. Even though Michellamine B may be too highly neurotoxic to be helpful directly as a drug, this discovery will have considerable value. Specifically, this compound does not work in the same way as AZT and other anti-HIV drugs, which affect the reverse transcriptases and proteases that have been found to play key roles in the viral replication. When its method of action is understood, it may well prove to be a useful molecular probe, and assist in the discovery of other drugs that will be effective against AIDS.

Against this background, it is easy to understand why the major pharmaceutical firms are expanding their programs of exploration for new, naturally occurring molecules with useful properties (Reid et al. 1993). In contrast, it is almost impossible to understand how any intelligent and reasonably sensitive individual could view the disappearance of a major proportion of the biodiversity on earth with equanimity, or why the world's nations have not already united in a major effort to explore and conserve the biodiversity on which so much of our common future will clearly depend.

In terms of plants generally, we are just starting to develop the techniques that will enable us to transfer genes and complexes of genes between unrelated kinds of organisms directly, and thus do a much more precise job of improving crops than has been possible earlier. Two important considerations relate to this skill. First, the world agricultural system is overly dependent on the close relatives of particular crops; the ability to interbreed strains will recede rapidly in importance as the years go by (Frankel and Hawkes 1974). Second, we must find ways to overcome our almost mystical distrust of genetically engineered organisms—especially when we are ready to use the same techniques immediately for medical purposes. By not doing so, we are foregoing a significant part of our ability to promote global sustainability for no perceptible gain.

Summarizing, the direct economic values of biodiversity are real and significant: values in crop production, medicines, and so forth. Prescott-Allen and Prescott-Allen (1986) carried out a detailed analysis of the contribution wild species of plants and animals made to the American economy, concluding that some 4.5 percent of the Gross Domestic Product (GDP) is attributable to wild species. The combined contribution to the GDP of wild harvested resources averaged some $87 billion per year over the period 1976 to 1980; the value undoubtedly has increased substantially since then. In general, the exact magnitude of these values is uncertain and new discoveries occur frequently, so that "option values" are considerable. Many values are not readily captured in the marketplace due to lack of property rights, public good problems, open access, and externalities (Pearce and Moran 1994). For these and other reasons, too little is invested in exploring and protecting biodiversity.

It is just 41 years since James Watson and Francis Crick first postulated a plausible structure for DNA, a postulate that led to the universal acceptance of this molecule as the genetic material that governs heredity. Although we have been applying molecular biology and its applications to the understanding of biological phenomena at all higher levels of complexity ever since, the hundreds of articles that appear monthly in journals throughout the world provide ample evidence that we still have much to learn. It is an even shorter time, only 21 years, since scientists first successfully transferred a gene from one unrelated kind of organism to another, giving rise to genetic engineering. We are just starting the task of sequencing entire genomes and beginning to understand them; at present, we cannot really estimate reliably the genetic difference between a corn plant and a human being, or what the differences signify. Clearly, we are at the dawn of a new era of biological understanding. Since we are just learning how to screen for the most useful properties of organisms to transfer, the technology for gene transfer, we must be deeply worried about the disappearance of a major proportion of our planet's organisms.

Sadly, many nonbiologists think about individual species as if they were simply packages of commodities on the supermarket shelf. As soon as an economic value is assigned to a species, it becomes a commodity that can be exploited. Economic arguments may therefore encourage consumption of nature. As Leopold (1949) has put it, "A system of conservation based solely on economic self-interest is hopelessly lopsided. It tends to ignore, and thus eventually eliminate, many elements in the land that lack commercial value but that are essential to its healthy functioning. It assumes that the economic parts of the biotic clock will function without the uneconomic parts." Even ignoring the ethical and aesthetic dimensions mentioned above, each species is in fact the unique product of more than 3.5 billion years of evolution, superbly adapted to its individual habitat, and with an expectable life span perhaps 10 times longer than our own species, *Homo sapiens*, has already experienced. Each species possesses a unique combination of genes—how many or in what proportion, we do not know—controlling all of its characteristics. These genes are now capable of being transferred from one system or species to another, but only if we face the problem of extinction and take action so that they still exist when we have developed the technology to use them intelligently. To paraphrase Leopold, the first rule of intelligent tinkering is to save all of the bits and pieces.

## Ecological Values

The third class of reasons that people should be concerned with the loss of biodiversity pertains to the array of essential services provided by natural ecosystems. Such services include protecting watersheds, regulating local climates, maintaining atmospheric quality, absorbing pollutants, and generating and maintaining soils (Schulze and Mooney 1993). Ecosystems, functioning properly, are responsible for the earth's ability to capture energy from the sun and transform it into chemical bonds, a form in which it is used to provide the energy necessary for the life processes of the fewer than 300,000 species of photosynthetic organisms and all the other 10 million or so species, including ourselves, which depend on them.

It could be argued that the existence or health of a forest or a population of fish depends on a system of ecological processes, relationships, and species; and that many, if not most of these, have no instrumental, aesthetic, or inherent value, at least as perceived by society at large. Perhaps one of the most important things that ecological economists can do is to identify how these ecosystem functions are, in fact, valuable to people. Norton (1988, 200–205) used the term "contributory value" to express the indirect benefits that species involved in predator-prey relationships essential to population stability of harvested species, and species diversity in general, confer to ecosystems. He argued that all species have contributory value, and that the loss of any species represents an incremental decrease in the overall utility value of ecosystems.

Ecosystems are also responsible for regulating the recycling of nutrients, derived from the weathering of minerals in the soil and from the atmosphere, and making them constantly available for the maintenance of life. The populations of organisms that control pests on adjacent crops are often maintained in natural or semi-natural ecosystems nearby, as are many of the insects and other animals that pollinate these crops, ensuring the production of fruits and seeds. People are not very aware of these ecosystem services, nor of the role of biodiversity in their lives, as can be illustrated by the "debate" about logging that has been going on in the Pacific Northwest, far western Canada, and the Alaska Panhandle for several decades. It is presented as a confrontation between loggers, who want to "preserve" their jobs by cutting down the remaining 10 percent of the irreplaceable forests that are still available to them on public lands—a process that would take less than five years; where would the jobs be then?—and the environmentalists' concern with a poorly differentiated race of a widespread western bird, the northern spotted owl. What of the beauty and majesty of these forests? What of the reasons that people choose to live in these regions, and enjoy them, and visit them as tourists? Tourism contributes more to the economy of the states involved than the lumber industry, but how do we deal with the direct negative relationship between these two economic forces? What of the role of the forests in protecting tens of thousands of species of other organisms, including the source of the only effective drug against breast cancer and ovarian cancer? How about the watersheds, the soils, the clear streams with their annual salmon runs—all dependent on the forests? We badly need a regional and national attitude that could really address the important questions, such as these, and not, by virtue of endless litigation, treat only a couple of its factors to reach a clumsy compromise that ultimately suits no one (Maser 1989).

This more comprehensive view is essentially the attitude that Secretary of the Interior Bruce Babbitt has espoused in many of the actions that the Department of the Interior has taken under his direction, including the formation of the National Biological Survey; the resulting controversy largely reflects society's incomplete understanding of sustainability, or perhaps our unwillingness to put what we know in our hearts to be true into practice for the benefit of our children and grandchildren. But the times are changing, and if those of us who understand something about the ecological systems that support us all are generous enough with our time and effort, they will change even more rapidly.

Cairns and Pratt (1990, 495–505) point out that simple communities are not capable of the same responses to stress as more complex ones. This may be a result of fewer redundant species in the species pool capable of exploiting changing conditions or of biological differences in the taxa found in early successional or immature comma than in later successional (more mature) stages. The result of these differences, the underlying biology of which is poorly known, is the inability of communities to disperse propagules to new habitat, to respond to toxic chemicals, or in the case of simple communities, to exclude invaders. Continual erosion of biological diversity may result in the loss of key species that regulate numbers of other taxa and allocation of resources to biomass.

Ecosystems of all kinds are what make the world look as it does, and function as it does, but we know next to nothing of the degree to which individual organisms can be substituted from one ecosystem into another—principles we shall need in order to utilize the survivors of the holocaust that we are now precipitating to restabilize the world once our numbers have become stable and "sustainability" becomes more than a kind of hopeful ideal, used to justify boundless expansionism. All of these inorganic and mechanical strategies tend to end in failure, because they do not have the self-sustaining qualities of the living systems into which our ancestors evolved, and which we are currently destroying so rapidly. As we modify ecosystems, we use ever-increasing amounts of pesticides, herbicides, fertilizers, chlorination, water control measures such as dams and irrigation, and other strategies that require large energy subsidies, based primarily on finite supplies of fossil fuels. Organisms are generally highly adapted to their roles in particular ecosystems, and in ways that we understand poorly. Clearly, much of the quality of ecosystem services will be lost if the present episode of extinction is allowed to run unbridled for much longer, and the rebuilding of these systems in which our descendants will necessarily be engaged is likely to be seriously impaired by our neglect.

## What Should Be Done?

The preservation of biodiversity can be accomplished only as part of an overall strategy to promote global stability. This would necessarily involve social equity issues, improved agricultural and forestry practices (for example, using cutover lands rather than mature forests for new enterprises), capping the activities that are leading to global warming and other drastic alterations of the earth's environment, and limiting overconsumption in industrialized countries to levels that the world could sustain (Woodwell 1990). Pollution, which poses problems for other organisms besides humans, is another factor that must be controlled if a reasonable sample of biodiversity is to be preserved. Since about 80 percent of the world's biodiversity is confined to tropical, developing countries, it is obvious that international assistance on a massive scale will be necessary to stabilize the common management of our planetary home and provide hope for the future of most of the world's animals, plants, fungi, and microorganisms. Building technical competence in nations around the world so that they can manage their own biodiversity for their own benefit, and thus for ours, is one of the best investments that we can make in our common future. It is only by saving organisms that we will be able to build the productive systems and ecological communities of the future, and the new Convention on Biological Diversity is an important international initiative toward this end (Glowka et al. 1994).

Concerning the fate of biodiversity, it is obvious that land conversion and habitat destruction are often carried too far as the result of subsidies to ranching, logging, farming, and other activities that cause biodiversity loss. Too little is spent to protect biodiversity and too much to destroy it, so that—from an economic perspective—what is happening is neither efficient nor "optimal." Much effort must be devoted to converting a business press that is largely ignorant of this reality, or chooses to ignore the facts, to one that addresses sound economic development as what it is: sound ecological development.

More specifically, and in the context indicated above, the preservation of selected natural and seminatural areas is the major strategy to ensure the greatest amount of biodiversity at the lowest cost (McNeely 1992). Managing ecosystems everywhere for maximum biological diversity, and limiting to the extent possible further human incursions into relatively undisturbed natural areas, would be important aspects of such a strategy. Human cultural diversity must also be taken seriously into account if biological diversity is to be preserved: the two are intimately connected. The use of primary forests should always be avoided when disrupted communities provide alternatives. In addition, many organisms will be preserved only if they are brought into cultivation, zoos, type culture centers, or similar facilities, where they can then be deep-frozen or otherwise preserved in a living condition outside their natural habitats. A world scheme for accomplishing the preservation of the maximum amount of biodiversity possible would be the most important single contribution that the people of our generation could make to the future (WRI, IUCN, and UNEP 1992).

At the same time, it is clearly important to encourage national and international efforts to learn more about biodiversity, so that it can be managed—in all senses of that word, including conservation—more effectively. The efforts of the National Biodiversity Institute in Costa Rica provide an outstanding example of the methods whereby a nation can take possession of its own biodiversity, conserve it, and then use it for its own benefit (Gámez 1991); analogous efforts have been initiated elsewhere, in Taiwan, Mexico, and the United States. Electronic data processing provides the tool necessary for the efficient handling of information about biodiversity, and should be utilized fully in this area of human knowledge.

In conclusion, quantifying the precise rate of extinction and determining a precise figure for the number of species on earth is of only minor relevance to the crisis facing the world today. Policymakers and the public may want to assess the magnitude of the extinction crisis, and thus the priority to be given to the issue, on the basis of an absolute rate, but the investment of time and effort in refining such predictions contributes little to tackling the root causes of the problem. Indeed, obsession with an absolute rate may give an unrealistically optimistic impression in that no allowance is made for the genetic impoverishment of the multitude of species brought to the verge of extinction through the progressive loss of discrete subpopulations.

Even if the growth in world population continues to decelerate from its historical peak of 1.9 percent in 1971 (it is now 1.7 percent, but with larger numbers of people being added each year because of the increase in the base population), the 1990s and the first few decades of the twenty-first century are likely to be the most tumultuous and destructive of the natural resources of our planet that we may ever face. We have less capacity to deal with the extra people now than we hope will be the case in the future. In light of this growing disaster, we can either decide to be conservative in dealing with biodiversity—since species are of such fundamental importance to us, and their loss is absolute and irreversible—or we can recklessly conclude, despite all evidence to the contrary, that what we are destroying is not important to us. In that case, we can continue to pursue our materialistic goals, which ultimately will be thwarted by the collapse of the ecological systems that we are not bothering to understand properly, and which are not necessarily the principles around which we wish to organize our lives and the societies within which those lives take place. One hopes that with population stabilization, we may come to the realization that we have only a single common resource, our planetary home. It is in our interest to preserve the organisms that allow our home to function, and on which we depend, as we have always done, rather than to assume tacitly, almost as if in a delirium, that they are irrelevant to us except as they may add to our bank accounts while we destroy them.

## Acknowledgments

We are especially grateful to Gordon Cragg, Herman E. Daly, Paul R. Ehrlich, Hugh Iltis, Stephen R. Kellert, Robert M. May, James S. Miller, Norman Myers, Stuart Pimm, Robert Repetto, Carl Sagan, Daniel Simberloff, Compton J. Tucker, and E.O. Wilson, all of whom assisted in the preparation of this paper or contributed materials for it.

## References

Akerele, O., V. Heywood, and H. Synge, eds. 1989. *The Conservation of Medicinal Plants*. Cambridge: Cambridge University Press.

Boyden, S. 1992. *Biohistory: The Interplace between Human Society and the Biosphere, Past and Present*. London: Pantheon.

Brown, L.R., C. Flavin, and H. Kane, eds. 1992. *Vital Signs: The Trends That Are Shaping Our Future*. New York: W.W. Norton.

Cairns, J., and J. Pratt. 1990. Biotic impoverishment: Effects of anthropogenic stress. In *The Earth in Transition: Patterns and Processes of Biotic Impoverishment*, edited by G. Woodwell. Cambridge: Cambridge University Press.

Darwin, C. 1936. *The Origin of Species by Means of Natural Selection or, the Preservation of Favored Means in the Struggle for Life and the Descent of Man and the Selection in Relation to Sex*. 1859. Reprint, New York: Modern Library.

Denevan, W.M. 1992. *The Native Population of the Americas in 1492*. 2nd ed. Madison: University of Wisconsin Press.

Ehrlich, P.R., and A.H. Ehrlich. 1981. *Extinction: The Causes and Consequences of the Disappearance of Species*. New York: Random House.

EPA (Environmental Protection Agency). 1987. *Protection of Stratospheric Ozone and an Assessment of the Risks of Stratospheric Modification*. Washington, D.C.: Environmental Protection Agency.

Erwin, T.L. 1982. Tropical forests: Their richness in coleoptera and other arthropod species. *Coleoptera Bulletin* 36:74–82.

FAO (Food and Agriculture Organization). 1995. The State of the World's Forests. Rome: Food and Agriculture Organization of the United Nations.

Farnsworth, M.R., and D.D. Soejarto. 1985. Potential consequences of plant extinction in the United States on the current and future availability of prescription drugs. *Economic Botany* 39 (2):231–40.

Frankel, O.H., and J.G. Hawkes, eds. 1974. *Plant Genetic Resources for Today and Tomorrow*. London: Cambridge University Press.

Gámez, R. 1991. Biodiversity conservation through facilitation of its sustainable use: Costa Rica's National Biodiversity Institute. *Trends in Ecology and Evolution* 6:377–78.

Glowka, L., F. Burhenne-Guilmin, H. Synge, J.A. McNeely, and L. Gündling. 1994. *A Guide to the Convention on Biological Diversity*. Gland, Switzerland: World Conservation Union.

Houghton, J.T., G.J. Jenkins, and J.J. Pephramus. 1990. *Climate Change: The IPCC Scientific Assessment.* New York: Cambridge University Press.

IUCN (World Conservation Union). 1993. *1994 IUCN Red List of Threatened Animals.* Gland, Switzerland: World Conservation Union.

Janzen, D.H. 1987. Insect diversity of a Costa Rican dry forest: Why keep it, and how? *Biological Journal of the Linnean Society* 30:343–56.

Jones, R., and T. Wigley, eds. 1989. *Ozone Depletion: Health and Environmental Consequences.* New York: John Wiley.

Labanderia, C.C., and J.J. Sepkoski Jr. 1993. Insect diversity in the fossil record. *Science* 261:310–15.

Leopold, A. 1949. *A Sand County Almanac and Sketches Here and There.* New York: Oxford University Press.

Lugo, Ariel. 1988a. Diversity of tropical species: Questions that elude answers. *Biology International* 19:1–37.

———. 1988b. The forest of the future: Ecosystem rehabilitation in the tropics. *Environment* 30:16–20, 41–45.

MacArthur, R.H., and E.O. Wilson. 1967. *The Theory of Island Biogeography.* Princeton, N.J.: Princeton University Press.

McNeely, J.A. 1992. *Parks for Life: The Proceedings of the Fourth World Congress on National Parks and Protected Areas.* Gland, Switzerland: World Conservation Union.

Martin, P.S., and R.G. Klein, eds. 1984. *Quaternary Extinctions: A Prehistoric Revolution.* Tucson: University of Arizona Press.

Maser, C. 1989. *Forest Primeval: The Natural History of an Ancient Forest.* San Francisco: Sierra Club.

May, R.M. 1990. How many species? *Philosophical Transactions of the Royal Society* B330:293–304.

Myers, N. 1983. *A Wealth of Wild Species.* Boulder, Colo.: Westview Press.

———. 1984. *The Primary Source: Tropical Forests and Our Future.* New York: W.W. Norton.

———. 1988. Threatened biotas: "Hotspots" in tropical forests. *Environmentalist* 8(3):1–20.

Nitecki, M.H. 1984. *Extinctions.* Chicago: University of Chicago Press.

Norton, B.G. 1988. Commodity, amenity, and morality: The limits of quantification in valuing biodiversity. In *Biodiversity*, edited by E.O. Wilson. Washington, D.C.: National Academy Press.

Olson, S. 1989. Extinction on islands: Man as a catastrophe. In *Conservation for the Twenty-first Century*, edited by D. Western and M.C. Pearl. New York: Oxford University Press.

Pearce, D., and D. Moran. 1994. *The Economic Value of Biodiversity.* London: Earthscan and World Conservation Union.

Pimental, D., ed. 1993. *World Soil Erosion and Conservation.* New York: Cambridge University Press.

Pimm, S.L. 1991. *The Balance of Nature? Ecological Issues in the Conservation of Species and Communities.* Chicago: University of Chicago Press.

Pinstrup-Andersen, P. 1993. *Socioeconomic and Policy Considerations for Sustainable Agricultural Development.* Washington, D.C.: International Food Policy Research Institute.

Prescott-Allen, C., and R. Prescott-Allen. 1986. *The First Resource: Wild Species in the North American Economy.* New Haven, Conn.: Yale University Press.

Principe, P. 1988. *The Economic Value of Biological Diversity among Medicinal Plants.* Paris: Organization for Economic Cooperation and Development.

Raup, D.M. 1986. Biological extinction in earth history. *Science* 231:1528–33.

Reid, W.V. 1992. How many species will there be? In *Tropical Deforestation and Species Extinction*, edited by T.C. Whitmore and J.A. Sayer. London: Chapman and Hall.

Reid, W.V., S.A. Laird, C.A. Meyer, R. Gámez, A. Sittenfeld, D.H. Janzen, M.A. Gollin, and C. Juma, eds. 1993. *Biodiversity Prospecting: Using Genetic Resources for Sustainable Development.* Washington, D.C.: World Resources Institute.

Sadik, N. 1992. *The State of World Population, 1992*. New York: United Nations Population Fund.

Schulze, E.D., and H.A. Mooney, eds. 1993. *Biodiversity and Ecosystem Function*. New York: Springer-Verlag.

Simberloff, D. 1992. Do species-area curves predict extinction in fragmented forest? In *Tropical Deforestation and Species Extinction*, edited by T.C. Whitmore and J.A. Sayer. London: Chapman and Hall.

Simon, J.L., and H. Kahn. 1984. *The Resourceful Earth: A Response to Global 2000.* London: Basil Blackwell.

Smith, F.D.M., R.M. May, R. Pellew, T.H. Johnson, and K.R. Walter. 1993. Estimating extinction rates. *Nature* 364:494–96.

Soulé, M.E., ed. 1987. *Viable Populations for Conservation*. Cambridge: Cambridge University Press.

United Nations. 1992. *Long-Range World Populations: Two Centuries of Population Growth, 1959–2150*. New York: United Nations.

Vitousek, P.M., P.R. Ehrlich, A.H. Ehrlich, and P.A. Matson. 1986. Human appropriation of the products of photosynthesis. *BioScience* 36:368–73.

WCMC (World Conservation Monitoring Centre). 1992. *Global Biodiversity: Status of the Earth's Living Resources*. London: Chapman and Hall.

Wilson, E.O. 1985. The biological diversity crisis. *BioScience* 35:700–706.

———, ed. 1988. The current state of biological diversity. In *Biodiversity*. Washington, D.C.: National Academy Press.

Woodwell, G.M., ed. 1990. *The Earth in Transition: Patterns and Processes of Biotic Impoverishment*. Cambridge: Cambridge University Press.

World Bank. 1992. *World Development Report, 1992*. New York: Oxford University Press.

WRI (World Resources Institute), IUCN (World Conservation Union), and UNEP (United Nations Environment Programme). 1992. *Global Biodiversity Strategy*. Washington, D.C.: World Resources Institute, World Conservation Union, and United Nations Environment Programme.

# Biodiversity and Public Policy:
# The Middle of the Road

by Ariel E. Lugo

If we want to ensure sustainability of development and a high quality of human life on earth, our policies must be driven by the need to conserve all biodiversity. This will enable us to maximize future options for development and increase our capacity to deal with the future. Tropical forests are especially important in view of their great species richness. Recent studies have shown that tropical forests are highly resilient and can absorb disturbances that change their structure and composition while maintaining their function and capacity to recover through succession. Thus, human activity is not incompatible with the maintenance of biodiversity. A significant number of species and ecosystems are likely to survive in any case, including those in areas that are not easily accessible to humans, those occupying extreme environmental conditions, and those closely associated with conditions selected by humans. It is clear, however, that the kind and intensity of future human activities will affect global biodiversity, as humans continue to adapt to environmental change; technology will never be a complete substitute for the free goods and services that people derive from biodiversity. Yet people are selecting, sometimes inadvertently, ecosystems that have lower biomass, smaller organisms, faster rates of turnover, and higher net primary productivity; exotic species can sometimes predominate in these new environments. Strategies for coping with the future include opportunism, coupling of science and management, approaching management at the ecosystem level, emulating natural ecosystem managers, managing succession, rehabilitating land, enabling ecosystems to design themselves, multiple seeding, and monitoring and feedback.

*

The public must decide whether or not public policy should be based on predictions of species extinction rates. One of the possible responses to this question would be to avoid basing public policy on predicted losses of species, since such predictions are exaggerated and a focus on biodiversity underestimates the ability of humans to solve resource development problems with technology. To the contrary, another response would be to base public policy on predictions of rates, because they are either correct or underestimated and the unprecedented extinction event will reduce the planet's capacity to sustain human activity. The fundamental issue underpinning the question facing the public is that of the role of accurate scientific information as a driver of public policy. If science shows that species will go extinct at faster and faster rates as a result of human activity, the sense that some fundamental life on earth is dramatically changing presumably provides urgency for its conservation. For some sectors of society, however, science's inability to provide accurate information about extinction rates removes the urgency to conserve species and ecosystems. There is a balance between these two contrasting views.

Although the biodiversity debate usually focuses on species extinction, it must be recognized that the term *biodiversity* means more than the variety of species. Biodiversity encompasses the whole expression of life on earth, ranging from genes to life zones *sensu* (Holdridge 1967, 206). Defining it has been the subject of much discussion (see, for example, Wilson 1992), and it is not my objective to define it further. The point is that biodiversity is a primary asset of this planet, equal in importance to, and integrated with, the flows of water, oxygen, carbon, essential nutrients, or solar energy. Survival of any life-form, including humans, in the absence of even one of these planetary assets is impossible. All these materials as well as the potential energy from the sun are kept in circulation by the world's biota, and together with the biota, form the complex interaction that we call "life."

The need to conserve and value the full spectrum of biodiversity transcends the ability of scientists to demonstrate with scientific certainty its status, rate of change, ecological function, or value. Scientific knowledge about issues of biodiversity is, of course, inadequate. All we can do is document the state of ignorance, while suggesting steps to improve knowledge and understanding. Scientists have been studying, classifying, and cataloging organisms at the species level for over 200 years. Yet the work is only about 20 percent complete (NRC 1980). Studies of biodiversity at the level of genetic material is just about 100 years old, and at the larger and more complex scales of ecosystems and life zones, scientific study started less than 50 years ago.

Lack of information and adequate knowledge is at the root of why estimates of species extinction rates are not sound. We don't have sufficient knowledge to resolve the models used to estimate species loss, and the models that we do have underestimate the resiliency of natural ecosystems (Lugo 1988a, 58–70). For example, there are many limitations to the species-area model as a tool for estimating the rate of extinction of species in the tropics. Among these are the assumptions that deforestation results in a complete loss of species, that ecosystems have little regenerative capacity, and that human intervention to rehabilitate and restore species to damaged sites is impossible.

As better ecosystem models are developed and sufficient data are gathered to resolve the models, species loss estimates are likely to be lower than those normally given today (Whitmore and Sayer 1992a, 1–14; Reid 1992, 55–73; Simberloff 1992,

75−89; Heywood and Stuart 1992, 91−117; Brown and Brown 1992, 119−42). Nevertheless, it will be a while before science can predict with accuracy the rate of species extinction attributable to human activities (Johns 1992, 15−53).

Some of the obstacles facing scientists trying to achieve this task include (cf. Whitmore and Sayer 1992b): an incomplete inventory of species; a poor understanding of critical taxonomic groups, such as invertebrates and microbes; the lack of information on natural rates of extinction and on how species numbers change as a function of natural or human disturbances; and an inadequate understanding of the relationship between land use change and species abundance. In short, ecologists have not developed a body of quantitative science to predict changes in species composition or explain the way in which species composition regulates ecosystem function. The study of biodiversity as an ecosystem parameter is just now developing (Schulze and Mooney 1993). Until this field has time to mature, it will be impossible to make predictions of any kind.

## The Resilient Tropical Forest

The issue of whether tropical forests are fragile or resilient is of fundamental importance for understanding the current debate about their future (Lugo 1995a, 3−17). From the 1960s and up to the 1980s, the prevailing view was that tropical forests were fragile since they were thought to have been undisturbed by the glacials that shaped the temperate and boreal forests of the world. Because tropical forests were believed to have developed under stable conditions, they were not expected to adapt to the changing environments caused by human activities.

Recent studies, however, clearly demonstrate that the tropical zone has been quite dynamic in the past (Bush et al. 1992; Bush and Colinvaux 1994) and that tropical forests evolved under changing conditions. In fact, many tropical tree species require disturbances for their regeneration (Denslow 1987). Moreover, tropical forests believed to be undisturbed have, in reality, been shaped by human activity for centuries (Lamb 1966, 220; Gomez Pompa, Flores, and Sosa 1987; Bush and Colinvaux 1994). Forest responses to disturbances are currently being studied, and results suggest adaptations to periodic catastrophes (Lugo and Waide 1993). Thus, tropical forests are now viewed as resilient ecosystems. A resilient ecosystem is one that can absorb disturbances that change its structure and composition, but can still maintain its function and recover through succession.

This analysis leads to the conclusion that human activity is not incompatible with the maintenance of biodiversity. Indeed, under proper management and care, human activity can coexist with a significant, but at this moment unknown, fraction of the world's biodiversity. By understanding the conditions that lead to the survival of this fraction, we may learn to minimize the extinction of the world's most vulnerable ecosystems.

## Surviving Global Change

At least three circumstances favor the survival of biodiversity under global change:

1. *Impeded Accessibility to Humans.* In many countries of the world, rugged topography or remoteness act as barriers to human intrusion, helping to protect biodiversity. In the Caribbean, for example, Lugo, Schmidt, and Brown (1981) found a direct relationship between the maximum elevation of mountains and the fraction of forests surviving on islands with high human population densities (about 1,000 people per square mile). Most of the biodiversity of these islands occur on mountain forests.

2. *Extreme Environmental Conditions.* Locations with extreme temperatures or precipitation, or that are hostile to human activity, are refugia for biodiversity. Examples are the deep ocean, alpine areas, hot springs, rock outcrops, deserts, wet and rain forests (with rainfall from four to and over eight meters per year), or anaerobic environments. These extreme habitats either are not used by people or support a low intensity of human activity. As a result, they conserve their natural biodiversity.

3. *Association to Conditions Selected by Humans.* This refers to the biodiversity favored by people, including natural reserves and ecosystems that are managed, or indirectly sustained, by human activity. As humans increase their influence over the planet, they will have a greater effect on the type of biota that survives. Today, the focus on the effects of human activity on biodiversity is its simplification of complex ecosystems. But humans also preserve complex ecosystems, such as protected natural areas, and create new environmental conditions that select for new combinations of species that form elaborate ecosystems. The complex Central American forests that resulted from the activities of the Mayans are an example (Lamb 1966, 220; Gomez Pompa, Flores, and Sosa 1987; Bush and Colinvaux 1994). Human-dominated landscapes can also be quite diverse and intricate (Burgess and Sharpe 1981).

The number of species and ecosystems that fulfill these three criteria is significant if one considers their combined geographic extension. Based on their analysis, Pimentel et al. (1992) found that a large fraction of the world's biodiversity resides in managed landscapes.

## Biodiversity, Human Emotion, and Population Size

Human activity cannot be divorced from biodiversity. We must obtain products and services from biotic systems, we compete with these systems for space and resources such as water, and our waste products can stress biotic systems. Therefore, the kind and intensity of future human activity will affect global biodiversity. While some people believe that a reduction in human activity will be required to minimize its effects on biodiversity, the direction of change in human activity—expansion or contraction—does not make a difference with regard to our relationship to biodiversity. We will impact biodiversity whether the fossil fuel energy base that supports our activities expands or contracts.

A declining fossil fuel energy base does make people more directly dependent on biodiversity for food, fiber, energy, and other products that in an energy-rich situation would be produced artificially. Therefore, should fossil fuel energy consumption decrease rapidly, the initial impact of an inflated human population on natural resources may be quite considerable before the population reaches a new steady state adjustment with available natural resources. The alternative trend, an increasing fossil fuel energy base, will also impact biodiversity because a larger population and more human activity can be sustained. Increases in population and human activity demands a greater amount of resources, and thus more pressure is placed on the world's biodiversity to keep up.

Yet no matter how clever we are, technology will not be able to completely substitute the free goods and services that we derive from biodiversity. Any technological improvement or increase in human activity will have to be matched with a product or service from the biodiversity sector. For example, we would need more freshwater and more space to accommodate waste. That means that the success of our future technological development depends, in part, on how successfully we manage the world's biodiversity. Therefore, our views on biodiversity will have to change soon, particularly as we enter the next century and face new circumstances in our struggle for survival.

A future of continuing environmental change is the prognosis for our civilization. This is inevitable since we have a fixed amount of space, a fixed natural resource base, and expanding energy consumption. An expanding consumption of energy means an expanding human population, which increases pressure on those resources that are fixed. Controlling the human population without controlling the expansion of energy and resource use is not a solution to our future problems: Strict control of human numbers has proven impossible without dramatically changing the socioeconomic conditions of people, and as long as consumption of energy and resources expands, pressure on natural resources will also increase. Further, the number of people may be irrelevant because not all humans consume resources at the same rate. Many see the increase in population in terms of the need to clothe and feed more people. But this is only one aspect of the problem. Another is the actual rate and magnitude of resource consumption, since the relationship between people and the environment varies in different social and economic contexts.

Faced with the issue of saving or losing species, our society has reacted with emotion. The range of responses vary widely from those that advocate maintaining the landscape in its natural state even if there is no room for people, to those who advocate using natural resources without allowing room for preservation. Some propose saving endangered, threatened, endemic, or keystone species; others prefer saving particular types of ecosystems. Still other people favor preserving useful species while declaring war on exotics, parasites, weeds, and the like. Given a choice, some people will favor preserving a job for a single person over saving the last population of any species, a native species over an exotic species, a mammal over a reptile, a bird over a tree, or an animal over an algae. Such is the world of emotion, and labeling of species and groups of species, that rules our views and management of biodiversity. These emotional responses can be powerful forces for advancing particular points of view, but they can also cloud the debate and obstruct the consideration of nonconventional approaches to the conservation of biodiversity.

## Key Questions

As we know from the past, the world's biota has not been constant in composition or abundance (Donovan 1989). Change in living systems and the level of biodiversity is as prevalent a phenomena as is change in the abiotic environment (Velikovsky 1955). It is naive, even irresponsible, to pretend that environmental change can be stopped or prevented, or to say that what is happening today is good or bad for $X$ or $Y$ reasons, without also suggesting realistic solutions. A more useful set of questions to help us deal with the future may be the following: How can we best manage the environmental changes that we are causing as well as those changes being imposed on us by natural phenomena? What kinds of ecosystems are we selecting for? What kind of biodiversity will survive the changes that are taking place? What tools do we have for managing biodiversity? What ecosystem attributes offer the best hope for successful biodiversity management? Is it possible to design ecosystems for specific purposes? Under which conditions should biodiversity be maximized? What actions are most effective for protecting biodiversity? How should society prepare itself for an unprecedented period of environmental change?

## Future Trends

The emerging trends for the next century involve greater use of fossil fuels and a growth in cities, where human activity is concentrated (Lugo 1991). These trends are already in motion. While less land is being used to supply the needs of people, more concentrated pollution is being generated and, consequently, more land is being damaged (Grainger 1988). We are converting primary old growth into secondary forests (Brown and Lugo 1990). The landscape is being fragmented, causing a de-coupling of ecosystems and damaging lands and ecosystems (Cutler 1991; Saunders, Hobbs, and Margules 1991). We are selecting for ecosystems that have lower bio-mass, smaller organisms, faster rates of turnover, and higher net primary produc-tivity. Exotic species can be common and, at times, predominate in these new envi-ronments (Burgess and Sharpe 1981; Crosby 1986; Lugo 1994, 218–20). With the spread of human activity and, therefore, exotic species, there is a tendency for fewer global weeds to spread about, thus homogenizing the world's genome. This is also an enrichment process, however, because local biodiversity is itself diversified, cre-ating new combinations of species, communities, and ecosystems. It will be inter-esting to observe the evolution of these new species combinations and the emer-gence of taxa, communities, and ecosystems adapted to human disturbances.

The potential extinction of endemic species unable to survive new environmen-tal conditions is an unmitigated consequence of all human activity. The conserva-tion of these species offers the most difficult challenge facing conservation biolo-gists, governments, and land managers. Sometimes, these species survive as exotics elsewhere; *Delonix regia*, for instance, which is endangered in its native Madagas-car, grows as a weed in Puerto Rico (Little and Wadsworth 1964, 249). Taking ad-vantage of such a situation may become an important conservation tool in the fu-ture (Conant et al. 1992).

These trends are already being realized today on tropical islands. The tropical is-lands of the Pacific and the Caribbean present an environmental situation that is likely to become more prevalent on the world's continents (Lugo 1994, 218–20; Lugo and Brown 1996, 280–95). These islands, with high population densities and ex-tremely fragmented landscapes, have experienced significant extinction events, and their landscapes are now enriched by the addition of exotic plants and animals that form new combinations of communities and ecosystems (Lugo and Brown 1996, 280–95). This island experience highlights tropical forest resiliency and suggests strategies for coping with the future in continental areas (Lugo 1995a, 1995b).

## Strategies for Coping

The following nine activities are emerging as sound strategies for managing bio-diversity in an era of continuous environmental change, globalization of human impacts on biodiversity, and increasing constraints on the availability of resources to managing agencies.

*Opportunism.* Our strategy for the future has to be one of opportunism: recognizing when to intervene and doing so effectively, and recognizing when not to do any-thing, to avoid wasting time, energy, and resources (Holling 1973; 1986, 292–317; 1993).

*Coupling of Science and Management.* More than ever, scientific research that is tightly integrated to resource management activities is critical to our survival and quality of life because we will have to resolve situations that are new to our experience. The management of the landscape has to expand in scope to include all lands (NRC 1993) and has to be done in an ecosystem context.

*Ecosystem Management.* We need to use holistic analysis to manage lands and water for products, services, and conservation of biodiversity. The success of ecosystem management depends on the type of analysis and the level of people involvement that occurs *before* execution of management actions. The influence of science in ecosystem management is through the development of paradigms that guide our thinking, planning, and execution of actions in the field. Four main areas with rapidly evolving paradigms of thought are driving ecosystem management: the perception of a quickly changing world; the notion of spatial and temporal hierarchies; the resiliency of ecosystems; and the human dimension of management, including people as agents of environmental change.

*Emulating Natural Ecosystem Managers.* Ecosystem management is necessary for our future success on this planet, but it can be expensive because it involves the whole landscape. To remedy this problem, human ecosystem managers must emulate natural ecosystem managers—such as beavers, alligators, elephants, or termites—who regulate complex landscapes with small energy expenditures. Basically, these organisms nudge the system in subtle ways, allowing the natural energies and functions of the ecosystem to perform most of the work, the results of which eventually favor the ecosystem manager.

*Managing Succession.* What we learn from observing natural ecosystem managers is that the first rule of ecosystem management is to allow natural processes to perform as much of the work required to accomplish our goals as possible with the least interference by the manager. This minimizes cost as well as risk since natural ecosystems are adapted to the variations of the environment. Natural succession, native species, and natural ecosystems are powerful tools to be used for accomplishing our landscape management goals. Their use must be mastered and always considered as our first option when intervening with a landscape.

*Land Rehabilitation.* Because land and water have a limited area, they must be protected from degradation, and when this happens, they must be rehabilitated. Rehabilitation of ecosystems will be a prevailing activity in the future, particularly in the proximity of urban areas, where damaged lands are so prevalent (Brown and Lugo 1994; Lugo 1988b, 1991; Grainger 1988). Public policy should support these resource-use strategies and adapt to changing circumstances.

There are times and conditions when the first option for ecosystem management (the use of native systems and natural succession) is either not available or not sufficient to accomplish our management goals. For example, sometimes natural ecosystems have low net productivity and, thus, have limitations in situations where maximizing a net yield is necessary. Also, when lands or waters are excessively damaged due to careless human activity, natural successions and native systems, when left alone, may not be effective for rehabilitation because native species grow slowly and succession is arrested. Under these conditions, it may be necessary to import genetic material from other geographic areas to accelerate the land or water healing process (Lugo 1992a, 247–55; Lugo, Parrotta, and Brown 1993; Parrotta 1992; Parotta 1993, 63–73). This involves the use of exotic or native species under carefully controlled management.

*Ecosystem Self-Design.* An ecosystem property that is relevant in this context is self-design, or self-organization, which in combination with multiple seeding, provides an approach to landscape-scale management of biodiversity. The following quote from Beyers and Odum (1993, 7) describes the two processes and their synergy (substitute *ecosystem managers* for *investigators*, and *landscape* for *microcosms*):

> Some of the most successful investigators have emulated nature by fostering waves of immigration into microcosms. This is usually termed multiple seeding. Organisms from similar, but geographically different sources are placed in the microcosm, and the system is allowed to establish its own web of interactions. This principle is called self organization [self-design]. Self organization allows the developing microecosystem to establish feedback pathways which reinforce those processes that contribute most, according to the maximum power principle.

Self-organization, or self-design, is a natural phenomena that can be observed today with increasing frequency throughout any landscape. It refers to the emergence of new combinations of plants and animals after natural communities have been disrupted or damaged by human activity. These new ecosystems are generally not valued by ecologists or ardent conservationists, but they are becoming dominant in the landscape, perform valuable ecological functions and services, maintain a significant fraction of the biodiversity of the landscape, are the ecosystems of the future, and are a necessary stage in the recovery of more mature native ecosystems (Brown and Lugo 1990; Lugo 1988b). It behooves us to study, value, and learn to manage, use, and conserve these types of ecosystems. These new ecosystems can be termed interface ecosystems because they occur at the interface between human activity and undisturbed landscapes.

*Multiple Seeding.* As a way to achieve or accelerate self-design of ecosystems, multiple seeding is different from establishing plantations in that seeds or propagules of many species (as opposed to a single one) are broadcasted and allowed to compete and regenerate on a site. Multiple seeding and self-design become highly relevant under two future scenarios.

First, given different sets of unique environmental conditions, we may have to design specific types of ecosystems to accomplish specific environmental goals that cannot be achieved through natural succession. For example, we can design microbial systems to absorb oil spills, wetlands to recycle sewage, forests to rehabilitate lost soil fertility, or forested wetlands to absorb floods. These are instances of humans managing biodiversity to solve problems and serve people. A historic example of this is the Mayan food forests, whose human-dominated species composition confused botanists for centuries (Gomez Pompa, Flores, and Sosa 1987).

Second, it is necessary to be able to anticipate and mitigate global climate change (Lugo 1992b, 336–44). A universal worry among ecologists and conservationists is the possible inability of the dispersal mechanism of species to adjust to rapidly changing climates. This assumes that the biota follows an orderly movement (north or south) to establish a new set of life zones analogous to the previous ones, yet displaced geographically along some gradient. This is unrealistic. More likely, the new climate will create a new mix of climatic and edaphic conditions requiring a complex response with species moving in all directions at the same time, or not moving at all. Self-design will prevail as new mixes of species occur. Moreover, people can use multiple seeding techniques to move genomes (species) to specific locations in advance of the expected change, and thus accelerate the response or take advantage of the process of climate change.

*Monitoring and Feedback.* Successful policymaking requires continuous feedback from field-level resource management activities. This is accomplished by monitoring ecosystem structures and processes so that the results of previous management actions can be compared with the expectations of the plans that led to the actions. Results from monitoring programs must be made available to planners, managers, policymakers, and scientists so that they can adjust plans, management actions, policies, and research programs (respectively). A loop is created between field actions, measurements for monitoring, checking against expectations, adjusting future actions, and conducting modified field actions. This loop is called adaptive management because each reiteration of activity is based on past experience.

## Conclusion

The genetic codes embedded in the world's biodiversity are a buffer for or a stabilizer of a world undergoing change. Within this vast pool of genetic information lies the secret for anticipating and coping with environmental change and uncertainty. The biota is our only true renewable resource, a weapon to combat, cope with, and anticipate environmental change. Wise use of this resource can be achieved through natural succession and, where required, multiple seeding followed by self-design. Conservation of biodiversity is thus at the core of our future survival as a civilization. By conserving biodiversity, we maximize future options for development and increase our capacity to deal with the future.

Predictions of species extinctions should not be used to drive public policy because they are too uncertain and may even be irrelevant to our future well-being. If we want sustainable development and a high quality of human life on earth, our policies must be driven by the need to conserve all biodiversity (from genes to life zones). What this means is that we need to manage all the waters and lands of our planet professionally and holistically, with a view to anticipating and adapting to environmental change. Flexible and enlightened use of biodiversity is the main card we have to play in the game of anticipating an uncertain future. The motto to guide public policy should be: manage and conserve the whole earth.

## Acknowledgments

This study was done in cooperation with the University of Puerto Rico. I would like to thank W. Silver, F.N. Scatena, J. Parrotta, A. Gonzalez, and the book editors for reviewing the manuscript.

## References

Beyers, R.J., and H.T. Odum. 1993. *Ecological Microcosms*. New York: Springer-Verlag.

Brown, K.S., and G.C. Brown. 1992. Habitat alteration and species loss in Brazilian forests. In *Tropical Deforestation and Species Extinction*, edited by T.C. Whitmore and J.A. Sayer. London: Chapman and Hall.

Brown, S., and A.E. Lugo. 1990. Tropical secondary forests. *Journal of Tropical Ecology* 6:1–32.

———. 1994. Rehabilitation of tropical lands: A key to sustaining development. *Restoration Ecology* 2:97–111.

Burgess, R.L., and D.M. Sharpe, eds. 1981. Forest Island Dynamics in Man-Dominated Landscapes. *Ecological Studies* 41. New York: Springer-Verlag.

Bush, M.B., and P.A. Colinvaux. 1994. Tropical forest disturbance: Paleoecological records from Darien, Panama. *Ecology* 75:1761–68.

Bush, M.B., D.R. Piperno, P.A. Colinvaux, P.E. DeOliveira, L.A. Krissek, M.C. Miller, and W.E. Rowe. 1992. A 14,300-year paleoecological profile of a lowland tropical lake in Panama. *Ecological Monographs* 62:251–76.

Conant, S., R.C. Fleicher, M.P. Morin, and C.L. Tarr. 1992. When endangered species are aliens: Some thoughts on the conservation of rare species. *Pacific Science* 46:401–2.

Crosby, A.W. 1986. *Ecological Imperialism: The Biological Expansion of Europe, 900–1900*. Cambridge: Cambridge University Press.

Cutler, A. 1991. Nested faunas and extinction in fragmented habitats. *Conservation Biology* 5:496–505.

Denslow, J.S. 1987. Tropical rainforest gaps and tree species diversity. *Annual Review of Ecology and Systematics* 18:431–51.

Donovan, S.K., ed. 1989. *Mass Extinctions: Process and Evidence*. New York: Columbia University Press.

Gomez Pompa, A., J.S. Flores, and V. Sosa. 1987. The "pet kot": A man-made tropical forest of the Maya. *Interciencia* 12:10–15.

Grainger, A. 1988. Estimating areas of degraded tropical lands requiring replenishment of forest cover. *International Tree Crops Journal* 5:31–61.

Heywood, V.H., and S.N. Stuart. 1992. Species extinctions in tropical forests. In *Tropical Deforestation and Species Extinction*, edited by T.C. Whitmore and J.A. Sayer. London: Chapman and Hall.

Holdridge, L.R. 1967. *Life Zone Ecology*. San Jose, Costa Rica: Tropical Science Center.

Holling, C.S. 1973. Resiliency and stability of ecological systems. 1973. *Annual Review of Ecology and Systematics* 4:1–23.

———. 1986. The resilience of terrestrial ecosystems: Local surprise and global change. In *Sustainable Development of the Biosphere*, edited by W.C. Clark and R.E. Munn. Cambridge: Cambridge University Press.

———. 1993. Cross scale morphology, geometry, and dynamics of ecosystems. *Ecological Monographs* 62:447–502.

Johns, A.D. 1992. Species conservation in managed tropical forests. In *Tropical Deforestation and Species Extinction*, edited by T.C. Whitmore and J.A. Sayer. London: Chapman and Hall.

Lamb, F.B. 1966. *Mahogany in Tropical America*. Ann Arbor: University of Michigan Press.

Little, E.L., and F.H. Wadsworth. 1964. Common trees of Puerto Rico and the Virgin Islands. In *Agriculture Handbook*. Washington, D.C.: U.S. Department of Agriculture.

Lugo, A.E. 1988a. Estimating reductions in the diversity of tropical forest species. In *Biodiversity*, edited by E.O. Wilson and F.M. Peter. Washington, D.C.: National Academy Press.

———. 1988b. The forest of the future: Ecosystem rehabilitation in the tropics. *Environment* 30:16–20, 41–45.

———. 1991. Cities in the sustainable development of tropical landscapes. *Nature and Resources* 27:27–35.

———. 1992a. Tree plantations for rehabilitating damaged forest lands in the tropics. In vol. 2, *Environmental Rehabilitation*, edited by M.K. Wali. Netherlands: SPB Academic Publishing.

———. 1992b. Managing tropical forests in a time of climate change. In *Forests in a Changing Climate*, edited by A. Qureshi. Washington, D.C.: Climate Institute.

———. 1994. Maintaining an open mind on exotic species. In *Principles of Conservation Biology*, edited by G.K. Meffe and C.R. Carroll. Sunderland, Mass.: Sinauer Associates.

———. 1995a. Tropical forests: Their future and our future. In *Tropical Forests: Management and Ecology*, edited by A.E. Lugo and C. Lowe. New York: Springer-Verlag.

———. 1995b. Management of Tropical Biodiversity. *Ecological Applications* 5:956–61.

Lugo, A.E., and S. Brown. 1996. Management of land and species richness in the tropics. In *Biodiversity in Managed Landscapes*, edited by R. Szaro and D. Johnston. New York: Oxford University Press.

Lugo, A.E., J. Parrotta, and S. Brown. 1993. Reduction of species due to tropical deforestation and their recovery through management. *Ambio* 22:106–9.

Lugo, A.E., R. Schmidt, and S. Brown. 1981. Tropical forests in the Caribbean. *Ambio* 10:318–24.

Lugo, A.E., and R.B. Waide. 1993. Catastrophic and background disturbance of tropical ecosystems at the Luquillo Experimental Forest. *Journal of Biosciences* 4:475–81.

NRC (National Research Council). 1980. *Research Priorities in Tropical Biology*. Washington, D.C.: National Academy Press.

———. 1993. *Sustainable Agriculture and the Environment in the Humid Tropics*. Washington, D.C.: National Academy Press.

Parrotta, J.A. 1992. The role of plantation forests in rehabilitating degraded tropical ecosystems. *Agriculture, Ecosystems, and the Environment* 41:115–33.

———. 1993. Secondary forest regeneration on degraded lands: The role of "foster ecosystems." In *Restoration of Tropical Forest Ecosystems*, edited by H. Lieth and M. Lohmann. Dortrecht: Kluwer Academic Publishers.

Pimentel, D., U. Stachow, D.A. Takacs, H.W. Brubaker, A.R. Dumas, J.J. Meaney, J.A.S. O'Neil, D.E. Onsi, and D.B. Corzilius. 1992. Conserving biological diversity in agricultural/forestry systems. *BioScience* 42:354–62.

Reid, W.V. 1992. How many species will there be? In *Tropical Deforestation and Species Extinction*, edited by T.C. Whitmore and J.A. Sayer. London: Chapman and Hall.

Saunders, D.A., R.J. Hobbs, and C.R. Margules. 1991. Biological consequences of ecosystem fragmentation: A review. *Conservation Biology* 5:18–32.

Schulze, E.D., and H.A. Mooney, eds. 1993. *Biodiversity and Ecosystem Function*. New York: Springer-Verlag.

Simberloff, D. 1992. Do species-area curves predict extinction in fragmented forest? In *Tropical Deforestation and Species Extinction*, edited by T.C. Whitmore and J.A. Sayer. London: Chapman and Hall.

Velikovsky, I. 1955. *Earth in Upheaval*. New York: Dell Publishing.

Whitmore, T.C., and J.A. Sayer, eds. 1992a. Deforestation and species extinction in tropical moist forests. In *Tropical Deforestation and Species Extinction*. London: Chapman and Hall.

———, eds. 1992b. *Tropical Deforestation and Species Extinction*. London: Chapman and Hall.

Wilson, E.O. 1992. *The Diversity of Life*. Cambridge, Mass.: Belknap Press of Harvard University Press.

# II Scientific Responses

# Biotechnology Can Help Reduce the Loss of Biodiversity

by Robert B. Horsch and Robert T. Fraley

Biotechnology offers new opportunities for significantly increasing the productivity of agriculture, reducing the cost of food production and decreasing the environmental damage of agricultural practices. When coupled with education, health care, and economic development, we can expect that the resulting increase in standard of living and quality of life will contribute to continuing declines in birthrates and greater stewardship of the environment. The key contributions of biotechnology will be severalfold:

- Producing more food on the same area of land, thus reducing pressure to expand into wilderness, rain forest, or marginal lands
- Reducing postharvest loss of food while improving the quality of fresh and processed foods, thus boosting the "realized nutritional yield" per acre
- Displacing resource- and energy-intensive inputs, such as fuel, fertilizers, or pesticides, thus reducing unintended impacts on the environment and freeing those resources to be used for other purposes or conserved for the future
- Encouraging reduction of environmentally damaging agricultural practices and adoption of sustainable ones such as conservation tillage and integrated pest management

These genetic improvements will also contribute to sustainable economic growth and development worldwide because they provide real gains in productivity for growers and processors and more value for consumers. Information-intensive industries, such as biotechnology, can continue to grow indefinitely and substitute for materials and energy-intensive industries that are limited by the earth's capacity to provide resources and tolerate wastes.

\*

The loss of habitat to nonsustainable and/or low productivity agriculture is the single largest threat to our planet's rich sources of biodiversity. The pressure to clear forests, erodible hillsides, or marginal lands for growing food or grazing animals is driven by the ever-increasing number of people on earth, over 90 million new people each year. Compounding this is the need to open new lands for cultivation as existing farmlands are degraded by nonsustainable practices. Pollution, overharvesting, and competition from nonnative species also contribute to the loss of biodiversity (Wilson 1992). Most of the sources of biodiversity on earth are concentrated in tropical countries with relatively low standards of living, high birthrates, and rapid destruction of natural habitat (Mittermeier and Bowles 1993). To adequately feed the world's growing population over the next 40 to 50 years, significant breakthroughs in agricultural productivity in developed and developing countries alike will be required (Bongaarts 1994a; Doeoes 1994). Even greater improvement will be needed to ensure an adequate diet for all people without plowing up the rest of our planet.

Many countries have set aside large blocks of wilderness as parks or preserves, and have moved toward more sustainable harvesting of natural resources from public lands (WCMC 1993; McNeely 1994). In the United States, over 50 million acres of farmland have been idled, much of which has been placed under conservation programs. Many countries have also begun to regulate pollution of the environment from industrial and other human activities. Economic hardships, however, resulting from restrictions on land use or practices that negatively impact the environment can be a strong deterrent to enacting and enforcing these protective measures. Technologies that increase net productivity or reduce pollution without loss of productivity will be an important part of developing and maintaining the social will and the financial ability to protect natural habitat and the organisms living there (CAST 1994).

During the second half of this century, the birthrate in many countries has declined as the standard of living has increased. This correlation offers some hope that the human population can be stabilized at a sustainable level with a good quality of life for everyone. To do so will require multifaceted programs and significant international cooperation on issues such as education, trade, medical care, and agricultural technology (Bongaarts 1994b). One critical component will be continued investment in new, more efficient, and more sustainable technologies for food production to both enable a decent quality of life and provide incentives for families to choose to have fewer children. For example, children have been a valuable source of labor in rural, agriculturally based societies. Agricultural technologies that improve productivity and reduce backbreaking, labor-intensive practices should help to encourage families to have fewer children, and give them the means to make a bigger investment in those children's future. Indeed, agricultural biotechnology may be seen as one of several essential factors to help stabilize the number of people on earth in a humane and voluntary way.

TABLE 1  Commercial Plantings of Transgenic Crops from Monsanto Company (1,000 acres)

| Product | U.S.A. | Canada | Mexico | Argentina | Australia |
|---|---|---|---|---|---|
| Roundup Ready Soybeans | 9,000 | 3 | – | 3,500 | – |
| Bollgard/Ingard Cotton (B.t.) | 2,400 | – | 40 | – | 150 |
| Roundup Ready Cotton | 600 | – | 10 | – | – |
| YieldGard Corn (B.t.) | 3,000 | 10 | – | – | – |
| Roundup Ready Canola | – | 500 | – | – | – |
| NewLeaf Potatoes (B.t.) | 25 | 5 | – | – | – |
| Laurate Canola (oil modification) | 70 | – | – | – | – |
| BXN Cotton (herbicide resistant) | 250 | – | – | – | – |
| Approximate Totals | 15,345 | 518 | 50 | 3,500 | 150 |

## Genetic Productivity

Genetic improvements, developed through breeding or biotechnology, provide great leverage for increasing the agricultural wealth of humankind. These improvements will change the design and efficiency of agricultural production systems, rather than just the magnitude or energy intensity. The key design changes are the substitution of information for materials, energy, and associated wastes, and a focus on increasing services provided without increasing material throughput.

For example, the application of biotechnology for incorporating new insect-resistant genes into crop plants will eliminate some of the need to convert petroleum resources and energy into pesticides, packaging, and distribution of the chemical product, as well as disposal of the wastes generated in the process (see table 1). In addition, the farmer will not need to drive a tractor or fly an airplane over the field multiple times during the growing season to apply pesticides. Since only the insects that eat the improved crops will be affected, beneficial insects that would have been killed by a broad spectrum insecticide will survive. In some cases, these beneficial insects may provide control of other insect pests, resulting in a double benefit. It is likely that the genetic improvements will provide better-than-average pest control and reach more acres than current products, thus increasing the overall yield per acre of the crop as well. All of these benefits are possible without ongoing capital investments or resource consumption for manufacturing, since they will come built into the seed without significant additional costs of production (see figure 1).

FIGURE 1    Insect- and Virus-Protection Options for Growing Potatoes

Monsanto estimate for control of  Colorado potato beetle and leaf roll virus

NewLeaf Plus potatoes, currently in Monsanto's pipeline, will add a virus-protection trait to NewLeaf insect-protected potatoes. These potatoes, which are expected to be commercialized within the next few years, will let growers reduce the amount of resources necessary to produce Russet Burbank potatoes in the United States.

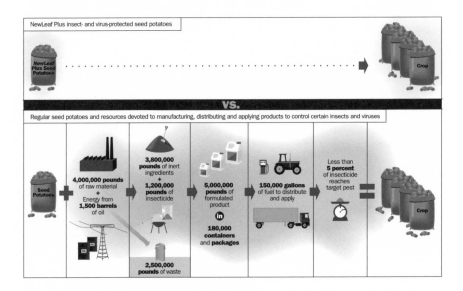

## Insect Control

One example of a gene for insect resistance is the protein gene from *Bacillus thuringiensis*. This microbe produces a special protein that is active against a narrow spectrum of insects yet nontoxic to other organisms (MacIntosh et al. 1990). Different strains of *Bacillus* produce different proteins that have different insect-control activities. The most commonly used *Bacillus* strain produces a protein with activity against many caterpillars and has been available commercially for over 30 years as a dry powder of bacterial spores. When dusted on the surface of plant leaves, caterpillars that eat a sufficient quantity will be killed or controlled. Unfortunately, the utility of this product is severely constrained by both its rapid degradation and because it cannot control caterpillars that burrow into the plant rather than eat the leaves where the powder can be applied. By engineering the gene that specifies the insect-control protein so that it expresses within the cells of a crop, the usefulness of this natural pesticide can be greatly expanded by providing continuous protection to the entire plant (Fischhoff et al. 1987).

In one example, insect-protected cotton has been developed using the insect-control protein gene that kills caterpillars, such as the cotton bollworm and bud-worm (Perlak et al. 1990). These voracious pests infest as much as 80 percent of all cotton acres. In the United States, almost $180 millon is spent annually on several million pounds of insecticides to control them. Even with the best chemical control practices available today, losses caused by these insects exceed $250 million a year. Worldwide insecticide use and loss of cotton harvest are four to five times greater than the U.S. estimates. Improved cotton varieties that protect themselves against these serious pests were planted on about 1.8 million acres across the cotton-growing regions of the United States in 1996 (Perlak and Fischhoff 1993, 199–211). We estimate that plantings of Bollgard cotton eliminated the use of about 25,000 gallons of insecticide in the United States during 1996, and resulted in an average yield increase of 56 pounds of lint per acre—all as a result of reduced pest damage attributable to increased pest resistance.

A different strain of *Bacillus* produces a protein that controls many beetles. The gene for this beetle-control protein has been engineered to protect potatoes against the Colorado potato beetle. Almost a million pounds of insecticides, costing close to $100 million annually, are currently used each year in the United States to protect potatoes from this pest. Insect-protected potatoes are presently in extensive field-tests across the United States in each of the major potato-growing regions, and show excellent control of the pest (Perlak et al. 1993). These improved potatoes promise to reduce the need for insecticides and should provide comparable or better control of the Colorado potato beetle than the best insecticides available today. Just like the protein that protects cotton, this protein is narrow in its spectrum of activity yet nontoxic to other organisms, able to be harmlessly digested within seconds if consumed. It also biodegrades rapidly in the environment without posing a threat to earthworms, honey bees, or other forms of life (Fuchs, Berberich, and Serdy 1993, 393–407). These improved potatoes, marketed under the trade name NewLeaf, were commercialized in 1995.

These and other insect-control proteins will be useful in a number of other crops as well (Fraley 1992). Corn has already been improved to be resistant to the European corn borer, another caterpillar-type pest, and was commercialized in 1996. Tomatoes, apples, and other fruits and vegetables suffer from a variety of caterpillar pests that could be controlled with these insect-control genes.

### Virus Resistance

Another major use of insecticides is to control pests that, by themselves, would not cause significant crop damage, yet acting as vectors to transmit plant viruses, lead to severe agricultural losses. Since there are few means to control viruses directly, insecticides control the insects that spread the viruses. It has been possible to confer genetic resistance to viruses by incorporating a gene from the virus into crops

(Sanders et al. 1992). For example, viral genes that produce the protein coat that surrounds viral particles have been used to confer resistance to a variety of viruses. Plants that make a small amount of the viral-coat protein are resistant to infection by the virus.

This virus-resistance trait was developed from studies of a natural phenomenon known as viral cross protection. Although it has long been observed that a plant infected with one strain of a virus is resistant to infection by a second strain of the same virus, scientists recently discovered that the coat protein is able to confer the resistance by itself without causing any of the viral disease symptoms or loss of yield (Abel et al. 1986). Viruses cause major losses of crop yield around the world. Engineering virus resistance is one of the top priorities of many international agricultural research programs aimed at developing countries. A few examples include potatoes in Mexico that are resistant to potato virus X and potato virus Y (Kaniewski et al. 1990), squash in Costa Rica with resistance to both watermelon mosaic virus-2 and zucchini yellow mosaic virus (Altman and James 1993), and research toward developing virus-resistant sweet potatoes for Africa (Horsch 1993). These and other virus-resistant crops under development should both displace significant amounts of insecticides and boost yields of crops that formerly suffered from chronic viral infections.

## Conservation Tillage and Herbicide Tolerance

Plowing is primarily a method of weed control, but since the soil is disturbed along with the weeds, it is exposed to wind and water erosion on a significant scale. The mandate to adopt conservation tillage practices is mainly aimed at reducing the massive soil erosion that results from plowing. Loss of topsoil, which reduces the productivity and sustainability of crop production, is considered to be one of the most serious threats to the sustainability of the world food supply (Kendall and Pimentel 1994; NRC 1993). Loss of soil carbon, soil compaction, water-absorbing soil structure, and surface water, as well as the creation of damaging runoff and silted waterways, are also accelerated by conventional tillage methods. When crop residues are left on the land with their roots in place, erosion is greatly reduced. Herbicides provide a means to control weeds without massive displacement of soil, and mechanical drills allow seeds to be planted in unplowed ground. As no-till and other conservation tillage practices expand, a larger portion of organic matter from crop residues have been shown to accumulate in the soil. The combination of higher organic matter and undisturbed soil encourages activity of earthworms, while also creating a better soil structure with greatly improved water-holding capacity and aeration.

Conventional weed-control practices rely on a combination of conventional tillage and incorporated herbicides, resulting in a significant expenditure of fossil fuels and release of carbon to the atmosphere from the diesel fuel used by the tractor to pull the plow, as well as from soil-stored carbon released after plowing, not to mention the fuel consumed and carbon released during the manufacture and application of herbicides. Kern and Johnson (1993) estimate that current weed-control and planting practices in the United States alone will release over 200 million tons of carbon to the atmosphere by the year 2020 if they remain unchanged. However, if conservation-tillage and no-till practices are expanded from one-quarter to three-quarters of U.S. farmland by 2020, there will be a dramatic drop in the release of carbon dioxide and a corresponding rise in the storage of carbon in soil organic matter, resulting in a net sink of over 240 million tons of carbon in the soil of U.S. farms. Energy consumption will be reduced as well, since herbicide use with minimum-tillage operations consumes only about half the total energy of conventional plowing.

Herbicide-resistant crops will help to accelerate the adoption of no-till practices by providing better methods for postemergence weed control. This will further reduce the cost of weed control by shifting the spectrum of herbicides used to better and more cost-effective chemistry with lower use rates as well as lower toxicity and more benign environmental properties. Energy consumption will also decline both from reduced fuel usage and the shift to lower use rate chemistry. Farmers will gain the flexibility to use the best postemergence herbicides only where and when they are needed, and at the minimal dose rate that will control the weeds that are present. Finally, herbicide-resistant crops will solve some tough problems caused by weed control (Gressel 1992).

Monsanto, for instance, is pursuing tolerance to the herbicide Roundup (Shah et al. 1986). The active ingredient in Roundup, glyphosate, is an excellent example of a new generation of herbicides that are extremely effective, user friendly, and environmentally friendly. Glyphosate is an efficient inhibitor of a single enzyme in aromatic amino acid biosynthesis, a pathway that humans and many animals do not have, but one that is essential for plants. Once applied to a field, glyphosate binds to soil particles and doesn't leach into water. It biodegrades into simple elements in the soil with a half-life of only a few weeks.

While glyphosate is quite effective at controlling weeds, it also damages crops, so its use today is restricted to preplant applications. Soybeans, corn, cotton, and other crops that have been engineered to be tolerant to glyphosate will offer farmers a more flexible, more effective, and lower-cost postemergence weed-control option (Padgette et al. 1994).

Glyphosate-tolerant soybeans will be compatible with all current weed-control practices and make new practices possible. Farmers will choose how to best incorporate the new soybean varieties into their favored planting, tillage, and herbicide regimes in ways that conserve their soil and give them an economic return on their investment of time, land, and money. The tolerant crops do not require the use of the herbicide, but rather, allow the herbicide to be used if and when weeds must be controlled, before the crop canopy closes. This flexibility fits well with the trend toward more careful management of crop production factors to reduce costs, boost yields, and conserve the soil. The tolerant beans are also compatible with any crop rotation pattern, since the herbicide has no residual activity and biodegrades between growing seasons.

Roundup ready soybeans were grown commercially on about 1 million acres in the United States in 1996 and on about 9 million in 1997. Contrary to the claims of critics who warned of the dangers of increasing herbicide use in connection with herbicide tolerant crops, Monsanto surveys indicate decreases in total herbicide use ranging from 9 percent in the eastern United States to 39 percent in the southeast. Thus, increases in Roundup use were more than offset by decreases in more expensive and more persistent herbicides.

### Postharvest Losses

Estimates of postharvest losses of food vary widely, ranging from 10 percent to 40 percent or more, and are difficult to accurately measure (Kendall and Pimentel 1994). Using a moderate figure like 20 percent still highlights the tremendous net loss of agricultural productivity to spoilage, inefficient handling, inefficient processing, or destruction by insects or rodents. A significant portion of these losses may be reduced by genetic improvements in the crops themselves. The same insect-control proteins that protect the crop in the field may also protect stored grains from insect pests. Calgene commercialized their Flavr Savr tomatoes, which have a longer shelf-life, in 1994. Genes that slow down ripening, and hence the post-ripening senescence and rotting processes, can lead to a longer shelf life for fruits and vegetables (Klee et al. 1993). This technology should also reduce losses of important staple foods, such as starchy bananas and plantains, which suffer huge postharvest losses today. Finally, genetic improvements in processing efficiency may reduce losses and energy consumption in a variety of ways (Stark, Barry, and Kishore 1993).

### Bioprospecting and Valuing Diversity

One of the key resources for genetic improvement of plants is the diversity of organisms that can be used as sources of genes for new, beneficial traits. Plant breeders have long recognized the value of genes found in landraces or wild relatives of cultivated crops, seeking to collect, preserve, and utilize these genetic resources in breeding programs. They have gradually expanded the pool of genes available for breeding by both wider collections within a species or from sexually compatible species, or by devising new means to breed with related species that were otherwise incompatible. The ability to isolate, improve, and transfer genes from a greater number of different plants or from other organisms into crops essentially gives breeders an even broader gene pool from which to produce the next generation of genetically improved crop varieties. Our capacity to study, understand, and find the most useful new genes from nature is still very limited, but growing steadily each year. Even as we are just beginning to tap this huge and valuable resource, it is being destroyed by loss of natural habitat and environmental destruction. The terrible irony is that effective use of the biodiversity on earth is one of the single most powerful means to protect it, if we act before it is lost.

As the public comes to recognize that the loss of species through extinction is also a loss of a potentially useful and irreplaceable resource for the future, the motivation and political will to conserve biodiversity should increase. There are a number of well-known examples of bioprospecting for medicinal activities now under way. Looking for agriculturally useful activities is still at an early stage and not as well funded as pharmaceutical searches. The importance of bioprospecting for crop improvement should increase as successful products are developed and brought to market.

## Other Genetic Improvements

As the knowledge base in plant biology continues to be expanded through a variety of gene-tagging and genome-mapping techniques, we can expect to see biotechnology solutions for an assortment of other pest problems, including nematode and fungal disease control. The recent isolation of several key genes involved in the plant's natural defense response to fungal pathogens is particularly encouraging (Moffit 1994). Advances in understanding carbon and nitrogen pathways may provide the basis for direct enhancement of crop yield. Clearly, an important target for the future is the ability to modify production of secondary products to produce novel fatty acids, oils, and other biomaterials that have industrial applications. The recent demonstration of the production of polyhydroxybutyrate, a biodegradable plastic polymer, in plants highlights their potential in the production of specialty materials.

## Animal Productivity and Feed Efficiency

Stiffer competition, declining resources for government support programs, and growing pressure to minimize the environmental impacts of farming all demand that producers get the most out of their land and livestock in the safest, most efficient manner possible. Dairy farmers have made particularly impressive gains in enhancing the productivity of dairy cows over recent decades, due in large part to their adoption of a variety of efficiency-boosting advances such as artificial insemination, better genetics, and improved nutrition. The use of supplemental bovine somatotropin (BST) is another technology that will allow farmers to substantially improve the efficiency of their operations. BST competes in the open market as a cost-effective dairy management tool that is equally available and beneficial to large and small dairies alike. It is a continuation of a long legacy of efficiency enhancing technologies that have boosted supply, reduced cost, and helped to maintain the international competitiveness of U.S. agriculture.

Utilization of this management tool, which was approved for use by the U.S. Food and Drug Administration (FDA) in November 1993, allows dairy farmers the opportunity to maintain current milk production levels with fewer cows, thus reducing the resources associated with the maintenance of larger herds and improving overall profitability. Over the last four decades, milk production has increased while the total number of dairy cows has dropped from 21 to 11 million, made possible by significant increases in milk production per cow. From a more global perspective, animal waste runoff and methane production have significant environmental impact. The use of supplemental BST, which safely and consistently improves milk output in dairy cattle, will have an important place in meeting the challenges of providing sources of high quality protein to consumers while reducing impact on the environment.

## Public Acceptance

Public acceptance is one of the last major hurdles for the first wave of products of agricultural biotechnology. As with any new technology, and especially with food products, people are naturally cautious about change. Several groups or regulatory agencies have explored the scientific issues at great length over the past 10 years. The FDA published their analysis and recommended procedures several years ago (Kessler et al. 1992). Issues under consideration, arguments being formulated and debated, and requests for regulatory approvals are published in the *Federal Register* to allow public comment. Recent actions include approval of the Flavr Savr tomato developed by Calgene and its initial sale in the United States in the spring of 1994.

The scientific demonstration of food safety and review by government agencies will prove critical in gaining public acceptance for products of this science. Other factors will be crucial as well. Among these are the role that credible experts will play in communicating the issues and results of the tests to the public. University and government scientists will be seen as more credible than individuals who work for the companies that will profit from the new products or who work for advocacy groups that raise their money by opposing new technology.

Clear and understandable consumer information is also an important part of the acceptance process. This is, unfortunately, often confused with the issue of labels on food products themselves. Labeling is one of several ways to provide information to consumers, but it would effectively block many uses of the technology because the cost of separate storage, transport, processing, distribution, and marketing would be higher than the savings that would otherwise accrue. Labels have traditionally been used to educate consumers about nutritional content or ingredients known to be a problem for certain people. Since the products under development today do not change the nutritional content or wholesomeness of food, it should not be necessary to impose a label requirement. Rather, the information can be made available to consumers through brochures or other educational means that still protect the public's right to know.

There are several other potential barriers to the full utilization of this important and necessary new technology for the benefit of humankind. These include restrictions on international trade; inadequate incentives for investments in the research, development, and marketing of new products; unnecessary delays in launching products; and misinformed public debate resulting in the loss of public confidence.

One of the consequences of focusing attention and debate on unimportant issues is that other issues of much greater importance are forgotten or under-resourced. For example, biotechnology has been criticized as a potential threat to biodiversity through unexpected environmental impacts. The products of biotechnology, however, are much more predictable and controllable than existing species that are introduced into new habitats. Improved crops developed by biotechnology retain the same characteristics as the original crop with one or two specific modifications that can readily be observed for changes in adaptive, competitive properties before release to the environment. The potential for environmental risk from these small changes can be analyzed by comparison to the characteristics of known weeds or pests, as well as by the study of traits already bred into modern crop varieties. A wide range of disease- and pest-resistance traits have been introduced into almost all crops grown today without significantly changing the competitiveness of those crops in the wild or causing harmful gene flow to sexually compatible weeds.

In contrast, aggressive wild organisms invading new habitats is the second greatest factor causing extinctions today (Wilson 1992). Geographically limited species may become competitive weeds or pests when released in a new environment, rapidly displacing native species and sometimes leading to extinction or severe economic losses. Rather than focusing excessive attention on products of biotechnology that have small and well-characterized changes, more effort should be directed toward understanding and managing the very real and often unpredictable risks of transporting regionally limited species, which are often poorly characterized, across natural boundaries.

## Recent Regulatory Approvals

A growing number of transgenic crops have received final regulatory approvals for commercial sale. In the United States, approval is needed from the U.S. Department of Agriculture (USDA) before the crop can be grown. This has been given in the form of a "determination of nonregulated status" when there is adequate evidence that the crop does not have the potential to be a plant pest. The FDA has responsibility for ensuring the safety of food and feed. Foods derived from transgenic crops are covered under their "statement of policy for foods derived from new plant varieties," which also covers varieties developed by traditional plant breeding. This is done by consulting with the FDA concerning the safety of the food, resulting in their concurrence that it is substantially equivalent to existing food. Approval may be needed from the Environmental Protection Agency (EPA) if the crop contains a new trait for some form of pest control, such as insect resistance. This is given in the form of a registration of the pesticidal component. In addition, the active component, as well as any inert ones, must have an exemption from the requirement of a tolerance. Approvals to date are listed in table 2. In 1997, over 25 million acres of crops enhanced with biotechnology were planted in the United States, including over 19 million acres of crops with traits developed by the Monsanto Company (see table 2).

## Technology Transfer to Developing Countries

A number of public and private foundations and institutions have targeted biotechnology as a key technology for solving food production problems in developing countries, particularly for subsistence farmers. Some important features of biotechnology are the low cost of goods and ease of distribution and use, since the products are genetically built into the seeds, something all farmers know how to use. The Rockefeller Foundation has funded a major program in rice biotechnology. The World Bank, USAID, UNESCO, the French Scientific Institute for Development Through Cooperation (ORSTOM), and others have also launched large funding programs to build institutional capacity for biotechnologies around the world. They have recognized the natural applicability of this technology for use by resource-poor farmers. Persley (1990) provides a good discussion of the compatibility of biotechnology with international institutions, programs, and needs.

Modern technology for agriculture is highly productive and, hence, greatly desired by farmers worldwide. It is also rapidly evolving toward being sustainable, environmentally compatible, and more energy efficient. Biotechnology will support and enhance these trends toward both continuing to increase productivity and protect the environment. Genetic improvements are also usable without high-input agricultural practices that may not be affordable or wise for developing

TABLE 2    At Least 35 Plant Biotechnology Products
Are Approved in at Least One Country

| Company | Trait | Company | Trait |
|---|---|---|---|
| AgrEvo Canada, Inc. | Glufosinate-tolerant canola<br>Glufosinate-tolerant corn | Du Pont | Sulfonylurea-tolerant cotton |
|  | Glufosinate-tolerant soybean | Florigene | Carnations with increased vase life<br>Carnations with modified flower color |
| Agritope, Inc. | Modified-fruit-ripening tomato | Monsanto Co. | Glyphosate-tolerant soybean<br>Improved-ripening tomato<br>Insect-protected potato |
| Asgrow Seed Co. | Virus-resistant squash I | | Insect-protected cotton |
|  | Virus-resistant squash II | | Glyphosate-tolerant cotton<br>Glyphosate-tolerant canola |
| Bejo-Baden | Male sterility/glufos.-tol. chicory | | Insect-protected corn<br>Glyphosate-tolerant corn |
| Calgene, Inc. | Flavr Savr Tomato<br>Bromoxynil-tolerant cotton | Mycogen | Insect-protected corn |
|  | Laurate canola | Northrup King | Insect-protected corn |
| China | Virus-resistant tomato | Plant Genetic Systems | Male sterile oilseed rape |
|  | Virus-resistant tobacco | | Male sterility/glufos.-tol. corn |
| Ciba Seeds | Insect-protected corn | Seita | Bromoxynil-tolerant tobacco |
| Cornell U./U. of Hawai'i | Virus-resistant papaya | U. of Saskatchewan | Sulfonylurea-tolerant flax |
| DeKalb Genetics Corp. | Glufosinate-tolerant corn | Zeneca/Petoseed | Improved-ripening tomato |
|  | Insect-protected corn | | |
| DNA Plant Technology | Improved-ripening tomato | | |

countries. Thus, it can be adapted to all situations in all world areas. Biotechnology is not a fundamentally different type of solution to agricultural problems than has been available in the past—rather, it is a logical extension of plant breeding. It does, however, provide a new gene pool from which to breed. This broader gene pool can be used in a way that is responsive to today's issues—concerns about environmental protection, long-term sustainability, and changes in the economics of farming and food production that were formerly dependent on cheap energy and lots of land.

Several private companies are also involved in technology transfer projects, such as the three-way collaboration between Monsanto, USAID, and the Kenyan Agricultural Research Institute to develop virus-resistant sweet potatoes for Africa. In many developing countries, decreases in root and tuber crops due to viral infection can reduce potential production by 20 to 80 percent, depending on the country. The objectives of this project are to begin developing virus-resistant sweet potatoes and then transfer the technology to an African institution. Training of African scientists includes: basic training in plant cell and tissue culture as well as in plant molecular biology; research to improve regeneration technology for the sweet potato, with emphasis on African varieties; development of *Agrobacterium* transformation techniques for sweet potato tissues; development of vectors with genes for resistance to sweet potato feathery mottle virus (SPFMV); and production of transgenic sweet potatoes with SPFMV vectors. Associated training includes assays to monitor the genes and virus resistance, as well as knowledge of the regulatory science and process in the United States. African varieties of sweet potato with genes for SPFMV resistance are now available for field testing in Africa during 1998.

A recent program launched by USAID at Michigan State University has incorporated collaborations involving companies such as DNAP (Sondahl 1993) and ICI Seeds (Wilson 1992) with public and private institutions in Indonesia and Costa Rica. A new nonprofit foundation, International Service for Acquisition of Agribiotech Applications (ISAAA), has facilitated programs between Monsanto and Mexico to develop local potato varieties that are resistant to potato virus X and potato virus Y, and between Asgrow and Costa Rica to develop virus-resistant melons (Altman and James 1993).

The four-year partnership between Monsanto and Mexico recently was expanded to allow Mexico to share the results of their efforts with other developing countries, since potato viruses X and Y can lead to crop losses of up to 20 percent. Alpha, the original target variety, is the most widely grown potato strain in Mexico, but is not resistant to late blight, the disease that caused the Irish potato famine. Two years ago, the program was extended to the nortena and rosita varieties, which are resistant to late blight and are grown for local consumption in smaller farm operations.

The Cornell University–based ISAAA, which helps transfer biotechnology and its applications to developing countries, arranged the joint project. The work was sponsored by the Rockefeller Foundation, which also supported the development by classical genetics of late-blight varieties in Mexico beginning in the late 1940s. The project began with the training of Mexican scientists at Monsanto's Life Sciences Research Center near St. Louis, Mo., and then at CINVESTAV, Mexico's leading biotechnology research center. The Kenyan Agricultural Research Institute has expressed strong interest in receiving the improved potato varieties for in-country use.

The agreement to transfer the technology does not extend to commercial varieties such as the atlantic, which is used by processors in Mexico. In this way, the agreement makes the technology available to those who cannot afford to pay for it, separate and apart from any efforts to develop commercial markets for new and improved varieties of potatoes.

Technology transfer has occurred in a number of other ways, and will continue to do so. The primary transfer vehicle will be business collaborations and marketplace sales of improved seeds. Publications, patents, and educational programs are also important mechanisms for technology transfer. Through the actions of public, private, and business institutions, the products of biotechnology will become available to farmers around the world in the next decade.

### Conclusion

Protecting our environment, while stabilizing our population and adequately feeding the people who will share the earth in the next generation, is the largest challenge facing humankind today. How this challenge can best be met is the subject of much scientific concern and popular debate. The projected growth of world population into the next century has triggered complex and often conflicting interactions between those interests concerned with global food production and those involved with environmental stewardship. The lack of common understanding and consensus has not only precluded the development of appropriate policies for addressing the critical issues surrounding population growth and food production, but it has thwarted the effective mobilization of public and private sector resources that are needed to solve this problem. Whether the challenge is viewed as adequately feeding the next 8 to 10 billion people over the next 40 to 50 years or as reclaiming the earth's environment for a smaller population base in the distant future, significant breakthroughs in food production will be required to ensure sufficient protection of the rich diversity and quality of life on earth.

Biotechnology applications to increase crop and animal productivity in ways that will lower costs, improve quality and abundance, and provide for protection of the environment and maintenance of natural habitats are critically needed. In combination with decisive actions to control population growth and reduce pollution, investments in new agricultural technologies form the basis of an action plan for the future. Only by development of clear policy initiatives that bring public and private sector resources together can we begin to mobilize the massive effort that will be needed to increase food supply, reduce population growth, and protect our land, water, and biological resources.

## Acknowledgments

The authors thank Barbara Rhodes for her editorial comments and assistance in preparing the manuscript.

## References

Abel, P.P., R.S. Nelson, B. De, N. Hoffman, S.G. Rogers, R.T. Fraley, R.N. Beachy. 1986. Delay of disease development in transgenic plants that express the tobacco mosaic virus coat protein gene. *Science* 232:738–43.

Altman, D.W., and C. James. 1993. Public and private sector partnership through ISAAA for the transfer of plant biotechnology applications. Annals of the *New York Academy of Sciences* 700:261.

Bongaarts, J. 1994a. Population policy options in the developing world. *Science* 263:771–76.

———. 1994b. Can the growing human population feed itself? *Scientific American* 270:36–42.

CAST (Council for Agricultural Science and Technology). 1994. *How Much Land Can Ten Billion People Spare for Nature?* Task force report 121:1–64.

Doeoes, B.R. 1994. Environmental degradation, global food production, and risk for large-scale migrations. *Ambio* 23:124–30.

Fischhoff, D.A., K.S. Bowdish, and F.J. Perlak. 1987. Insect tolerant transgenic tomato plants. *BioTechnology* 5:807–13.

Fraley, R.T. 1992. Sustaining the food supply. *BioTechnology* 10:40–43.

Fuchs, R.L., S.A. Berberich, and F.S. Serdy. 1993. Regulatory considerations for pesticidal plants: Insect-resistant cotton as a case study. In *Advanced Engineered Pesticides*, edited by Leo Kim. New York: Marcel Dekker.

Gressel, J. 1992. Genetically-engineered herbicide-resistant crops: A moral imperative for world food production. *Agro-Food-Industry Hi-Tech*, 3–7.

Horsch, R.B. 1993. Commercialization of genetically engineered crops. *Philosophical Transactions of the Royal Society of London Biological Sciences* 342:287–91.

Kaniewski, W., C. Lawson, B. Sammons, and et al. 1990. Field resistance of transgenic russet burbank potato to effects of infection by potato virus X and potato virus Y. *BioTechnology* 8:750–54.

Kendall, H.W., and D. Pimentel. 1994. Constraints on the expansion of the global food supply. *Ambio* 23 (May): 198–205.

Kern, J.S., and M.G. Johnson. 1993. Conservation tillage impacts on national soil and atmospheric carbon levels. *Soil Science Society of America Journal* 57:200–210.

Kessler, D.A., M.R. Taylor, J.H. Maryanski, E.L. Flamm, and L.S. Kahl. 1992. The safety of foods developed by biotechnology. Science 256:1747–49, 1832.

Klee, H.J., M.B. Hayford, K.A. Kretzmer, G.F. Barry, and G.M. Kishore. 1993. Control of ethylene synthesis by expression of a bacterial enzyme in transgenic tomato plants. *Plant Cell* 3:1187–93.

MacIntosh, S.C., T.B. Stone, S.R. Sims, P.L. Hunst, J.T. Greenplate, P.G. Marrone, F.J. Perlak, D.A. Fischhoff, and R.L. Fuchs. 1990. Specificity and efficacy of purified *Bacillus thuringiensis proteins* against agronomically important insects. *Journal of Invertebrate Pathology* 56:258–66.

McNeely, J.A. 1994. Lessons from the past: Forests and biodiversity. *Biodiversity and Conservation* 3:3–20.

Mittermeier, R.A., and I.A. Bowles. 1993. The GEF and biodiversity conservation: Lessons to date and recommendations for future action. *International Biodiversity Policy Program* (May): 1–19.

Moffit, A.S. 1994. Mapping the sequence of disease resistance. *Science* 265:1804–5.

NRC (National Research Council). 1993. Soil and Water Quality—An Agenda for Agriculture. Washington, D.C.: National Academy Press.

Padgette, S.R., D.B. Re, and G.F. Barry. 1994. New weed control opportunities: Development of soybeans with a Roundup ready gene. In *Herbicide-Resistant Crops: Agricultural, Economic, Environmental, Regulatory, and Technological Aspects*, edited by S.O. Duke. Boca Raton, Fla.: CRC Press.

Perlak, F.J., R.W. Deaton, T.A. Armstrong, R.L. Fuchs, S.R. Sims, J.T. Greenplate, and J.T. Fischoff. 1990. Insect resistant cotton plants. *BioTechnology* 8:939–43.

Perlak, F.J., and D.A. Fischhoff. 1993. Insect-resistant cotton: From the laboratory to the marketplace. In *Advanced Engineered Pesticides*, edited by L. Kim. New York: Marcel Dekker.

Perlak, F.J., T.B. Stone, Y.M Muskopf, L.J. Peterson, G.B. Parker, S.A. McPherson, J.Wyman, S. Love, G. Reed, and D. Biever. 1993. Genetically improved potatoes: Protection from damage by Colorado potato beetles. *Plant Molecular Biology* 22:313–21.

Persley, G.J. 1990. *Beyond Mendel's Garden: Biotechnology in the Service of World Agriculture*. Wallingford, Conn.: CAB International.

Sanders, P.R., B.B. Sammons, W. Kaniewski, L. Haley, L. Layton, B.J. LaValle, X. Delannay, and N.F. Tumer. 1992. Field resistance of transgenic tomatoes expressing the tobacco mosaic virus or tomato mosaic virus coat protein genes. *Phytopathology* 82:683–90.

Shah, D.M., R.B. Horsch, H.J. Klee, G.M. Kishore, J.A. Winter, N.E. Tumer, C.M. Hironaka, P.R. Sanders, and C.S. Gasser. 1986. Engineering herbicide tolerance in transgenic plants. *Science*. 233:478–81.

Sondahl, M. 1993. Newsletter of the Agricultural Biotechnology for Sustainable Productivity Project, Michigan State University, East Lansing. *Biolink* 1:2.

Stark, D.M., G. F. Barry, and G.M. Kishore. 1993. Impact of plant biotechnology on food and food ingredient production. In *Science for the Food Industry of the Twenty-first Century: Biotechnology, Supercritical Fluids, Membranes, and other Advanced Technologies for Low Calorie, Healthy Food Alternatives*, edited by M. Yalpani. Shrewsbury, Mass.: ATL Press.

WCMC (World Conservation Monitoring Centre) and IUCN (World Conservation Union) Commission on National Parks and Protected Areas. 1993. *United Nations List of National Parks and Protected Areas*. Washington, D.C.: Island Press.

Wilson, E.O. 1992. *The Diversity of Life*. Cambridge, Mass.: Belknap Press of Harvard University Press.

Wilson, M. 1993. Newsletter of the Agricultural Biotechnology for Sustainable Productivity Project, Michigan State University, East Lansing. *Biolink* 1:2.

# Agricultural Industrialization and the Loss of Biodiversity

by Laura L. Jackson

The claims of industry that biotechnology products will assist in species and natural resource conservation are false. This assessment rests on three erroneous assumptions: that biotechnology is different from other forms of agricultural industrialization; that industrialized agriculture is productive and, therefore, desirable everywhere; and that agriculture should be conceived of and managed separately from nature and nature reserves. The unintended, multiscale costs of continued industrialization through the application of biotechnology are explored using two case studies, herbicide-resistant soybeans and bovine growth hormone. Herbicide-resistant soybeans are in demand ultimately because crop rotations were abandoned in the 1950s, leading to massive soil erosion and the need for herbicide-intensive, minimum-tillage methods. Bovine growth hormone, when compared to a low-input, management-intensive grazing technique, is likely to increase both the environmental and social costs of dairy farming. Industrialization of agriculture in the North American corn belt has eliminated natural refugia necessary to species conservation, while degrading land and rural communities. It is, ultimately, the loss of more people from the land due to agricultural industrialization—of which biotechnology is one part—that will endanger the remaining biodiversity, and eliminate the hope of natural area restoration in agricultural landscapes.

*

Representatives of Monsanto, a company that sells agrochemicals and seeds, claim that their newest genetically engineered products will contribute to the conservation of biodiversity (Horsch and Fraley, this volume). Evoking projected increases in the human population, they follow Avery (1995) in reasoning that more production will be required to meet human needs, and that genetically engineered, high-yielding crops will come to the rescue by releasing some of the pressure to clear undisturbed tropical habitats. Toenniessen (this volume) accepts this argument, and then discusses how to overcome the barriers to access of biotechnology by poor countries, where improved crop and livestock varieties may become increasingly inaccessible due to the expanding exercise of intellectual property rights.

Monsanto representatives also claim that biotechnology products will make agriculture more sustainable by reducing agricultural inputs and energy use. They contend that incorporating pest and disease resistance into crops through genetic engineering could reduce pesticide use. Herbicide-resistant soybeans, by speeding the adoption of minimum-tillage practices in the corn belt, could reduce soil erosion and increase carbon storage in the soil. Finally, they assert that Monsanto's bovine growth hormone, Posilac, will add to the general efficiency of farming by reducing the number of cows needed to produce the same amount of milk, thereby reducing problems of manure disposal.

These claims are logical within the context of an industrializing agriculture increasingly dominated by vertically integrated corporations. Critical consideration of the effects of biotechnology, however, must examine not only the solutions themselves, but also the context in which agricultural problems—and the scientific research to address them—are framed. Indeed, there are three false assumptions underlying their claims, which call for a more thorough analysis of agricultural intensification and continued industrialization. This analysis should reach beyond the crop and field to encompass the cropping system, agricultural landscape, regional ecosystem, rural economy, and culture itself. Two of the products described by Horsch and Fraley, herbicide-resistant soybeans and bovine growth hormone, will be examined on these several scales. Iowa will then be used to make the case that industrialization of agriculture has destroyed not only the once-rich natural landscape embedded in the rural countryside, but also the human culture capable of enjoying, defending, and restoring that natural diversity.

## Three False Assumptions

In order to accept the idea that biotech products will benefit biodiversity by increasing yields and thus sparing nature elsewhere, one must accept the following assumptions:

*1. Biotechnology Products Are Environmentally Unique.* Actually, Horsch and Fraley assert that biotech products are *not* fundamentally different from their predecessors in the agricultural seed and input market—except that they will benefit the environment. Essentially, this is having it both ways. Despite their novel origins, we are assured that these new organisms are quite ordinary. Advocates of biotechnology are accustomed by now to calming the fears of their critics, who worry about whether novel genetic material will escape and persist in wild populations, or whether gene products derived from these new techniques will pose health risks. The authors would also like us to believe, however, that this new generation of agricultural genetics and inputs is different from its predecessors when it comes to the environment. This claim capitalizes on the popular perception that biotech

TABLE 1   Characteristics of "Family" and "Industrial" Farms

| Family farming | Industrial farming |
|---|---|
| Owner operated | Separate owner, investor, manager, worker |
| Decisionmaking vested in the farmer | Absentee ownership, decisions made elsewhere |
| Use of internal resources, minimum purchased inputs | High purchased inputs and dependence on outside capital |
| Family centered, a way of life | Just a business |
| Smaller | Larger |
| Dispersed land ownership | Land concentration |
| Diversified | Specialized monoculture |
| Resource conservation based on biology | Costs to the environment are externalized |
| Community centered | Community has no value |

Note: No one farm exhibits all of these characteristics; typically, they are mixtures of elements from both extremes. With few exceptions, however, farms have tended to move toward the industrial end of the spectrum (Strange, 1988).

products are miracle organisms, made to order according to our most practical or fanciful wishes.

Yet biotechnology products may be problematic, not because of the new methods by which they are created, but because they are so *similar* to past innovations. They are part of the trend toward greater industrialization of agriculture. *Industrial agriculture* has been defined by Strange (1988) as one end of a spectrum of farming practices and social organization (table 1). At the farm or landscape level, industrial agriculture tends toward greater regional- and farm-level specialization in one or two crop species; increasingly large fields and livestock facilities; absentee ownership; increasing reliance on purchased inputs rather than whole farm management for fertility, as well as for pest and disease control; and spatial separation of feed grain production from livestock production. Vertical integration of food production is now virtually complete for poultry, where a few companies control every step of production from feed grain to livestock genetics to packinghouses to frozen dinners (Heffernan 1994). Agriculture as an industry may thrive while agriculture as a way of life, a culture, and a creative engagement with nature shrivels, and while the biodiversity of the rural landscape becomes increasingly impoverished.

Biotechnology has been conceived, born, and raised within the industrial model of agriculture. Can biotech products benefit subsistence farms or family-sized farms that sell their food locally? Can they enhance the cultural and biological diversity of rural, agricultural areas? While some may do so in certain situations more or less by coincidence, none will be specifically designed to solve problems in nonindustrial contexts, and thus will often provide answers to the wrong set of questions. Meanwhile, the worldwide trends toward greater industrialization and biological impoverishment will continue.

*2. Industrialized Agriculture in Developed Countries Is Highly Productive and Should Be Adopted Worldwide.* Since biotechnology promises to boost the productivity and efficiency of industrial agriculture, it is widely assumed that growing developing countries must adopt biotechnology and industrialize their agriculture in order to feed themselves (Avery 1995; Toenniessen, this volume).

However, the units of productivity and efficiency need to be carefully defined. In terms of human food produced per unit area, the productivity of North American agriculture is routinely overestimated. In order to make fair comparisons between the productivity of the North American corn-and-feedlot belt and Asian rice fields, for instance, it is necessary to divide corn yield by at least two. North American feed grains must first pass through livestock before they become food for people, and this involves a 50 to 90 percent reduction in their original caloric value, depending on whether the animal is a chicken, hog, or cow. So, with respect to feeding the world's growing population, North American maize fields apparently yielding 150 bushels to the acre are actually yielding 15 to 75 bushels-worth of human food per acre.

How long can such productivity be sustained? Industrial agriculture is extremely dependent on fossil energy and other nonrenewable resources. In 1983, the food energy return on a kcal of fossil energy for maize stood at 2.5 kcal, down from 3.0 in 1964 (Pimentel and Dazhong 1990, 147–64). Once transportation, processing, and preparation costs were figured in, the energy balance was 9.8 kcal of fossil energy per kcal of food energy (Lovins, Berry, and Bender 1984). The trend toward locating livestock in high-density, climate-controlled confinement facilities distant from sources of feed will continue to diminish the energy return on grain production. Without even beginning to address the social and political problems of poverty and food distribution, it is clear that in the United States, high yield per unit area has little to do with productivity in the absolute (energy) sense.

The productivity of American agriculture is further diminished by soil loss. The National Research Council (NRC 1993) estimates that one-quarter of U.S. soils are subject to sheet and rill erosion greater than the soil replacement rates, and a significant but difficult to quantify area also experiences wind, gully, and ephemeral gully erosion. The costs of soil erosion and the consequent ground and surface water degradation are difficult to estimate, and methods of doing so are controversial, but off-site costs have been estimated at between $2 billion and $8 billion per year (USDA 1987), while on-site costs of lost productivity may be anywhere from $1 billion to $18 billion annually (NRC 1986; Pimentel 1987, 217–41; USDA 1987).

Further reducing agricultural productivity is declining soil quality, or "the potential utility of soils . . . resulting from the natural combination of soil chemical, physical and biological attributes" (Johnson et al. 1992, 72–78; NRC 1993). Soil quality is deteriorating as a result of compaction, salinization, declining levels of organic matter and degradation of the soil's physical structure, and the consequent loss of water-holding capacity and biological activity (NRC 1993; Larson and Pierce 1991, 175–203). Yields on the experimental plots of the International Rice Research Institute (IRRI) have declined under conventional, intensive production, despite the replacement of old varieties with new ones. Researchers suspect this is due to a decline in soil quality (Pingali 1994, 384–401). Permanent abandonment of agricultural lands due to soil degradation and groundwater decline masks some of the degradation. About 47,000 hectares of former cotton land now lie idle in the Santa Cruz valley of Arizona (Jackson and Comus in press), another invisible casualty of "productive" American agriculture.

In Mexico, agricultural industrialization has not necessarily resulted in the ability to feed more people in that country. Intensive vegetable production in the Culiacan valley provides blemish-free, winter vegetables for American tables while endangering the health of migrant laborers (Wright 1990). Ancient cactus forests in the Mexican state of Baja California are being bulldozed to grow irrigated tomatoes for U.S. markets (Bashan 1992), not to feed the hungry. Meanwhile, over 50,000 hectares of irrigated wheatland, cleared between the 1950s and 1970s, has been abandoned in the Hermosillo coastal plain, a crucible of Mexico's green revolution. Overpumping and an upstream dam in this part of the Sonoran Desert caused saline water invasion into the fragile aquifer (Jackson and Comus in press). These regions may be modern and industrialized, but no one in Mexico is eating better as a result.

Hewett and Smith (1995) have reviewed the ecological impacts of chemically based, intensive agricultural systems and found evidence of declining productivity of agricultural lands, with a corresponding loss of function and productivity in nearby terrestrial and aquatic ecosystems. The losses in productivity, in estuarine ecosystems in particular, offset the projected gains in food production due to increased inputs.

There are certainly many examples where industrialization has at least temporarily increased the essential human food output of a region. Yet experiences in the United States and Mexico suggest that industrialization does not aim to enhance the long-term ability of the land to produce food for the people who need it most. On the contrary, market forces and governmental decisions drive these systems without respect for basic human needs or ecological health.

*3. Agricultural Landscapes Are Artificial Ecosystems.* By characterizing farming and nature conservation as purely separate activities, advocates of industrialization conveniently segregate nature in space and thought from what occurs in fields and on farms. This makes it possible to ignore the consequences of industrialization for the biodiversity located within agricultural landscapes.

Some agricultural ecosystems are nearly as old as the North American tallgrass prairie, born after the Pleistocene glaciers retreated 10,000 years ago. Agriculture is a complex mutualism between *Homo sapiens* and the animals, plants, fungi, bacteria, viruses, and protists evolving to take advantage of the habitats we purposefully or accidentally provide. This is most apparent when we look back on traditional subsistence agricultural systems, such as those of the Hidatsa of midwestern North America. These and other Native American systems were based on small, extended-family plots; use and cultivation of a full range of domesticated crops and crop varieties as well as weedy and wild species; and dependence on the culture of soil and natural flood events for nutrient cycling (Wilson 1987). The wild nature of agriculture is just as evident in "modern" fields, however, where hundreds of species of weeds and insects have evolved resistance to pesticides (Georghiou 1986).

Agribusiness separates nature from agriculture and relegates it to the nostalgic, nice-but-regrettably-impractical-now past, in order to sell products. Purchased fertilizers, herbicides, and genetics are, in the analysis of Kloppenburg (1988), substitutes for the labor of people, who once managed ecological and evolutionary processes to make a living. Services and products once provided by the agricultural ecosystem, combined with the skills and wisdom maintained by culture and practiced by individuals, have been "commodified."

This relegation of nature to remote, publicly owned, pristine preserves is also a denial of the needs of rural people for pleasant work. Many farmers in the Midwest will attest that a chief attraction of farming was the freedom to work outside with nature—not the pleasure of keeping steady company with a piece of roaring machinery (Logsdon 1994; Kline 1990). Agricultural landscapes contain significant wildlife habitat in the form of farm lanes and ponds, hedgerows, woodlots, creek bottoms, pastures, and "waste" places, as well as the lively soils that produce crops (Bunce and Howard 1990).

It is often overlooked that in many places farming is done by whole families, including women accompanied by their small children. The "production agriculture" that has developed in the midwestern United States is often too dangerous for children, and the once central role of women in producing food has shrunk considerably. It is understood in Iowa that real, serious, commercial agriculture is performed solely by men and machines. As gardening, canning, the home chicken flock, and other features of the domestic economy performed by women have been marginalized, the natural and domestic biodiversity around farmsteads has been replaced with uniform lawns. It is likely that this would occur in other industrializing agricultural economies as well.

## Beans and Cows: A Technology Analysis

The claim that biotechnology will increase gross production per unit area can be verified at the scale of the plant and field. The prediction that high yield per unit area will feed the world and protect global biodiversity (Avery 1995; Horsch and Fraley, this volume) is a field-level analysis inappropriately extrapolated to the global scale. Several scales in between have been skipped. The most obvious and immediate problem of people's ability to pay for the food that is produced has been covered well elsewhere and will not be repeated here (e.g., Pinstrup-Anderson 1993).

New technologies can cause changes in land use that impact not only the target plant or animal to which they are applied, but farming systems, landscapes, regional ecosystems, economies, and even cultures as well (several examples are reviewed in Hewett and Smith 1995). Yet we rarely try to predict the full impact of these changes ahead of time. Some of the gaps in the analysis of the effects of biotechnology on biodiversity can be filled in using the two examples given by Horsch and Fraley (this volume): herbicide-resistant soybeans and bovine growth hormone. A third example, incorporation of *Bacillus thuriengensis* (Bt) genes into crops for insect control, is briefly discussed.

In order to evaluate the effect of herbicide-resistant soybeans at several scales, it is necessary to consult the history of agricultural development in the midwestern United States. Why is herbicide resistance required in the first place? This new technology is marketed as part of the solution to soil erosion, in combination with minimum-tillage practices.

Soybeans are most often grown in rotation with corn in the upper midwestern United States. Between the early 1950s and 1980s, weeds were managed with a combination of tillage, herbicides, and human labor. Repeated tillage in the spring eliminated already-germinated weeds, and herbicide was applied for control of grasses. One or two cultivations controlled weeds between rows, and farm children "walked beans" to pull the broadleaf weeds coming up within the rows. Corn, likewise, was cultivated several times during the season to control weeds. Often the moldboard plow was used in the fall to turn over the soil, partially burying weed seeds that had accumulated on the surface.

Since the 1980s, conservation, minimum-tillage, and no-till methods have been adopted by many farmers in order to lessen exposure of the soil to erosion. New planters designed to plant seeds and fertilize into a preexisting sod have made it possible to leave the stubble from previous years' crops in place over the winter. In place of tillage, herbicides are used to "burn down" the weeds in the spring before planting. The benefits of minimum-tillage methods depend on the soil type, slope, and crop and management factors, but in general, soil organic matter increases in the top few centimeters, and soil erosion is reduced compared to full-tillage systems. The problem is that soybeans are sensitive to the herbicides used on corn to kill broadleaf weeds.

Herbicide-resistant soybeans appear to be the answer for better soil conservation, but only if the past is forgotten and certain aspects of the present are downplayed. Before the late 1950s, Iowa farmers regularly rotated their crops. Densely planted "smother crops," such as oats and barley, were planted in early spring and interseeded with a "catch crop" of clover or alfalfa. After the grains were harvested in mid-July, the legumes took over, holding the soil all fall and winter. The following year required no tillage, and the ground was covered by a nitrogen-fixing legume sod that was cut for animal fodder or grazed. In late spring of the third year, a moldboard plow, disk, and harrow were used to till before planting corn, which was then cultivated several times to remove weeds. (In a three-year rotation, corn would be the final crop; in a four-year rotation, the corn would be followed by soybeans; and in a five-year rotation, a second year of corn would follow the soybeans before returning to small grains.)

The rotation had many complementary effects. Nitrogen was provided by the legumes as well as the application of animal manures. Because the sequence of crops never optimized growing conditions for any one weed species, and controlled input of weed seeds into the soil, weed populations were reliably managed by mechanical means (Liebman and Dyck 1993). In a sense, weed population dynamics were managed by the rotation itself. Insects, such as the corn rootworm, were not a problem due to the interruption of the insect's life cycle. Heavy tillage, including cultivation, occurred in one out of every three years (three out of five in the five-year rotation).

The effectiveness of traditional rotations in saving soil likewise depended on soil type, slope, and individual management. In general, however, rotations contributed to erosion resistance by using smaller fields (shortening the length of the slopes) and limiting the area devoted to erosion-prone soybeans and corn. Due to the prevalence of sod crops and manure application, traditional crop rotations helped to maintain the size and abundance of water-stable soil aggregates and earthworms. These features of soil quality increase the water infiltration rate, reduce runoff, and help the soil to withstand raindrop impact (Russell 1973). In 1957, the *United States Yearbook of Agriculture*, entitled *Soils and Men*, acknowledged these advantages to crop rotation when it considered the new move to monocropping, and warned that use of one or two crops would require extra herbicide, fertilizer, pesticide, and erosion prevention. In that same year, the close correspondence between the area planted to row crops (corn and soybeans) versus small grains and hay in Iowa began to disintegrate (figure 1) and has never recovered.

In a crop rotation system, livestock were needed to provide manure and to use the small grain and hay crops produced, but with monocropping, farmers could sell the livestock and, for once, take a real winter vacation. Newly reliant on income

FIGURE 1   Crop Rotation in Iowa, 1840–1994

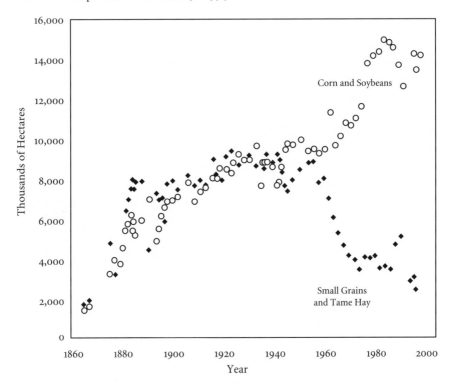

from cash grains instead of livestock, and responding to the "fencerow to fencerow" mandate of then–Secretary of Agriculture Earl Butz, farmers tilled up about 1.6 million hectares, or 36 percent of the existing permanent pastures in Iowa (figure 2). Some of these pastures, particularly in western Iowa, would have included remnant native prairies. Pastures all over the state maintained wild plant species along fences, hedges, and hard-to-graze places. Wet native hay meadows that had been grazed by dairy cows were drained and tilled as well.

Pastures had been on marginal land, either too steep or too wet. Plowing up the steep land laid some of the most highly erodible land open and vulnerable for the first time. The USDA-sponsored National Resources Inventory in 1977 reported national average soil erosion rates of 6.8 tons of soil per acre per year (1.8 tons *above* the official but scientifically suspect "tolerable" level of 5 tons per acre). By 1982, the estimate was increased to 8 tons per acre (Soule and Piper 1992); 1992 estimates based primarily on remote sensing show we are still losing 5.5 tons per acre per year nationwide (Kellog, TeSelle, and Goebel 1994). Siltation caused by erosion impacts streams and wetlands, endangering freshwater mollusks and fish that require

FIGURE 2    Decline of Pasture after 1940

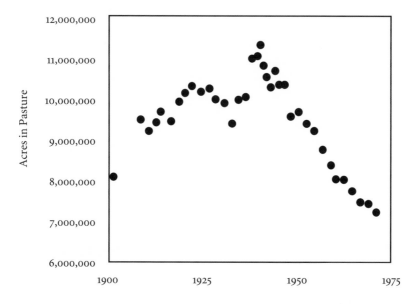

clear water. The fertilizer that came with the soil changed stream life and simplified species composition of streamside plants. Herbicide runoff contributed to the decline of prairie species in roadside refugia. Insecticides killed native pollinators as well as domestic bees. Artifical drainage and stream channelization shortened the distance that each raindrop travels to the sea, increasing the intensity of floods.

Another effect of this simplification of cropping systems has been the decline of grassland-nesting, Neotropical migratory bird species, such as the bobolink and the upland plover (Herkert 1991). This did not occur while prairies were being converted to agriculture (1820s to 1920s across the upper Midwest), or even in the 30 years following, as one might expect. Instead, it began in the late 1950s (Warner 1994).

Land use change—the abandonment of rotations and livestock for monocropping, and the conversion of marginal pasture land to row crops—precipitated the crisis in soil erosion that demanded conversion to minimum-tillage methods. Other contributing factors were the increasing size of farms, and the loss of farm numbers and people on farms. Between 1940 and 1987, Iowa dropped from 213,000 to 107,000 farms, while the average farm size increased from 65 to 127 hectares (or 160 to 313 acres; Iowa Agricultural Statistics 1988). Minimum-tillage methods save time and trips across the field.

Minimum-tillage practices also require more herbicide, particularly Round-up, which netted Monsanto close to $1 billion in 1992, according to the 8 September

1993 *Wall Street Journal*. Additionally, according to farmers interviewed in the summer of 1994, minimum-tillage practices cause shifts in the composition of weed populations. This has led to the demand for different herbicide formulations.

Despite claims that minimum-till methods reduce soil erosion and the agricultural chemicals that go with it, heavy rains and flooding in the Midwest in 1993 resulted in extraordinarily high levels of silt and agricultural chemicals in the Mississippi River, still detectable months later in the Gulf of Mexico. Over half the estimated reductions in soil erosion in the last 10 years have been achieved by taking the most highly erodible land out of production, rather than by improved farming methods (Kellog, TeSelle, and Goebel 1994). In 1991, the U.S. Geological Survey found the herbicide atrazine in each of 146 water samples collected throughout the Mississippi drainage basin; 75 percent also contained alachlor, metolachlor, or cyanazine (Goolsby, Coupe, and Markovchick 1991). An independent study by the Environmental Working Group, a Washington, D.C. nonprofit organization, estimated that 14 million Midwesterners are exposed to pesticides in their drinking water that are known carcinogens at levels above federal maximum standards (EWG 1994). At the field scale, as reported in the 8 September 1993 *Wall Street Journal*, some farmers observe that in no-till systems there is a risk that rain right after the application of chemicals will produce silt-free but chemical-laden runoff, due to poor water infiltration and the presence of chemical-soaked plant debris on the soil surface.

Therefore, although minimum-tillage techniques have been touted as progressive and soil-saving, the benefits are only in comparison to full tillage, without crop rotation, on land that should never have been tilled in the first place. Herbicide-resistant soybeans, by making minimum tillage work better, *enable* a bad system to persist, in the same way that the spouse of an alcoholic may enable the addict to function in society. Meanwhile, alternative "no-tillage" methods—such as grazing, hay production, cover crops, and agroforestry—have been overlooked.

Bovine growth hormone (BGH) is a product that has been thoroughly studied by its manufacturers at the level of the cow, but questions raised at the level of the farm, farm landscape, and rural economy have been left to others to sort out. Even at the cow level, the product is disturbing. The veterinary sheet that comes with a dose of Posilac contains the following warnings:

- BGH almost doubles the time during which a cow undergoes catabolic stress, metabolizing her own body tissues to recover from pregnancy and produce more milk. This increases her need for high-quality feed, increases susceptibility to disease, and may cause her to run an elevated temperature.
- There is an increased risk of both clinical (milk visibly abnormal) and subclinical (milk not visibly abnormal) mastitis and of somatic cell counts in the milk (indicative of poor milk quality).

- Use of Posilac is associated with increased digestive disorders, loss of appetite, permanent swelling at the injection site, and swelling and lesions of the knee and foot.
- The cow may experience reduced pregnancy rates, increases in cystic ovaries and disorders of the uterus, decreases in gestation length and birth weight, and a higher incidence of retained placenta. The effect of BGH on bull calves born to cows on the drug has not been established.

The product's ability to stimulate milk production by an average of 12 percent is better appreciated when one considers that a first-lactation heifer is not only producing milk, but simultaneously adding 70 to 100 pounds to her own body weight while pregnant with her second calf.

Liebhardt (1993) has recently taken a powerful analytic approach to the evaluation of BGH as a new technology. Contributors to this interdisciplinary study evaluated the effects of BGH on the dairy business compared to an alternative rotational grazing. It is commonplace to evaluate new technologies such as BGH versus the status quo and find them superior. This method is biased, in part, by the speculative nature of the new technology's benefits. While the side effects, particularly at larger scales of analysis, of the new technology are only partially known, the pitfalls of the current system are thoroughly understood. Another reason new technologies tend to "win" in the typical analysis is that they are designed to take the existing system one step further, and thus they compete under the existing structural framework or rules of the game, however flawed. By comparing BGH to a truly alternative system that "changes the rules" of dairy production toward higher resource efficiency instead of higher per-cow yield, the consequences of these alternative structural frameworks are better evaluated.

In Liebhardt's study, the effects of BGH and rotational grazing are evaluated for consumer health risks, the economic impacts, the economic and social viability of rural communities, and the quality of life of the farmer. While both technologies showed the potential to increase profit and decrease feed costs, the systems differed greatly in other respects. The consumer health analysis focused on mastitis and the effects of different hormones in the milk. The researchers concluded that the increased mastitis rates associated with BGH will hasten the evolution of antibiotic resistance in the bacteria that cause mastitis, making it increasingly difficult and expensive to monitor for the unapproved antibiotics that this microbial arms race will generate. Another concern was the insulin-like growth factor (IGF–1), which exists in elevated concentrations in the milk of cows treated with BGH. Its effects on humans are poorly known and require more testing. In contrast, rotational grazing of cows on grass and clover had no additional health questions associated with them, and mastitis rates were lower than in conventional grain-based dairy systems (Liebhardt 1993).

At the farm level, rotational grazing was increasingly competitive with BGH when milk prices were low, feed costs were high, and interest rates and capital costs were high. The two techniques were observed to use different economic strategies: BGH aims to increase productivity faster than it increases feed and drug costs; rotational grazing aims to increase profitability by reducing feed and drug costs while improving cow health and maintaining productivity. Case studies showed that rotational grazing decreased feed costs by up to 36 percent per 100 pounds of milk; decreased energy costs by up to 75 percent; increased the grazing season by up to 100 days; increased milk protein; and reduced labor for feeding hay, spreading manure, and storing forage (Liebhardt 1993).

A review of eight studies on the economic and social impacts of BGH at the community or regional level concluded that larger farms will be most likely to benefit from the early adoption of BGH, and that this will precipitate a decline in the number of midsized dairies (Liebhardt 1993). Because midsized farms are crucial to the social and economic health of rural communities, BGH is not likely to help them and will probably hurt in the long run. In contrast, rotational grazing helped to maintain the profitability of small to midsized family dairies, preserving community jobs while encouraging the children of dairy farmers to stay in the area and take up the business themselves. Because rotational grazing requires human skill and reliance on natural resources instead of purchased inputs, capital stays in the community. As one northeast Iowa dairy farmer explained, "Only so much milk will be consumed in this country and only so much money can be made off of that milk. The question is whether you want to give some of that money to Monsanto" (Stewart 1994).

Several consequences of BGH are predictable at the scale of the farm landscape and regional ecosystem. The farm landscape—a mixture of pastures, native hay meadows, forages, and row crops—is linked to the number of dairies because of the profitable use of pasture and forage by cows, as well as the need for manure by row crops. In northeast Iowa in the mid-1930s, when crops were rotated, there were more than five creameries per county, and the number of dairies was not even reported, apparently because they were so numerous (ISDA 1936). Today, all these creameries have been abandoned along with the crop rotation, and the decline in dairies continues. Between 1984 and 1994, according to the *Waterloo Courier* of 13 November 1994, Iowa lost 44 percent of its dairy farms, from 8,356 to 4,653. Total milk production went down only 8 percent because the existing dairies expanded.

From the point of view of regional biodiversity, BGH promises to increase the amount of land in row crops relative to other land uses. Because of the high feed quality required by treated cows, even less pasture and hay is likely to be needed. Corn and soybeans contribute more nitrate, herbicide, and pesticide to the watershed than do pasture and hay; are responsible for more soil erosion; and result in the loss of more soil organic matter. Pasture and hay, on the other hand, can store organic matter in the soil, acting as a sink for atmospheric carbon dioxide ($CO_2$), a greenhouse gas. To the extent that BGH will reduce the number and increase the size of dairies, and increase reliance on row crops, it will have negative impacts on regional ecosystems.

In contrast, some dairy farmers in the upper Midwest have converted most of their row crops to permanent, intensively managed pastures (Stewart 1994; Chan-Muehlbauer, Dansingburg, and Gunnink 1994). In addition, rotational pastures can incorporate native plants from the prairies for forage. The plant diversity of a pasture is a small fraction of its total biodiversity, which includes birds and invertebrates depending exclusively on grassland habitat.

Beyond the questions of ecosystem and economic health is the issue of quality of life for the farmer. Quality of life is a major reason some dairy farmers are switching to rotational grazing. Instead of investing in the equipment and working long hours to raise the grain and silage, feed the cows, and spread the manure, farmers let the cow do everything except the milking itself. Low-cost electric fencing and water lines have made these innovations relatively simple to adopt; farmers' focus is now on getting the most out of pastures through paddock design and grazing management (Liebhardt 1994; interviews with Iowa farmers, summer 1994). In contrast, BGH obligates farmers to work harder to prevent or fight disease and exhaustion in their treated cows, yet none of their other work is predicted to decrease.

An innovative, interdisciplinary study coordinated by the Land Stewardship Project in Minnesota has begun to quantify the effects of rotational grazing on animal and human health, farm- and community-level economics, farm family quality of life, ecological sustainability, and regional biodiversity. Farmers, in cooperation with university scientists and others, are monitoring soil quality and the diversity of invertebrates, amphibians, and birds, as well as more standard measures of farm profitability and production (Jackson 1994). The monitoring has already begun to change farming practices. For instance, farmers in the program have expressed great enthusiasm for hands-on measures of earthworm density, and want to know how to manage the land to increase their earthworm populations (Dorsey, 1994).

When BGH and herbicide-resistant soybeans are analyzed in terms of alternatives that lie outside of the trend toward industrialization of farms, their benefits are one-sided. They help to simplify the landscape toward even more corn and soybeans, and eventually drive more family farms off the land, further undermining rural economies and cultures along the way. Those who are left behind are consigned to deal with animals pushed beyond natural limits and, thus, always in danger of becoming sick; greater dependence on seed, fertilizer, herbicide, and insecticide; and declining economic conditions and decaying rural infrastructures. By comparison, techniques such as management-intensive rotational grazing—

which depend on the health of agricultural ecosystems, biological diversity, and the farmers' own management skills and resources—appear to have the opposite effect.

Incorporation of Bt genes into crops for insect control would appear to have fewer multiscale impacts than the previous two examples of biotechnology in industrial agricultural systems. Its widespread use, however, in agroecosystems would almost certainly lead to the evolution of resistant insect strains. This will deprive organic producers, from the farm to the family gardener, of an important tool for managing pest populations without chemical pesticides. What is essentially a public resource—pest susceptibility to Bt—could be squandered in a few years, depending on the spatial distribution and number of seasons that the Bt-augmented strains are in use.

### Agricultural Landscapes and Refugia

About 10 percent of terrestrial ecosystems are agricultural. Rather than treat agricultural landscapes as biotic sacrifice areas, it is important to learn what kind of agriculture will harmonize food production with good and satisfying work, a healthy rural culture, and the diversity of both wild and domestic plants and animals. A key concept in biological conservation that has proved durable in many parts of the world is that of the biotic refuge. Much biodiversity can be preserved in agricultural landscapes if natural refugia of the proper size exist. The legacy of agricultural industrialization in Iowa, however, has been to remove those refugia. The roots of the industrial model go back well before World War II and the loss of crop rotations. The values of efficiency, machinery, size, science, and specialization were expressed in the rhetoric of what was to become the USDA at least since the 1850s (U.S. Patent Office 1856). The use of the categories "unimproved pasture" and "waste places" to describe natural refugia on farms in the U.S. agricultural census (for example, U.S. Bureau of the Census 1954) is indicative of prevailing attitudes about the future of the agricultural landscape. Our task was to get things cleaned up and improved as soon as possible.

As a direct result, nature reserves are rare or absent on arable land. In Iowa, where tallgrass prairie once covered 12.2 million hectares, less than 0.1 percent remains (Smith 1992, 195–99). Remaining protected remnants are tiny—the largest state preserve is 100 hectares and the median size is 13.5 hectares. However, species refugia can and do exist in and around farms in even smaller refugia. Traditionally, these have been located on nonarable land, such as streams, marshes, steep hillsides, and rocky or dry uplands. Often hunted, grazed, hayed, logged, or put to some other less-intensive purpose than row-crop agriculture, these places nonetheless provide the only refuge for many wild species. Refugia may also occur along fencerows, roadsides, and railroad tracks, as well as in temporary and permanent pastures, woodlots, and graveyards. The occasional landowner has managed refuge habitat for the sake of curiosity or enjoyment.

While few globally rare species may persist in these small refuges, the local area is enriched by them. Local species extirpation has the same effect as complete extinction on local population processes. For instance, the absence of an important insect pollinator may cause low fertilization rates and inbreeding in some plant species. The absence of species-rich refugia in a landscape also impoverishes the human experience. Traveling to another state or 60 to even 100 miles to see a scrap of natural area with the species that once were common throughout Iowa is no substitute for rambling through such a place every summer as a child. In the first two instances, the natural area is a museum; a destination of academic interest. In the last instance, it is home—a part of the landscape and a birthright.

Intensive and productive farming has coexisted with these refugia. Ninety-seven percent of Iowa was developed for agriculture by 1910 (ISDA, 1911); total land in farms has actually declined slightly since then (Iowa Agricultural Statistics 1988). One might conclude that the battle over Iowa's prairie biodiversity had been lost by 1910, but in fact, a great deal of diversity and ecosystem resilience remained. Shimek (1917) noted that prairies persisted in strips near roads and along fences in row crops, and that a constant heavy rain of weed seeds was unable to invade the intact prairie sod. Not only could remnant prairie hold its ground in 1917, but it could invade disturbed areas. Shimek documented rapid reinvasion of old fields, disturbed road cuts, and railroad right-of-ways. Seed sources were "amply sufficient," in Shimek's judgment, to reseed all suitable areas.

During this era, prairie hay was a significant resource for farmers. Three hundred and fifty thousand hectares of prairie hay were cut in 1910 statewide, in some counties as much as 11,000 (ISDA 1911). In addition, there were 3.9 million hectares of pasture, roughly equal to the area of corn production at the time. The abundant seed sources were disappearing, however. The native flora along roadsides was rapidly disappearing with the widening of the main roads and indiscriminate enforcement of weed laws on the secondary roads (Smith 1992, 195–99). As early as 1913, Wilhelm Miller of the University of Illinois Agricultural Extension Service advocated the replanting of prairie roadsides that "relieve the cast iron system that laid out the country in absolute, unvarying squares" (Egan and Harrington 1992, 147–51). Apparently, Illinois' roadside prairie remnants were nearly gone. Prairie hay continued to decline at a constant 7 percent per year; by 1970, less than 5,000 acres were harvested for wild hay statewide, after which the government stopped keeping records (Iowa Agricultural Statistics 1988).

Today, the Iowa tallgrass prairie is no longer resilient. The rarity and small size of remnants, the lack of additional seed sources, and the ubiquitous presence of aggressive exotic species such as *Bromus inermis* (smooth brome) make prairie reconstruction difficult and natural recovery of disturbances next to impossible.

Alert and vocal witnesses to the impoverishment of the Iowa landscape were apparently few. In 1928, Aldo Leopold began to gather evidence that fencerows, borders, woodlots, remnant prairies, and wetlands were disappearing due to agricultural intensification, and that the wildlife they supported were disappearing as well (Meine 1987, 39–52). In *A Sand County Almanac* (1949), Leopold lamented the narrowing of the prairie sod fencerows by the most forward-looking, industrious farmers to "a few feet." Even Leopold, decrying the relative simplification and sterilization of the Iowa landscape compared to his boyhood days around the turn of the century, did not predict the complete loss of those fences, or the pastures, crop rotations, and animals that made them necessary.

Ecologists and historians bent on restoration of this lost landscape can partially deduce its species composition and distribution based on clues from the soil and the few remaining natural refugia in state preserves and on farms. Restoration of the once diverse countryside will require more than just natural refugia, however. It will require human refugia—witnesses to a loss so complete that when these people are gone, no one will know how to value what has been lost. An Iowa naturalist tells me that a wetland restoration program has been successful partly because there are semiretired farmers who still remember the pleasures of fishing and hunting in the undrained prairie potholes of their youth. It is not so much the information on the biology of what grew there, as appreciation and affection that we are quickly losing.

The number of farms in Iowa has been cut in half since 1940, and continues to drop (Iowa Agricultural Statistics 1988). Communities struggle to maintain the critical mass of people to keep the volunteer fire departments and health clinics, schools, and other services running. The aging farmers that are left barely have time, much less the interest or money, to restore habitat or preserve genetic diversity in their crops and livestock, even if they remember it with some fondness (many do not). Why should they? Their children and grandchildren won't come back to farm. Even worse, there will soon be no one left on the land who remembers anything but corn and beans, and the high-input ways of growing them. Conservation of biodiversity, like agriculture, is an expression of culture. When culture erodes, so does the potential for local appreciation and tending of biodiversity. It required nearly 100 years for Iowa and Wisconsin to produce one Leopold who was capable of interpreting the landscape and the history of the region as both a landowner/user and a conservationist. Today, there are probably very few children who are being exposed, as Leopold was, to the rich agricultural and natural beauty of their region. Most of the landscape is no longer beautiful, and there are far fewer children left in the countryside to experience it. Without people who possess a sense of place, and the political and economic enfranchisement to do something about it, conservation will be left to remote government regulators of abstract "resources."

There are small human refugia among the Amish and Mennonite communities that continue to thrive and expand, even while modern farms go under around them. Their strategy has been to partially isolate, rather than open, their economy to global demands. Their approach to technology has been rational and deliberate: they determine how it will affect their farms, their families, and their communities before deciding whether to adopt it (Logsdon 1994). As a result, they were some of the first people in the country to adopt solar electric fence chargers. At least one Amish family (Kline 1990) enjoys a rich diversity of bird and insect life around their farm, attributing it to the proper scale and diverse habitats afforded by their farm.

There are also small numbers of innovative farmers who rely more on farm and ecosystem health than brute force technologies to make their living. A recent study on the profitability of four sustainable farms in Minnesota found that they had similar strategies in addition to their common stewardship ethic: little or no use of chemical fertilizer or pesticide; employment of numerous soil-building strategies; minimization of excessive capital costs; and an emphasis on net return rather than gross return. Their net return per acre was three times the average for their regions, and the diversity of products raised on the farms made them more flexible and resilient to changes in the market (Chan-Muehlbauer, Dansingburg, and Grunnink 1994). As these success stories become more widely known among farmers, the idea that stewardship and successful farming can go together may begin to take hold once again.

Organizations such as the Practical Farmers of Iowa and the Innovative Farmers of Ohio foster on-farm research and link farmers who have already begun to cut chemical use with those who are interested in trying, but need information and support. Nonprofit organizations devoted to research, education, and policy analysis—such as the Center for Rural Affairs, the Land Institute, the Kansas Rural Center, the Land Stewardship Project, and the Wisconsin Rural Center—promote family-sized farms, and they call attention to the relationships between the structure of agriculture, the sustainability of farming, and the health of natural ecosystems.

These efforts are all overshadowed by the rise of the livestock confinement industry. While innovative farmers can still raise hogs and dairy cattle more cheaply and with fewer environmental impacts than the high-density livestock facility, they suffer as their neighbors go out of business and the infrastructure and markets for livestock crumble, as has happened in the poultry industry years ago. Without a market to sell their animals, even the most practical, conscientious, and sustainable operations—including those of the Amish and Mennonites—are in danger of disappearing. When the minds responsible for these farms have left the countryside, replaced by minimum-wage labor in factory-style facilities, so will the potential to conserve and improve the agricultural landscape.

Products of biotechnology in the service of industrial agriculture will continue to erode the biodiversity and cultural viability of the American Midwest. If this were the sacrifice necessary to preserve the wonderful, species-rich tropical rain forests in South America, a majority of Midwesterners might generously conclude that elimination of our own wonderful, species-rich prairie ecosystem was worth it. However, industrialization here has roots that go back before the turn of the century, long before the current biodiversity crisis was apparent. The impoverishment of this rich country has not prevented the deforestation of our poorer neighbors; and only a hopeless Pollyana would believe that it will do so now. In order to assess the multiscale effects of the biotech industry and the brave new world of biodiversity prospecting, we need to understand the ecological history of the American Midwest, where the industrial model of agriculture has had its fullest expression.

## References

Avery, D. 1995. *Saving the Planet with Pesticides and Plastic: The Environmental Triumph of High-yield Farming.* Indianapolis: Hudson Institute.

Bashan, Y. 1992. Telephone convsersation with author. Director, Division of Experimental Biology, Centro de Investigaciones Biologicas de Baja California Sur, Apdo Postal 128, La Paz, B.C.S., Mexico (682) 53633.

Bunce, R.G.H., and D.C. Howard. 1990. *Species Dispersal in Agricultural Habitats.* London: Belhaven Press.

Chan-Muehlbauer, C., J. Dansingburg, and D. Gunnink. 1994. *An Agriculture That Makes Sense: Profitability of Four Sustainable Farms in Minnesota.* Land Stewardship Project, 14758 Ostlund Trail North, Marine on St. Croix, MN 55047; (612) 433-2770.

Dorseg, J. 1994. Personal communication.

Egan, D., and J.A. Harrington. 1992. Use of native vegetation in roadside landscaping: A historical review. In *Proceedings of the Twelfth North American Prairie Conference*, edited by D.D. Smith and C.A. Jacobs. Cedar Falls: University of Northern Iowa.

EWG (Environmental Working Group). 1994. *Tap Water Blues.* Environmental Working Group, 1718 Connecticut Avenue, NW, #600, Washington, D.C. 20009; (202) 667-6982.

Georghiou, G. 1986. The magnitude of the resistance problem. In *Pesticide Persistence: Strategies and Tactics for Management*, edited by the National Resource Council. Washington, D.C.: National Academy Press.

Goolsby, D.A., R.C. Coupe, and D.J. Markovchick. 1991. *Distribution of Selected Herbicides and Nitrate in the Mississippi River and Its Major Tributaries, April through June 1991.* Water resources investigations report 92–4163. Denver, Colo.: U.S. Geological Survey.

Heffernan, W.D. 1994. Agricultural profits: Who gets them now, and who will in the future. *Leopold Center for Sustainable Agriculture 1994 Conference Proceedings.* Ames, Iowa.

Herkert, J.R. 1991. Prairie birds of Illinois: Population response to two centuries of habitat change. *Illinois Natural History Survey Bulletin* 34:393–99.

Hewett, T.I., and K.R. Smith. 1995. *Intensive Agriculture and Environmental Quality: Examining the Newest Agricultural Myth.* Greenbelt, Md.: Henry A. Wallace Institute for Alternative Agriculture.

Iowa Agricultural Statistics. 1988. *1988 Agricultural Statistics.* Des Moines, Iowa: Iowa Department of
  Agriculture and Land Stewardship, and U.S. Department of Agriculture.
ISDA (Iowa State Department of Agriculture). 1911. *Twelfth Annual Iowa Year Book of Agriculture for
  1910.* Des Moines: State of Iowa.
————. 1936. *Thirty-seventh Annual Iowa Year Book of Agriculture for 1936.* Des Moines: State of Iowa.
Jackson, D. 1994. Personal communication.
Jackson, L.L., and P.W. Comus. In press. Ecological consequences of agricultural development in a
  Sonoran Desert Valley. In *Ecology and Conservation of Sonoran Desert Plants: Attribute to the
  Desert Laboratory,* edited by R. Robichaux. Tucson: University of Arizona Press.
Johnson, M.G., D.A. Lammers, C.P. Andersen, P.T. Rygiewcz, and J.S. Kern. 1992. Sustaining soil quality
  by protecting the soil resource. In *Proceedings of the Soil Quality Standards Symposium.* (San An-
  tonio, Tex., 23 October 1990). Watershed and Air Management report no. WO-WSA-2. Washing-
  ton, D.C.: U.S. Department of Agriculture, U.S. Forest Service.
Kellog, R.L., G.W. TeSelle, and J.J. Goebel. 1994. Highlights from the 1992 National Resources Inventory.
  *Journal of Soil and Water Conservation* 49:521–27.
Kline, D. 1990. *Great Possessions: An Amish Farmer's Journal.* San Francisco: North Point Press.
Kloppenburg, K. 1988. *First the Seed: The Political Economy of Plant Biotechnology, 1492–2000.* Boston:
  Cambridge University Press.
Larson, W.E., and F.J. Pierce. 1991. Conservation and the enhancement of soil quality. In *Technical Pa-
  pers.* Vol. 2 of *Evaluation for Sustainable Land Management in the Developing World.* Bangkok: In-
  ternational Board for Soil Research and Management.
Leopold, A.C. 1949. *A Sand County Almanac and Sketches Here and There.* New York: Ballantine Books.
Liebhardt, W.C. 1993. *The Dairy Debate: Consequences of Bovine Growth Hormone and Rotational Graz-
  ing Technologies.* Davis: University of California Sustainable Agriculture Research and Education
  Program.
Liebman, M., and E. Dyck. 1993. Crop rotation and intercropping strategies for weed management. *Eco-
  logical Applications* 3:92–122.
Logsdon, G. 1994. *At Nature's Pace.* New York: Pantheon.
Lovins, A.B., L.H. Lovins, and M. Bender. 1984. Energy and agriculture. In *Meeting the Expectations of
  the Land: Essays in Sustainable Agriculture and Stewardship,* edited by W. Jackson, W. Berry, and
  B. Colman. San Francisco: North Point Press.
Meine, C. 1987. The farmer as conservationist: Leopold on agriculture. In *Aldo Leopold: The Man and
  His Legacy,* edited by T. Tanner. Ankey, Iowa: Soil Conservation Society of America.
NRC (National Research Council). 1986. *Soil Conservation: Assessing the National Resources Inventory,*
  *Vols. 1–2.* Washington, D.C.: National Academy Press.
————. 1993. *Soil and Water Quality: An Agenda for Agriculture.* Committee on Long-range Soil and
  Water Conservation, Board of Agriculture. Washington, D.C.: National Academy Press.
Pimentel, D. 1987. Soil erosion effects on farm economics. In *Agricultural Soil Loss: Processes, Policies,
  and Prospects,* edited by J.M. Harlin and A. Hawkins. Boulder, Colo.: Westview Press.
Pimentel, D., and W. Dazhong. 1990. In *Agroecology,* edited by C.R. Carrol, J.H. Vandermeer, and P.M.
  Rossett. New York: McGraw Hill.
Pingali, P. 1994. Technological prospects for reversing the declining trend in Asia's rice productivity. In
  *Agricultural Technology: Policy Issues for the International Community,* edited by J.R. Anderson.
  Wallingford, Conn.: CAB International.
Pinstrup-Anderson, P. 1993. *World Food Trends and How They May Be Modified.* Washington, D.C.:
  CGIAR International Centers Week.
Russell, E.W. 1973. *Soil Conditions and Plant Growth.* 10th ed. London: Longman.
Shimek, B. 1917. The persistence of the prairie. *University of Iowa Studies in Natural History* 11:3–24.

Smith, D.D. 1992. Tallgrass prairie settlement: Prelude to the demise of the tallgrass ecosystem. In *Proceedings of the Twelfth North American Prairie Conference*, edited by D.D. Smith and C.A. Jacobs. Cedar Falls: University of Northern Iowa.

Soule, J.D., and J.K. Piper. 1992. *Farming in Nature's Image: An Ecological Approach to Agriculture*. Washington, D.C.: Island Press.

Stewart, M. 1994. Interview by author. Oelwein, Iowa. August.

Strange, M. 1988. *Family Farming: A New Economic Vision*. Lincoln: University of Nebraska Press.

U.S. Bureau of the Census. 1954. *Counties and State Economic Areas*, Vol. 1, part 30 of U.S. Census of Agriculture. Washington, D.C.: Government Printing Office.

USDA (U.S. Department of Agriculture). 1957. *Soil: The 1957 Yearbook of Agriculture*. Washington, D.C.: Government Printing Office.

———. 1987. *Agricultural Resources—Cropland, Water, and Conservation—Situation and Outlook Report*. AR-8. Washington, D.C.: Economic Research Service.

U.S. Patent Office. 1856. *Report of the Commissioner of Patents for the Year 1855*. Washington, D.C.: Cornelius Wendell.

Warner, R.E. 1994. Agricultural land use and grassland habitat in Illinois: Future shock for midwestern birds? *Conservation Biology* 8:147–56.

Wilson, G.L. 1987. *Buffalo Bird Woman's Garden: Agriculture of the Hidatsa Indians*. St. Paul: Minnesota Historical Society Press.

Wright, A. 1990. *The Death of Ramon Gonzalez*. Austin: University of Texas Press.

# An *Ex Situ* "Library of Life" Strategy

by Gregory Benford

While *in situ* conservation is the ideal, it needs to be supplemented by *ex situ* approaches, such as captive breeding programs, zoos, seed banks, and botanical gardens. But even these may not be sufficient to meet the needs of future generations, when biotechnology is likely to be far beyond what is imagined today. This chapter, therefore, proposes a systematic program of cryopreservation. Freezing organisms will allow future biologists to extract DNA from frozen samples to study the exact genetic source of biodiversity, providing access to important information that may be unavailable unless such preservation steps are taken. Such a program would have numerous collateral benefits, such as enabling evolutionary biologists of the future to study evolution over perhaps thousands of insect generations. A global program to establish this frozen "library of life" is estimated to cost around $2 billion for a century, a modest investment to enable future biologists to benefit from the biodiversity found today.

\*

Intuitively, we prefer *in situ* preservation of endangered species. The broad public, and probably most biologists, think in these terms, with some appreciation for the conservation aspects of such places as zoos and aquaria. But two liabilities of *in situ* measures become apparent as our global crisis worsens: cost and reliability.

No one would propose simply buying up the tropical forests. The huge cost is the least of the difficulties; what will deflect the vast social pressure of a burgeoning human population? The hardworking farmer with a family and a chain saw is an often unknowing enemy of biodiversity. The farmer's modest economic ambition makes many *in situ* schemes too expensive, economically and politically. This pressure leads to long-term uncertainty in holding the line of even the best local programs, as in Costa Rica. The price of *in situ* conservation is eternal vigilance.

Inevitable defeats will bring the loss of uncounted species, which we protected for only a while. How can we hedge our bet?

## The Library of Life

A sweeping program of extreme *ex situ* preservation, the library of life (Benford 1992), is a necessary supplement to *in situ* approaches. It would aim to freeze species in threatened ecospheres, salvaging samples randomly for eventual use by future generations. Under such a scheme, *preservation* means keeping alive representatives of at least each genus—*in situ, in vivo* protection in reserves—while freezing representatives of species within the genus. A parallel program of limited *in situ* preservation of ecosystems is essential to allow later expression of frozen genomes in members of the same genus, as well as to eventually study and potentially resurrect frozen species.

Sampling without studying—no taxonomy—can lower costs dramatically. Local labor can do most of the gathering. Plausible costs of collecting and cryogenically suspending the tropical rain forest species, at a sampling fraction of $10^{-6}$, are about \$2 billion for a full century.

By simply saving whole organisms, much more information than species DNA will be saved. This could allow future biotechnology to derive high information content and perhaps even to resurrect then-extinct species.

This idea arises as the true dimensions of our biodiversity crisis emerge. We have only begun the elementary taxonomic description of the world biota; about 1.4 million species have been given scientific names, while estimates of the total number of species range up to roughly 30 million or higher (Wilson 1989). Other chapters in this volume demonstrate that systematists may very well not know the species diversity of the world fauna to the nearest order of magnitude.

The time to even catalog our living wealth is running out. This dire moment demands radical thinking. While *in situ* methods appeal deeply to our spirit, perhaps proposals strongly linking the *in situ* preservation community with the *ex situ* conservationists promise better long-range results. *In situ* measures are supported by economic, environmental, and aesthetic arguments. To preserve the genome of many species, however, *ex situ* methods may suffice and, with future technical advances, allow for intricate interplay between the two approaches. Considering this possibility serves to separate the kinds of arguments we make for conservation methods, including concepts of our moral debt to posterity.

## Space and Species

A globally systematic but locally random sampling might follow the criteria of Vernon Heywood (1992, 15–19):

1. The area is species rich.
2. The area is known to contain a large number of edaphic plant species.

3. Sites contain important gene pools of plants valuable, or of potential value, to humans.
4. Sites include a diverse range of habitat types.
5. Sites contain a significant proportion of species adapted to special edaphic conditions.
6. Sites are threatened or are under imminent threat of development.

Several salient questions about even such broad criteria immediately occur: Should we limit ourselves only to sites that are species rich? Should we assume we know what the future will deem important?

A humble recognition that biology is a young science, only beginning to explore its deep genetic fundamentals, suggests that we pursue strategies unencumbered by our narrow present concerns. Often, "species rich" does not count microbial densities (Dial and Marzluff 1988). The concept of species is itself a bit slippery, certain to be refined and altered in the future. Further, current programs, such as sampling of tropical trees by insecticidal fogs and active searching of the canopy, may be most useful if they do not have species-specific goals.

Teams trained to simply collect, without analyzing, require minimal supervision by research biologists. Freezing at the site can be done with ordinary ice or dry ice; liquid nitrogen suspension need only occur at the long-term repository. Extensive work by taxonomists enters only when samples are eventually studied and classified, perhaps a century from now.

Our primary aim should be to pass on to later generations the essentials of our immense biodiversity. Even bare information about the existence of a species is useful, for in the future, without a sample, one cannot be sure whether a given variation existed at all or simply became extinct without being observed. Detailing allelic variations within a population, or another variability (for example, conspecific, interpopulation, congeneric, or ecosystem), probably will not be worth the trouble.

It seems likely that captive breeding programs, protected areas, and zoos can preserve only a tiny fraction of the threatened species. Moreover, microbe preservation is neglected in most such programs, yet their species density may be very high.

To save the biosphere's genome heritage demands going beyond existing piecemeal strategies of seed banks, germ plasm and tissue culture collection, and cryopreservation of gametes, zygotes, and embryos; these programs mostly concentrate on saving traditional, domesticated varieties (Cavalli-Storza 1986). Our goal is a dense, random sample of threatened species.

## Against Taxonomy

Many prominent biologists have proposed bold strategies for analyzing organic diversity. Desire for a complete catalog of species is widespread, and the library of life project would appear to be an excellent opportunity to begin classification of all the world's species. Only by attempting this on a small scale can we establish the methods for a larger problem.

But complete classification is both economically unfeasible and logistically impossible. Wilson (1989), arguing for a complete identification of the world's species, calculated that 25,000 professional lifetimes would be needed to identify 10 million species. If his estimate is correct, and each lifetime cost only $1 million (a conservative figure), the total expense would be $25 billion. Clearly, taxonomy is far more expensive than frozen salvage. To be fair, Wilson argued that computerized identification methods could speed up the process, but at the same time, 10 million species might be a gross underestimate.

Even if classification were cheap, other problems make it impossible to provide comprehensive coverage. There are fewer than 1,500 professional systematists in the world with expertise in tropical species, and this number is rapidly declining. Taxonomists are moving to better-paid positions, and are not replaced when they retire or are laid off from museums and herbariums. Quite probably, even a much larger team could not keep up with the tremendous volume of material obtained from an aggressive sampling project, such as the library proposal. John Terborgh (1992) illustrates the problem:

> Even when specimens are in hand for each of the 600 to 800 trees that typically occupy a hectare, the job is far from done. The specimens must be taken back to a world-class herbarium and then sorted, first to family and then to genus. The specimens are then packaged and mailed out, genus by genus, or family by family, to perhaps the only human-being competent to identify them. Likely as not, the specialist is burdened with dozens of such requests, each of which must await its turn. It might be a year or two before the specimens come back with names sanctioned by the expert. Specimens lacking flowers or fruits often cannot be identified at all.

The problem is worse for insects. For instance, we do not know how many species of beetles there are, even to the nearest order of magnitude. Quite likely, many species will appear hundreds or thousands of times in the sampled collection. This is a virtue, in that we get a rough measure of the area and density distributions of species, but bare taxonomy of them would be pointless. Repeatedly classifying the same species pushes the cost of identification painfully high. Keying out hundreds of millions (perhaps more) of organisms could take centuries.

Then, too, the usefulness of such information is itself in doubt. To further the library analogy, would one rather have the card catalog or the books themselves?

Given the books, one can later catalog at leisure.

## Preserve the Genus, Freeze the Species

A combined strategy can blend the virtues of *in situ* and *ex situ* by preserving alive some fraction of each ecosystem type ("biome"), its population intact at the genus level; and by freezing representatives of as many species related to the preserved system as possible, often relying on random sampling.

At a minimum, this will allow future biologists to extract DNA from frozen samples and study the exact genetic source of biodiversity. An intact beetle carries much obvious information—a glance shows it's a beetle, without reading DNA—and even such small organisms carry mites, parasites, and microbial forms as well. Organisms are themselves pocket ecosystems.

In the long run, genes of interest could be expressed in living examples of the same genus, by systematic replacement of elements of the genetic code with information from the frozen DNA. Resurrecting mastodons through elephants might be possible. Obviously, the preserved genus is essential.

These techniques would open broad attacks on the problem of inbred species. A ravaged environment can constrict the genetic diversity of individual species. Reintroducing diverse traits from frozen tissue samples could help such a species blossom anew, increasing its resistance to disease and the random shocks of life.

Beyond this minimum—the DNA itself—future biologists will probably find great use for recovered cells in reexpressing a frozen genome. Cell use for mollusks, trees, insects, and other species is a cloudy, complicated issue. For mammals, uterine walls and elements of the sexual reproductive apparatus should prove essential, since placentation and the highly variable physiology of different taxa are crucial. Germ plasm cryobiology can store, thaw, and use sperm and embryos only with much collateral information. It seems highly unlikely that one can make appropriate placental and endometrial choices in the many steps from genome to newborn, merely from reading DNA.

As saviors of the library of life, we are at best marginally literate, hoping that our children will be better readers, and wiser ones. Many biotechnological feats will probably emerge within a few decades—many ways, let us say, of reading and using the same genetic "texts." But no advanced "reader" and "editor" can work on texts we have lost.

Selectively reintroducing biodiversity in the future could gradually recover lost ecosystems. Individual species can be resurrected from small numbers of survivors, as the nearly extinct California condor and black-footed ferret have proved. Frozen genes could later increase genetic diversity in such populations. Fidelity in reproducing a genome may not be perfect, of course. Many practical problems arise (such as placental environment and chemistry, for example) to complicate expression of a genotype. In any case, future generations may well wish to edit and shape genetically those species within an ecosystem as they repair it, for purposes we cannot now anticipate.

Loss of nearly all of an ecosystem would require a huge regrowth program, for which the library of life would prove essential. Suppose, though, that we manage to save a large fraction of a system. Then the species library will provide a genetic "snapshot" of biodiversity at a given time and place, which evolutionary biologists can compare with the system as it has evolved much later. This would be a new form of research tool.

Already a crash program to collect permanent cell lines, DNA, or both from vanishing human populations has excited attention (Cavalli-Storza 1986). This program maintains cell lines by continuous culture, a costly method that invites random mutation. Such records may allow a deeper understanding of our own origins and predispositions (Diamond 1991), but banking frozen tissues of endangered species is the only way to ensure that any genetic disease diagnosed in the future in small, closed populations (the "founder effect") can be mapped and managed (Ryder 1988). Long ago, Ehrlich (1964) suggested the creation of "artificial fossils" in such a fashion. The "frozen zoo" in San Diego, California, begun with this in mind, has immersed 2,400 mammal fibroblast cell cultures and 145 tissue pieces in liquid nitrogen—about 300 species in all. Cryonic mouse embryo banking for genetic studies is now routine (Glenister, Whittingham, and Wood 1990).

All *ex situ* programs have liabilities, arising from the brevity of human concerns. Programs from seed banks to systematically frozen samples have all failed because a career ends, funding dries up, freezer space gets scarce, or zoos lose interest. A deliberate, full-scale, and institutionalized attack can perhaps better survive the erosions of time.

The far larger prospect of eventually reading and using a library of life is difficult for us to imagine or anticipate, at the early stages of a profound, dramatic revolution in biological technology. Our situation may resemble that of the Wright brothers, had they tried to envision a moon landing within three generations.

## Tropical Tactics

As an example of a truly massive program, consider sampling the rain forests. This poses perhaps the most difficult, stratified problems. Ground-based sampling methods of terrestrial insects and animals are common and well-established. Litter-sifting techniques can be especially effective in capturing a large diversity of species. Tactically, this is ground well-covered. But big problems loom overhead.

Rain forest trees often have distinct levels of stratification, with the tallest attaining heights of 40 or 50 meters. Indeed, some trees in Southeast Asia can reach as high as 70 meters. Most photosynthetic activity occurs at the canopy, rich in leaves, fruits, and flowers. As well, the canopy hosts countless insects, epiphytes, and many higher order animals.

This complicates study, as there is no convenient means for reaching the canopy. We have easy direct access only to the bottom two meters of the forest, or about 5 percent by volume. Unfortunately, these two meters are often the least productive zone. Clearly, gaining access to the canopy is paramount in order to accurately sample the forests' diversity. Climbing trees to sample the uppermost layers is arduous, dangerous, and time consuming. Researchers have often spent months climbing stands of fewer than 100 trees, attempting to gain statistical information.

The library of life proposal does not seek this type of information. There would be no attempt at determining quantitative information, such as species density or diversity. Furthermore, resources do not exist for thorough sampling of every plot on a global scale. Nevertheless, the canopy must be sampled.

One method of sampling the canopy involves fogging it with biodegradable pyrethrin. Many insects and other invertebrates are overcome by the fog and fall onto plastic sheets, where they can be collected. This method led Erwin to his 1990 estimate of 30 million insect species. It is also effective in collecting a great variety of species (Ryder 1988). Other research focuses on accessing the canopy from the sky, via dirigibles or inflatable platforms that rest on the tops of the trees. These exotic methods are still experimental, but should be discussed; relatively small, successful investment here could greatly simplify the collecting task and lower costs. Now consider expense. After the initial investment for equipment and collecting, which could cost $100 million, the biggest expense will be replacing liquid nitrogen. This depends ultimately on how much organic matter is collected.

Details of how big the sampled areas should be, how many plots should be sampled, and so forth, need statistical discussion. For simple concreteness, consider the following method. Suppose we sample 9,000 plots of about 100 meters on a side. This is 1 for every 1,000 square kilometers of rain forest. If each plot costs $10,000 to sample (likely an overestimate), then $90 million for sampling would be needed. Assume 100 trees are sampled per plot, taking 10 kilograms from each tree and its surrounding area, including soil. This yields 9 million kilograms to be preserved. The greatest bulk of sampled matter, by volume, will be plant material, with a fraction of the density of water. Assume this fraction to be one for the moment, and assign it to the packing fraction, discussed shortly. This means we need 9,000 cubic meters of cold storage.

A variety of biomedical freezers and storage vessels abound, but most can merely hold small vials for sample storage. By far the most efficient are the largest, so-called "big foot" containers. These stainless steel, cylindrical dewars are vacuum sealed and internally wrapped 75 times with "super insulation." They have an effective capacity of 1,760 liters, with a liquid nitrogen boil-off rate of only 12 liters per day.

To allow for the wide range in packing methods, and sample densities ranging from tubular plants to topsoil, let us assume a volume packing fraction of 75 percent. Then we need about 6,800 big foot containers. With liquid nitrogen costs at current levels of about 20¢ per liter, the cost of keeping the samples frozen for 100 years comes to about $600 million.

The big foot dewars can be mass-produced at under $10,000 each, so containers will cost $68 million. The up-front costs for sampling, containers, and a $20 million facility come to $178 million. Even with $200 million for employee salaries and the $600 million for liquid nitrogen, it is difficult to imagine a total of greater than $1 billion. That comes out to an average cost of only $10 million per year over 100 years.

Conservative estimates have been made throughout, so the total may be considerably less. The most uncertain variable is the average density of samples, multiplied by the packing fraction. Assuming that we can cleverly pack materials at 0.75 the density of water may be optimistic. Since the largest term, nitrogen cost, is linear in this quantity, storage for a century could cost $6 billion if the product were 0.075, for example. Clearly, this is the critical parameter.

However, accurate numbers are not crucial at this stage. This exercise shows that the idea is, indeed, affordable, and in fact, quite inexpensive compared with buying and maintaining a reserve of the sampled area in the tropics.

This simple scenario uses existing technology, undoubtedly not the best possible. For example, larger and more efficient dewars should be developed, for substantial savings. Also, liquid nitrogen accounts for the greatest expense and probably can be more affordably made in large, dedicated production plants. Lowering the cost of liquid nitrogen by even a few cents per liter yields huge savings; about $30 million saved for every penny decrease in the nitrogen price per liter in a century. This would easily justify spending a few million for liquid nitrogen production facilities. But once salvaged, how much might we recover?

### Cryostress

Seeds can germinate after lengthy freezing and microbes can routinely sustain cryogenic temperatures. Simple cells, such as sperm and ova, survive liquid nitrogen preservation and function after warming. Generally, organs with large surface to volume ratios preserve well, such as skin and intestines.

Of course, more complex systems suffer great freezing damage, although ongoing research is attempting to minimize this. Several kinds of damage occur, and little is known about methods of reversing such injury. Biochemical and biophysical freezing injury arises from shrinking cell volume as freezing proceeds. Plants display extrusion of pure lipid species from the plasma membrane, as cells contract during freezing (Mazur 1984, 47–56). Such lipids do not spontaneously return to the plane of the membrane during volume expansion on thawing, so that restoration of approximate isotonic volume near the melting point causes cellular lysis due to inadequate membrane surface area. While osmotic injury can be reversed, there is a net loss of membrane proteins. Reorganization of membrane bilayer structure into cylindrical lipid tubes may be reversible with warming (Lyons 1972; Kynoova, Tenchov, and Quinn 1989; Steponkus and Lynch 1989). Structure of the cytoplasm may break down into blobs of proteinaceous matter (Jacobsen et al. 1988; Fahy 1981). Major fracturing of cells, axons, dendrites, capillaries, and other elements causes extensive damage at temperatures below the glass point (Fahy, Saur, and Williams 1990), suggesting that this be avoided. For some purposes, then, immersion in liquid nitrogen may be unacceptable.

Even then, all may not be lost. The problem of recovering cells from frozen samples is complex, but even low survival rates of one cell in a million are irrelevant if the survivor cells can produce descendants. We need not rely on present technology for the retrieval. Progress in biological recovery can open unsuspected pathways.

Recent advances underline this expectation. Techniques such as the polymerase chain reaction can amplify rare segments of DNA over a million-fold (Mullis 1987). Such methods have enabled resourceful biologists to recover specific segments from such seemingly unlikely sources as a 120-year-old museum specimen, which yielded mitochondrial DNA of a quagga, an extinct beast that looks like a cross between a horse and a zebra (Higuchi et al. 1984). A 5,000-year-old Egyptian mummy has yielded up its genetic secrets (Paabo 1985). Amplifiable DNA in old bone is beginning to open up the study of the bulk of surviving organic matter from the deep past (Hagelberg, Sykes, and Hedges 1989). The current record for bringing the past alive, in the genetic sense, is DNA extracted from a fossilized magnolia leaf between 17 and 20 million years old (Golenborg 1990). This feat defied the prediction from *in vitro* estimates of spontaneous hydrolysis rates, which held that DNA could not survive intact beyond about 10,000 years (Sykes 1991).

We should recognize that future biological technology will probably greatly surpass ours, perhaps exceeding even what we can plausibly imagine. Our attitude should resemble that of archaeology, in which a fraction of a site is deliberately not excavated, assuming that future archaeologists will be able to learn more from it than we can.

### Trade-offs

Sampling and freezing have little aesthetic appeal. To some, they will smack of fatalism; it may be merely realism. As well, freezing species does not offer the immediate benefits that *in situ* preservation yields. (Samples would probably be taken only from areas not already highly damaged.) More concretely, this proposal will not hasten benefits for new foods, medicines, or industrial goods. It will not alter the essential services an ecosphere provides to the maintenance of the biosphere. We should make very clear that this task is explicitly designed to benefit humanity as a whole, once this age of rampant species extinction is over.

*Ex situ* methods do not address many legitimate reasons for preserving ecospheres intact, and it should not be seen as opposing them. Indeed, only by preserving *in vivo* a wide cross section of biota can we plausibly use much of the genetic library frozen *in vitro*.

Habitat preservation may compete politically with a sampling and freezing program, but there is no intrinsic reason why this need be so. They are not logically part of a zero-sum game because they yield different benefits over different time scales. Of course, we would all prefer a world that preserves everything. But the emotional appeal of preservation should not disguise the simple fact that we are losing the battle, let alone argue against a prudent suspension strategy.

Sampling is far less expensive than *in situ* preservation—which is why it is more likely to succeed over the long run. Even competition for "debt swap" funds will not necessarily be of the same economic kind. Conservationists seek to buy land and set up reserves, putting funds into the hands of (often wealthy) landowners. A freezing program will more strongly spur local, largely unskilled employment, affecting a different economic faction.

Some will see in this idea a slippery slope: to undertake salvaging operations weakens arguments for *in situ* conservation. To avoid this, two parallel programs of *in situ* preservation and *ex situ* conservation freezing must be kept clear. Only when *in situ* measures collapse should the *ex situ* program stand alone. Though they should cooperate, in the real world, *ex situ* funds do not come directly from *in situ* programs. If the Topeka Zoo budget is cut, the city does not transfer funds to Zaire to save gorillas.

This two-pronged proposal, strongly linking *ex situ* with *in situ*, avoids the problem of deciding which species are of probable use to us, or are crucial to biodiversity. By sampling everything we can, we avoid some pitfalls of our present ignorance. Too often, preservation efforts focus on "charismatic vertebrates," neglecting the great bulk of diversity (Soulé 1991), and with no *ex situ* backup. In the long run, this could be fatal for many species, which may never be known to anyone, present or future.

### References

Benford, G. 1992. Saving the "library of life." *Proceedings of National Academy of Sciences*, USA 89:11098–101.

Cavalli-Storza, L.L., J.R. Kidd, K.K. Kidd, C. Bucci, A.M. Bowcock, B.S. Hewlett, and J.S. Friedlaender. 1986. DNA markers and genetic variation in the human species. *Cold Spring Harbor Symposia on Quantitative Biology* 51:411–18.

Dial, K.P., and John M. Marzluff. 1988. Are the smallest organisms the most diverse? *Ecology* 69:1620–24.

Diamond, J.M. 1991. A way to world knowledge. *Nature* 352:567.

Ehrlich, P. 1964. Some axioms of taxonomy. *Systematic Zoology* 13:109–23.

Erwin, T. 1990. *The Ground Beetles of Central America*. Washington, D.C.: Smithsonian Institution Press.

Fahy, G.M. 1981. Analysis of solution effects injury. *Cryobiology* 18:550–70.

Fahy, G.M., J. Saur, and R.J. Williams. 1990. Physical problems with the vitrification of large biological systems. *Cryobiology* 27:492–510.

Glenister, P.H., D.G. Whittingham, and M.J. Wood. 1990. Genome cryopreservation. *Genetical Research* 56:253–58.

Golenberg, E.M., D.E. Giannasi, M.T. Clegg, C.J. Smiley, M. Durbin, D. Henderson, and G. Zurawski. 1990. Chloroplast DNA sequence from a Miocene *Magnolia* species. *Nature* 344:650–58.

Hagelberg, E., B. Sykes, and R. Hedges. 1989. Ancient bone DNA amplified. *Nature* 342:485.

Heywood, V.H. 1992. Efforts to conserve tropical plants—a global perspective. In *Conservation of Plant Genes*, edited by R.P. Adams and J.E. Adams. San Diego: Academic Press.

Higuchi, R., B. Bowman, M. Freiberger, O.A. Ryder, and A.C. Wilson. 1984. DNA sequences from the quagga, an extinct member of the horse family. *Nature* 312:282–84.

Jacobsen, I.A., D.E. Pegg, H. Starklint, C.J. Hunt, P. Barfort, and M.P. Diaper. 1988. Introduction and removal of cryoprotective agents with rabbit kidneys: Assessment by transplantation. *Cryobiology* 25:285–99.

Koynova, R.D., B.G. Tenchov, and P.J. Quinn. 1989. Sugars favor formation of hexagonal ($H_{I}$I) phase at the expense of lamellar liquid-crystalline phase in hydrated phosphatidylethanolamines. *Biochim. Biophys. Acta* 980:377–80.

Lyons, J.M. 1972. Phase transitions and control of cellular metabolism at low temperatures. *Cryobiology* 9:341–50.

Mazur, P. 1984. Freezing of living cells: Mechanisms and implications. *American Journal of Physiology* 247:C125.

Mullis K.B., and F.A. Faloona. 1987. Specific synthesis of DNA *in vitro* via a polymerase-catalyzed chain reaction. *Methods in Enzymology* 155:335–50.

Paabo, S. 1985. Molecular cloning of ancient Egyptian mummy DNA. *Nature* 314:644–46.

Ryder, O. 1988. Founder effects and endangered species. *Nature* 331:396.

Soule, Michael E. 1991. Conservation: Tactics for a constant crisis. *Science* 253:744–50.

Steponkus, P.L., and D.V. Lynch. 1989. Freeze/thaw-induced destabilization of the plasma membrane and the effects of cold acclimation. *Journal of Bioenergetics Biomembranes* 21:21–41.

Sykes, B. 1991. The past comes alive. *Nature* 352:381.

Terborgh, J. 1992. *Diversity and the Tropical Rain Forest*. New York: W.H. Freeman.

Wilson, E.O. 1989. Threats to Biodiversity. *Scientific American* 261:108–14.

# III Economic Responses

# Sustainable Development and North-South Trade

by Graciela Chichilnisky

The present acceleration of environmental destruction can be linked to the economic trading strategies that came into vogue after World War II. The theory of comparative advantages of trade, which recommends that developing countries emphasize resource exports and exports of labor-intensive products, has proven devastating to both the economies and environments of Latin America and Africa. In contrast, the Asian Tigers approach based on external economies of scale, has generated knowledge-intensive products where benefits spread across whole industries and whole economies, leading to more economic growth with much less environmental degradation. Such an approach should be promoted throughout the world trading system instead of the resource-intensive patterns of growth that continue to threaten our global environment. This is particularly important because other resource-conserving strategies, such as green accounting and property rights regimes, remain politically unattainable.

*

## The Global Environment Today

Human beings, or their close genetical relatives, have lived on earth for several million years. Yet it is only recently that human activity has reached levels at which it can affect fundamental natural processes, such as the concentration of gases (chlorofluorocarbon, or CFC, and $CO_2$) in the atmosphere of the planet, and the complex web of species that constitute life on earth. There is considerable uncertainty about the magnitude and impact of the changes that humankind is causing, but it is known that industrial activity has, for the first time in history, reached levels at which it can alter the planet's atmosphere and destroy its biodiversity. Incidents of destruction of particular species are recorded; however, the overall destruction of biodiversity on the planet is not (Kalin, Raven, and Sarukhan 1992). Some fear that humanity's survival may be at stake.

The June 1992 Earth Summit in Rio de Janeiro underscored this concern. In the United Nation's Agenda 21,150 nations explicitly recommended new development patterns based on the satisfaction of basic needs, a concept created and developed within the Bariloche model (Chichilnisky 1977a; Herrera, Scolnik, and Chichilnisky 1976). The summit chose three major areas in which concerted international action is urgently needed: biodiversity, climate change, and sustainable development. Two Framework Conventions were assigned the task of designing international policies that can avert damage in these areas. The areas are so closely connected as to be inseparable in real as well as conceptual terms.

Biodiversity cannot be seen in isolation, and the threats to biodiversity must be contextualized within the patterns of economic development. Most biodiversity loss is due to habitat destruction, for example through deforestation. Deforestation, in turn, is mostly due to the transformation of forested areas into agricultural sites for growing cash crops, or for exploring and extracting resources such as petroleum. The driving force behind most biodiversity loss is economic development based on the exploitation and intensive use of natural resources.

Often the exploitation of resources is for trade in the international market. Petroleum exploitation in the Amazon basin and copper extraction in Chile are good examples, and so are most cash crops grown in Central and South America and in Africa, such as sugar, coffee, and soya and palm oil.

This chapter examines the problem from a North-South perspective, focusing on patterns of trade between industrial and developing nations. It suggests patterns of sustainable economic development that are in harmony with the earth's resources. In the last 50 years, industrial society has developed a voracious appetite for the earth's resources, and in the midst of its "golden age," it is starting to bite its own tail.

Major changes in our understanding of economic development are needed to ensure the preservation of the world's remaining resources and its biodiversity. Economic development is not about doing more with more, but rather, doing more with less.

## Major Concerns and Little Action

Three years have passed since the Framework Conventions were charged with their task at Rio's Earth Summit, but little action has been taken so far. Two major factors hinder the negotiations. The first is the differing perceptions of global environmental problems by industrial and developing countries. A second contributing factor is that our scientific knowledge of global environmental issues is poor.

Global environmental problems are relatively new; because they are global in nature, traditional physical sciences are ill-equipped for understanding them. However, scientific uncertainty about physical phenomena should not by itself detain the negotiations. After all, one often assumes decisions under uncertainty; in this case, under scientific uncertainty. It is the first problem—North-

TABLE 1   Share of Total World Carbon Dioxide Emissions, Population, and GNP
Industrial and Less-developed Countries 1986

| Countries | Cumulative $CO_2$ Emissions | Current $CO_2$ Emissions | Population | GNP |
|---|---|---|---|---|
| Industrial | 70% | 60% | 23% | 84% |
| Less developed | 30% | 40% | 77% | 16% |

*Sources*: Author's calculations are based on Susan Subak, *Accounting for Climate Change* (Stockholm: Stockholm Environmental Research Institute, 1990).

South differences—that hinders the negotiations. This point was underscored at a conference of the parties in Berlin in April 1995. The North-South differences go to the root of the global environmental problem, and lead us to question fundamental issues about economic development, international trade, and the distribution and use of resources on the planet.

## North-South Issues

Developing countries view global environmental issues in a historical perspective, and observe that most of the damage to the global environment originates in the industrial countries. The industrial countries, for obvious geographical reasons, are called the North. For the same reasons, the developing countries are called the South.

Greenhouse gas emissions are a prime example. Carbon emissions arise from the use of energy: the more energy that is used, the more carbon that is emitted. Energy consumption is proportional to the level of production, and most production takes place in the developed countries of the North, which house about 20 percent of the world's population. Table 1 shows that 70 percent of the world's $CO_2$ and most of the CFCs are emitted by the industrialized countries, which have also substantially altered their biomass. In per capita terms, the emission of carbon is 10 times greater in industrial counties than in developing countries. Therefore, the North consumes most of the energy produced in the world.

By contrast, most of the biodiversity and forests remaining on the planet are in the South, home to most of the planet's people as well. The developing countries account for four-fifths of the total world population. From the perspective of developing countries, the North is fundamentally responsible for the current situation, and nothing short of changing the North's pattern of development and environmental use can change matters. The problem, of course, is that restricting emissions requires restricting energy use and output. Therefore, the North must decrease its use of resources to make a dent in the problem.

Industrial countries have a different perspective on global environmental problems: they focus not on the past or present, but on a long-run future. They fear rapid population growth in developing countries and the damage that this could eventually produce on the environment. China and India, countries with enormously large populations and fast growth rates, also have large coal deposits. As their industrial growth proceeds, they are expected to burn coal and emit carbon into the atmosphere in substantial amounts. In the case of China, generally accepted projections indicate that it could approximate U.S. levels of carbon emissions in about half a century. At present, the United States emits around 25 percent of all carbon emissions and consumes approximately 25 percent of all petroleum produced, even though it contains less than 5 percent of the world's population. Replicating such a pattern could potentially predicate disaster.

Much rhetoric has focused on population growth as a source of environmental problems. In a sense, the issue is real: without humans, the problem that concerns us today would not exist. It is incorrect, however, to blame matters on population growth. Those regions of the world that have the lowest population growth, namely the North, account for most of the damage to the global environment. While this trend could, of course, be reversed in the future, global environmental damage has so far not been related to population growth.

### Overconsumption and Overproduction

The world's energy use, and the concomitant use of the atmosphere to absorb carbon emissions, is symptomatic of a larger problem: the use of natural resources as a whole. Most exhaustible resources, such as minerals, and most of the renewable ones, such as wood and the agricultural products obtained from fertile land, are consumed in the North. To a large extent, most of these resources are overproduced in and then exported from the South. In fact, two-thirds of all Latin American exports are resources.

The interrelated overconsumption and overproduction of fossil fuel, for example, can be described as a North-South process. Most carbon emissions originate from the burning of fossil fuel, which is generally exported by developing countries to industrialized ones. Further, industrialized countries consume 160 gigajoules of fossil fuel per person as compared to 17 gigajoules per person in developing countries. For aluminum, the figure is 16 to 2; for copper, 16 to 1; and for beef and veal, 27 to 4.5 (WRI, UNEP, and UNDP 1995).

The most prominent fossil fuel is petroleum, and its overproduction and overconsumption clearly implicates the international market. The general consensus is that the world's rapid rate of consumption of fossil fuels is linked to low petroleum prices: the lower the price, the higher the consumption. Petroleum prices in the United States are 2.5 to 3 times lower than those paid by the German and Japanese consumer. Corresponding to this, the United States uses petroleum much less efficiently than Japan and Germany: a unit of GDP in the United States has approximately a 40 percent higher petroleum content. The United States imports its petroleum from Latin America—Mexico, Venezuela, and Ecuador—despite the fact that about 20 percent of the recoverable petroleum deposits in the world are in North America.

While this North-South pattern of resource use has been known for some time, it has not been considered a problem until recently; in fact, many economists view this pattern as a manifestation of the efficient functioning of markets.

What is less clear is how this situation of overproduction and overconsumption beyond the point of sustainability evolved. Many of us have been considering this question for the last 20 years.[1] But it seems fair to say that many have not thought about this question often, because it does not fit easily with traditional economic views of the world, and in particular, the standard vision of economic development and international trade based on resource-intensive exports by developing countries.

## Economic Causes and Ecological Consequences

The environmental question is difficult because it falls in the gap joining several different disciplines: economics, biology, and earth sciences. In simple terms, the causes of the phenomenon that we face are economic: we destroy biodiversity and its habitat, the forests, for economic reasons. For example, 90 percent of the deforestation that takes place in the tropics, which house 60 to 70 percent of the world's biodiversity, is to the land to grow cash crops, most of which are for the international market. The Amazon forest is being cleared in Ecuador for the exploration and extraction of petroleum by U.S. companies, despite protests from indigenous people; oil accounts for 50 percent of Ecuador's exports. In Brazil, the Amazon forest is used as a source of wood, such as mahogany, of which 50 percent is exported to Britain; it is also cleared to grow coffee and soybeans, again for international market. The Korup forest between Cameroon and Nigeria—at 60 million years old, the oldest forest in Africa—is used as a source of palm oil, also sold in the international market. Therefore, although the causes are economic, the results are biological and physical.

This often means that one discipline—economics—observes the causes, and others—such as biology and geophysics—observe the results. Yet each misses a crucial side of the equation. One can only deal with the global environmental dilemma by looking at both sides of the equation. This requires interdisciplinary cooperation, a subject that does not come easily in our traditional university system.

## The Postwar World Order

Why is it that after millions of years on the planet, humans now perceive a global environmental crisis that could threaten the species survival? Have matters really changed in recent years?

Most scientists now agree that they have, and they give a date for the onset of today's global environmental problems. The last 50 years are seen as the period when most of the damage to the planet's biodiversity and the change in atmospheric concentration of greenhouse gases has occurred. There is wide agreement that the emission level of greenhouse gases in the past half decade has exceeded the extent of emissions in recorded history. Indeed, biologists claim that today's devastation is only comparable to four or five main incidents of global biodiversity destruction, such as the fall of a meteorite held to be responsible for the disappearance of the dinosaurs. What happened fifty years ago?

Fifty years ago, World War II was won by the Allies. For the first time, the United States of America, which led the victory, dominated the world economy, producing 40 percent of the world's products as a consequence of the war's destruction of the Japanese and European economies. Today, the United States is back to a much lower level, approximately 25 percent of the world economy measured in terms of the Gross National Product (GNP).

After the war was won, a new world order emerged with its norms punctuated by the creation of four major international organizations: the United Nations, the International Monetary Fund (IMF), the World Bank (WB), and the General Agreement on Trade and Tariffs (GATT). These organizations followed and implemented the leading country's vision of economic growth: an extremely resource-intensive growth corresponding to a rapidly expanding frontier economy, with an enormous consumption of resources, and the domination of nature through rapid technological change.

In the last 50 years since the end of the war, the world has grown at a rapid pace. Yet, international trade has grown much faster: in fact, three times faster than the overall growth of the world economy during the same period.

At the same time that these four major international organizations were created to implement a new world order, two major theories of trade and growth were developed and implemented. One was the neoclassical theory of optimal economic growth, originated in the United States, which views as a long-run steady state a path of development with exponential rates of population growth and correspondingly exponential increases in the use of resources. This theory ascribes to unlimited expansion in terms of the economy and of its use of resources, paralleling the pattern of development in the United States. The second economic theory was complementary to the first, and originated in Sweden, although it was also widely applied and developed in the United States. This is the theory of comparative advantages in international trade, which recommends that developing countries emphasize resource exports and exports of labor-intensive products, and in turn, trade these for capital- and technology-intensive products produced by the industrial countries. The vision of development that these theories advanced was one based on unlimited and inexpensive resources. Even today, this view is prevalent in the United States: inexpensive oil is seen as the basis for healthy economic growth, almost a birthright of its citizens, a right for which wars can be, and are, fought. Any rational attempt to redress this view meets with political failure.

These two theories of growth and trade, which have prevailed since the 1950s, have had major implications for the way we use and trade resources. These theories were implemented by international organizations such as the WB and IMF, both of which provided strong incentives to developing countries to follow resource-intensive development patterns, and recommended exporting more and more resource-intensive products as a precondition for loans and other important economic incentives. At least as important, economists and civil servants from developing countries under the influence of the United States were imbued with a sense of finality in the way things are: developing countries are only good for resources, cash crops, and cheap labor products. This proposition is largely uncontested today in Latin America and Africa, the two areas that have fallen behind in terms of economic growth over the past two decades. It underscores the heavily resource-intensive patterns of production and exports in these two regions.

Why do developing countries overproduce and overexport resource-intensive products such as cash crops, which require extensive land clearing, and minerals such as petroleum, which also affect the health of many forested areas? Why do developing countries export environment-intensive products at prices below the social costs? Is it true that developing countries have a comparative advantage in environment-intensive products, such as cash crops, minerals, and dirty industry, which use clean air intensively? If so, doesn't efficiency dictate that this comparative advantage be exploited, and shouldn't this lead to everyone's gain? In sum, is there a fundamental contradiction between economic gain and environmental preservation?

The answer to this latter question is "no," and it leads to a new theory of why countries trade (Chichilnisky 1994a, 159–69; 1994b). All this is found by analyzing the behavior of competitive international markets, taking into account important institutions, such as property rights for common property environmental resources, which are typically different in industrial and developing countries.

## Property Rights and International Trade

Before industrialization, many traditional societies had long managed their common property resources, such as fisheries and forests, using various forms of local governance. *Common property* is a term that refers to shared group ownership, rather than individual ownership. An example is Valencia's *Tribunal de las Aguas*, a local 1,000-year-old Spanish court, which still meets today on a weekly basis to administer costs and allocate the use of the region's water network. Other examples are the Iriaichi system of managing common lands in Japan, and Bahia's system of sea tenure in northeastern Brazil (Chichilnisky 1994b). These traditional systems, however, require a population that is both stable—in the sense that successive generations remain in the same area—and not too large—so that penalties from antisocial use of resources can be administered effectively and, if necessary, across generations (Stone, this volume).

Such systems of resource management tend to break down during periods of industrialization, when outsiders move into the common property area, and can easily move out to avoid penalties. During the process of industrialization, then, populations become large, mobile, and unaccountable. Well-managed common property is treated as unmanaged "open access" resources, which can be had for the taking. A "first come, first serve" system prevails.

In many of the now-industrialized countries, industrialization was preceded by the privatization of common property resources. For example, in the United Kingdom, industrialization was proceeded by a major change in property rights: the privatization of the commons. With large and mobile populations, private property regimes often work better in the conservation of local resources than do common property regimes; the privatization of oil in the United States, the "Hot Oil Act" of 1936, is an example. The United States extracts little oil compared with the levels of extraction in developing countries with less well-defined property rights on this resource, such as Mexico. This is true even though the United States has enormous deposits, while Mexico does not.[2] The fact that the United States uses its local oil resources more carefully, however, does not mean that it consumes less oil: the difference between production and consumption is made up by imports, and the United States economy is today the largest single importer of oil in the world.

Why do property rights matter? When a pool from which a resource is extracted, such as a forest or a lake, is treated as open access, only the cost of actually extracting the resource, such as a tree or fish, is computed. The responsibility for the cost of managing the system, which is often substantial, is not computed. In such cases, noncooperative systems of exploitation emerge: at each market price, more is extracted under open access regimes than under traditional managed systems or private property regimes. The resource is overextracted, and can dwindle or even disappear (Dasgupta and Heal 1979; Chichilnisky 1994b).

Private goods are goods whose consumption is rival, in the sense that what one person consumes, others cannot. Furthermore, the levels of consumption can be chosen independently by each person. Examples of private goods are eatable products. Public goods differ from private goods in that they are available to everyone in about the same amount, and within limits, are not rival in consumption—for example, a road, a bridge, or an army. Moreover, with public goods, one person's consumption need not detract from others'. A good example is knowledge: one may share knowledge with others without losing it oneself. (Of course, knowledge should not be identified with the financial gains that can be obtained from it.)

Classic public goods are supplied by governments. Biodiversity or greenhouse gas concentrations in the atmosphere are public goods, but not in the classical sense. They are produced by each individual in the economy, rather than by governments. For example, carbon emissions are "produced" by each person or firm privately when they drive their cars or burn fossil fuels to release energy. These are private activities that a government does not generally regulate.

The trading of private goods is very different from the trading of public goods. In markets with private goods, efficiency is divorced from equity in the sense that under any distribution of property rights, a competitive market with private goods achieves an efficient outcome at a market equilibrium. This is not true in markets with public goods. In such markets, there is a relation between efficiency and equity. A rigorous, general equilibrium treatment of markets in which some of the goods are private, and others are privately produced public goods, such as property rights on emission of carbon dioxide, is in Chichilnisky 1994a, Chichilnisky and Heal 1994, and Chichilnisky, Heal, and Starrett 1993. These papers show that certain property rights regimes on the use of global environmental goods are consistent with the efficient operation of competitive markets, and others are not. A certain "equity" is needed for environmental markets to operate efficiently.

It has been shown that if a traditional economy treats a pool from which a natural resource is extracted as open access, then at each market price it will offer more of the resource, leading to an *apparent* comparative advantage in resource-intensive products even where there is none. More resource-intensive products will be produced at each market price than is socially optimal. In particular, more is available for export at each price.

A typical example of this phenomenon emerges in societies where resources are unregulated, national property effectively treated as open access. Besides apparent comparative advantages, it also leads to apparent gains from trade, even in cases where they are losses. For example, Honduras exports mahogany to the United States even though it has no comparative advantage in wood products. Mexico exports petroleum to the United States even though it has small reserves.

Resource-intensive products are exported at prices that are below social costs. They are overconsumed by the countries with well-defined property rights and overproduced by those with ill-defined property rights. Moreover, export countries do not compute the cost of replacing the stock of trees harvested for wood or the cost of the depleted resource in a case of oil. Finally, the world economy as a whole consumes an inefficient quantity of resources, because it does not take into account the costs to the world economy of the resource overuse.

It is remarkable that all this happens even while markets are perfectly competitive. Of course, monopolistic practices may exist, but we need not invoke them in any way to explain these inefficient patterns of consumption, production, and trade. The overuse of resources, and the inefficient market solutions just described, do not derive from market imperfections. Rather, they derive from a defective system of property rights as the world economy moves away from traditional forms of resource management into industrial societies. The process of industrialization itself leads to the patterns of North-South trade that we observe, and that are at the core of the environmental dilemma today.

## Green Accounting

A proposal currently being considered by the United Nations would modify the system of national accounts followed by all countries to formally incorporate environmental costs. Green accounting is the practice of deducting environmental costs from the computation of the GDP. For example, the national accounts would depreciate the value of the stock of forests or minerals extracted, much in the same way that private individuals and firms depreciate the value of their own stock when reporting their personal or corporate income.

Green accounting can indeed help to reduce the overuse of resources, as well as excessive extraction and trade, by correcting the miscalculation described above. It can make a large difference in reporting economic performance based on GDP in resource-intensive countries. For example, the GDPs of Costa Rica and Mexico were recomputed using this practice, and dropped to a fraction of their former level as computed by standard practices.

Green accounting can only help indirectly, however, by deducting the depreciation of the stock of exhaustible resource or the cost of replacing the stock of renewable resources from the GDP. It is an indirect mechanism in that it can only induce growth-oriented politicians to follow more environmentally sound policies if they perceive their political target as the maximization of national economic growth. If a politician's reelection depends on the measure of national economic growth, and it often does, Green accounting could be helpful in reorienting environmental policy.

## The Asian Tigers, Latin America, and Africa

After disclosing the negative impact that North-South trade has had on the global environment, the natural question that arises is whether a solution can be found that does not conflict with free markets. Some policymakers, for example, advise import substitutions, tariffs, or other practices that restrict trade. It is true that in emergencies, a ban on the trade of certain species that are close to extinction may be necessary: examples include trade in elephant tusks, tigers, and more recently, in the United States, box turtles. Humans' irrational cruelty to animals admits on occasion to no other redress. The problem, however, goes beyond the restriction of animal trade: the overexploitation of minerals and forestlands inflicts damage not only to their producers, but also to their consumers.

Yet it is possible to reorient patterns of trade and development without hindering or interfering with international trade (Chichilnisky 1992). A positive example of such a strategy is offered by the Asian Tigers, while Latin America and Africa serve as negative examples. The former are export oriented, but have moved swiftly away from traditional comparative advantages—such as labor-intensive and resource-intensive products—into knowledge-intensive products. Such products include microprocessors, consumer electronics, financial products, and other sophisticated, technology-based products. The term *economies of scale* refers to the fact that such products are produced more efficiently at larger scale levels. An example is provided by the well-known concept of "learning by doing," where the more one produces, the more productive one becomes. This means that their prices drop as their supply increases: a typical case of economies of scale is the computer industry.

*External economies of scale* is a term used to characterize production processes where efficiency gains at larger scales are not restricted to one firm. They are, instead, distributed throughout the whole industry or economy. In other words: the more that is produced, the more productive all producers are. Such external economies of scale are typical of industries that require knowledge as an important input. For instance, a better-trained labor force benefits all firms, not just one of them. Knowledge diffuses across a whole industry. Examples include electronic products, hardware and software, biotechnology, and electronic-based services such as data communication, consumer electronics, and financial services, all of which are based on knowledge. These are the most dynamic sectors in the world economy today.

An important aspect of external economies of scale is that they are not connected with monopolistic behavior, require neither large capital outlays nor large-sized plants. To the contrary, external economies of scale typically occur in industries with many competitive firms.

The type of skilled labor required for industries with external economies of scale is available in many developing countries. Mexico is currently a producer of electronic products such as microchips and software; India is becoming one of the largest exporters of software in the world, having produced the software that manages the entire United Kingdom train system. Since software production is labor intensive, and does not require large capital outlays, it fits the Indian and Mexican economies quite well.

Many authors are concerned that educational conditions in developing countries such as the Caribbean would not allow the transition from resource-intensive production to knowledge-based production in the near future. However, the most

FIGURE 1    Composition of World Wealth (Percentage of Total)

Raw Material Exporters (4.6%)

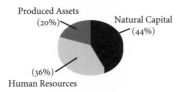

Other Developing Countries (15.9%)

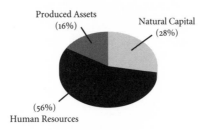

High-income Countries (79.6%)

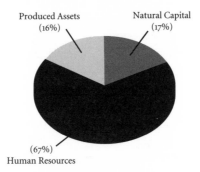

important aspect of knowledge-intensive forms of development, as far as this chapter is concerned, is its low impact on the environment. Knowledge-intensive growth is not only more successful in pure economic terms, it is also more compatible with a sustainable global environment. Recent empirical work at the Inter-American Development Bank in Washington, D.C., belies this view for the Caribbean region as well. Harris (1994) established that the initial conditions found in Caribbean countries 20 years ago, in terms of education and the general satisfaction of basic needs, matched those of the East Asian economies at the same period. Since then, the East Asian countries have moved rapidly toward technology-intensive practices with success. The Caribbean countries, and indeed the whole of Latin America, instead emphasized resource-intensive growth, losing ground. Today, Barbados is redressing this policy and attempting to make a swift transition toward an information age society (Fidler 1995). Mexico and India have active and outward-oriented computer hardware and software sectors.

Some say that the loss of biodiversity might be slowed for a while in countries following the path of the Asian Tigers, but eventual development—including high-class housing, more cars, and roads—may renew the threats to biodiversity while increasing other types of environmental degradation. This is not a foregone conclusion, however. Knowledge-intensive growth need not degenerate into resource-intensive growth. The idea that it must misses an essential point in the argument. Knowledge-intensive growth, if it occurs, is less dependent on resources than resource-intensive growth—initially and forever.

Indeed, the industries that use knowledge most intensively need never revert to resource-intensive production. The information revolution has led to forms of communication and production that decrease, in absolute terms, the need for roads and cars. Computers and communication networks have decreased the need to travel to a workplace. Heating and power managed by computerized systems is more efficient in the use of fuels. The products as well as the processes of the information society are much less intensive in the use of the atmosphere as a sink for emissions than is the industrial society, and of many other resources. The only resource whose demand has increased in recent years in industrial societies is paper, mostly due to the low price of trees for wood pulp.

### Can the Asian Model Be Generalized?

The answer is "yes." The whole world can become knowledge intensive. There are certainly skeptics to this view, and until it happens, this point cannot be proven. But a simple analogy explains how this can, and potentially will, happen.

In the last 100 years, the nations of the North shifted to industrial societies. At the beginning of this process, large parts of their economies were related to the agricultural sector, directly or indirectly. At that point, it would have seemed impossible that the agricultural sector would shrink simultaneously in all of them. But it happened. Today, for instance, the United States has at most 2.5 percent of its GDP in the agricultural sector, and most Organization for Economic Cooperation and Development (OECD) nations have similar proportions. The same process is occurring throughout the South today.

Where will all the food come from when the process of industrialization is completed in the whole world? Can the whole world turn into an industrial society? The answer is "yes, it can." Industrial societies will keep producing food. In fact, they produce even more food today than agricultural societies because they are much more efficient at it. Currently, the industrial countries of the OECD have a more or less permanent surplus of food, while the agricultural societies of the South have deficits. An industrial world need not have a food deficit. The problem with industrial countries is not lack of food; it is, as already discussed, overexploitation of the global environment.

The same argument can be used to understand the transformation process of industrial societies into knowledge-based society that will take place in the next few decades. Ultimately, time will tell.

In any case, there is no reason to believe that the South must follow the North's process of industrialization. The whole point of this chapter is to recommend that it should not. The South should move directly from an agricultural society to a knowledge-intensive society. Examples of agricultural societies that have bypassed heavy industry abound. The area surrounding Chicago is one; a wheat and beef economy was turned into financial products and services. Other examples include India, which is becoming a major software exporter, and Barbados, which intends to become an information society within a generation.

## NAFTA, EU, and WTO

External economies of scale can also play a positive role in reconciling potential conflicts between regional trade agreements (such as the North American Free Trade Agreement [NAFTA] and the European Union [EU]), and the liberalization of world trade (the World Trade Organization [WTO], formerly the General Agreement on Tariffs and Trade [GATT]) (Chichilnisky 1992).

These conflicts can be summarized as follows: a region with more market power has generally more incentives to raise tariffs on outsiders, because it is in a better position to win a "trade war." This means that a regional trading block such as NAFTA would typically conspire against the overall liberalization of the world's trade.

Yet this traditional proposition breaks down with external economies of scale, which take the wind out of the protectionists' sail (Chichilnisky 1992). This happens because anything that restricts trade, such as tariffs, also decreases efficiency and increases domestic costs of production when there are economies of scale. Therefore, a major incentive to raise tariff barriers is gone.

It is widely believed that the main rationale for the formation of the EU was the exploitation of economies of scale among European countries: the fact that each producer could be more efficient when producing for the whole European market than when planning production for their smaller, domestic market. Yet none of this has permeated the logic of NAFTA, which is solidly based, instead, on traditional comparative advantages.

The complete mobility of labor in the EU is a clear indication: it demonstrates that the EU trading block does not see its gains in trade of labor-intensive goods from labor-rich countries, such as Portugal, against capital-intensive goods from capital-rich countries, such as Germany. In contrast, NAFTA does, and again, the lack of mobility of labor within NAFTA is an indication of this fact.

Another negative aspect of regional trade blocks is that they can induce *diversion* of trade, which means that they may lead a country to purchase from a higher-cost producer within the region, rather than from a more efficient, lower-cost producer outside the region. This occurs simply because the outsiders have tariffs on their products, which makes them appear less competitive.

However, economies of scale can also defeat this problem. The higher-cost producer within the region may be so because it produced for a small market, and did not benefit from economies of scale: by offering an opportunity to produce in larger scale for the whole region, the regional trade block may change matters. The high-cost producer can turn into an efficient, low-cost producer. If so, there is no trade diversion.

### Conclusion

Biodiversity cannot be seen in isolation. It is inextricably connected with economic development, as most biodiversity destruction arises from the loss of habitat for economic purposes. A move away from patterns of development and North-South trade based on resource-intensive traditional comparative advantages toward knowledge-intensive patterns of trade and growth is essential. Resource-intensive growth has shown itself to be unsuccessful during the last 20 years, and threatens to destroy our global environment.

Economic growth and trade based on external economies of scale, and on knowledge-intensive products, has proven itself successful in the East Asian economies during the last 20 years, and is more consistent with the environment. Our international organizations (the IMF, WB, and WTO) must evolve to regard knowledge-intensive patterns as a major goal of development.

Biodiversity is hostage to our backward mentality about economic development. Only a fundamental and swift change in our vision, and corresponding changes in our practice of development and trade, can ensure the survival of the planet's remaining biodiversity.

## Acknowledgments

The author wishes to acknowledge the research support provided by the National Science Foundation grants 92–16028 and 94–08798, the Operational Research Department of Stanford University, and Stanford Institute for International Studies.

## Notes

1  For this purpose, I introduced and elaborated the idea of development based on the satisfaction of basic needs (Chichilnisky 1977a, 1977b, 1982) within the context of the Bariloche global model (Herrera, Scolnik, and Chichilnisky 1976), which provided an empirical examination of patterns of trade and development in five continents that could be consistent with environmental preservation. The concept of basic need recently became a cornerstone of the definition of sustainability in the Brundtland Report.

2  The United States has about 2 percent of all the world's known recoverable oil deposits (WRI 1994), while Mexico's resources are expected to be exhausted early in the coming century (Chichilnisky and Heal 1993).

## References

Bromley, D. W. 1991. *Environment and Economy: Property Rights and Public Policy.* Oxford: Basil Blackwell.

Chichilnisky, G. 1977a. Economic development and efficiency criteria in the satisfaction of basic needs. *Applied Mathematical Modeling* 1:290–97.

———. 1977b. Development patterns and the international order. *Journal of International Affairs.* 31:275–304.

———. 1982. Basic needs and resources in the North-South debate. *World Orders Models Project, Working Paper No. 21.* New York: Institute for World Order.

———. 1992. Traditional comparative advantages and economics of scale: NAFTA and the GATT United Nations project on trade liberalization in the Americas. *Journal of International Comparative Economics* 20:147–81.

———. 1994a. Comment: The abatement of carbon emissions in industrial and developing countries. In *The Economics of Climate Change*, edited by T Jones. Paris: Organization for Economic Cooperation and Development.

———. 1994b. North-South trade and the global environment: Technical report no. 31, Stanford Institute of Theoretical Economics, 1992. *American Economic Review* 84:851.

———. 1994c. Property rights on renewable resources and the dynamics of North-South trade. *Structural Change and Economic Dynamics.* Oxford: Oxford University Press.

Chichilnisky, G., and G. Heal. 1994. Who should abate carbon emissions? An international perspective. *Economics Letters* 32:443–49.

Chichilnisky, G., G. Heal, and D. Starrett. 1993. *Equity and Efficiency in International Markets with Emission Permits. Publication No. 81.* Stanford, Calif.: Stanford Center for Economic Policy Research.

Dasgupta, P., and G. Heal. 1979. *Economic Theory and Exhaustible Resources*. New York: Cambridge University Press.

Fidler, S. 1995. An information age society is booming. *Financial Times*, 26 April.

Harris, D. 1994. *Determinants of Aggregate Export Performance of Caribbean Countries: A Comparative Analysis of Barbados, Costa Rica, Dominican Republic, Jamaica, Trinidad, and Tobago*. Stanford, Calif.: Interamerican Development Bank and Department of Economics.

Herrera, A., R. Scolnik, and G. Chichilnisky. 1976. *Catastrophe or New Society? A Latin American World Model*. Ottawa: International Development Research Center.

Kalin, M.T., P.H. Raven, and J. Sarukhan. 1992. Biodiversity. In *An Agenda of Science for Environment and Development in the Twenty-first Century*, edited by J. Dodge, G. Goodman, and J. la Riviere. New York: Cambridge University Press.

WRI (World Resources Institute), UNEP (United Nations Environment Program), and UNDP (United Nations Development Program). 1994. *World Resources: People and the Environment, 1994–95. A Guide to the Global Environment*. New York: Oxford University Press.

———. 1995. *World Resources: The Urban Environment, 1995–96*. New York: Oxford University Press.

# Markets and Biodiversity

by Geoffrey M. Heal

This chapter addresses the issue of whether markets help or hinder the preservation of biodiversity. Biodiversity is valuable as a source of knowledge, in its contribution to life-support systems, and as an intrinsic good independently of any value to people. Four economic values crucial to answering the question posed are discussed. First, property rights are a critical element, though the issue is far from simple; even when biological resources are owned, their diversity is not a tradable commodity and, therefore, is difficult to value. Second, biodiversity is a public good, providing benefits to all humanity. But it is a privately provided public good, which also makes it difficult to value. An innovative economic mechanism for assigning value to biodiversity might be "tradable depletion permits," which would establish a clear economic incentive not to deplete. Third, the costs of a loss in biodiversity will be felt in full only over periods of at least 50 to 100 years, so the discount rate commonly used by economists ensures that biodiversity is given an extremely low value. Correcting this market failure will require a symmetric treatment of the present and the long-term future in such a way as to place a positive value on the very long run, along with explicit recognition of the intrinsic value of biodiversity. And fourth, because the future is uncertain, biodiversity also has an option value that relates to the possibility that it might be more valuable in the future than it is today. The chapter concludes that while the market cannot be relied on to preserve biodiversity, it can provide some incentives for its preservation.

*

## Markets for Biodiversity

Biodiversity is being destroyed at a rate paralleled only by the rates of destruction in historical episodes such as the extinction of the dinosaurs. Today, however, the driving forces are mainly human economic and social organizations. In particular, markets, and the incentives that they generate, are playing a central role. This observation leads one naturally to question whether there is a basic antipathy between market forces and biodiversity. Can markets, in fact, capture the social value that resides in biodiversity? Do they provide incentives to maintain and develop this value? Or do we need different forms of economic and social organization to maintain the earth's biological resources intact? This chapter presents a preliminary review of these issues, trying to classify biodiversity as an economic commodity, understand how it fits into the economic system, and assess the actual and potential adequacy of market economic systems as institutions for managing the resources of biodiversity.

The earth's biodiversity is a characteristic of its biological resources: that is, its total resources of naturally occurring living entities, such as plants, animals, fish, insects, birds, bacteria, and so forth. Biodiversity is a measure of the genetic variance contained within these biological resources; it increases with the number of distinct species. Many populations have critical masses or thresholds: their growth functions are such that once the population falls below a critical number, the species is doomed to extinction. So biodiversity is not related simply to the total quantity of biological resources, but to the distribution of these relative to the threshold levels of the populations of the species represented. If all species are well above these levels, an increase in biological resources does nothing for biodiversity. On the other hand, if some are at or near these threshold levels, an increase in biological resources focused on these populations will increase or maintain biodiversity. More forests, more cod, or more tigers are not *per se* beneficial to biodiversity; this depends on populations relative to thresholds.

The issue, then, is whether markets help or hinder the preservation of biodiversity. Certain cases come to mind immediately in which markets do, in a limited sense, place an economic value on biodiversity. Examples of these include:

- eco-tourism, and in general tourism, based on environmental assets such as coral reefs, unspoiled scenery, unique environments, and so on;
- biological prospecting—INBio, Shaman Pharmaceuticals, and other attempts to exploit traditional knowledge of the medical properties of plants and insects; and
- fisheries and forestry.

Each of these is an example of a market in which one trades goods or services produced from biodiversity. But do these markets provide adequate incentives for the preservation of biodiversity? Do the incentives that they provide cover all important aspects of biodiversity?

The answers to these questions are probably "no" and "no." Nevertheless, markets do provide some incentive to preserve biodiversity, and more intelligent use of market forces can surely be of value in the quest for conservation.

In order to understand both the potential and limitation of market forces in the area of biodiversity, we need to understand two sets of issues. What is the value that resides in biodiversity? And how do these sources of value fit within the general framework of resource allocation theory? A general review of the relevant aspects of economic theory can be found in Dasgupta and Heal (1979) and references there.

## The Value of Biodiversity

There are (at least) four ways in which biodiversity is a source of value.

1. It may be valuable as a source of knowledge. This is the source that is tapped in biological prospecting and the famous Merck-INBio deal.[1]
2. Biodiversity has value in life-support systems. For example, green plants produce oxygen, and bacteria clean water and fertilize soil. All of these activities are crucial to the maintenance of human life.
3. Biodiversity—such as animals, plants, and even landscapes—may have an intrinsic value, a value in and of themselves independently of their anthropocentric value, and similarly, they may have a right to exist independently of their value to humanity.[2]
4. Biodiversity may have cultural and aesthetic values. For example, in certain societies animals or plants have great symbolic value, as with elephants for Hinduism, the bald eagle for the United States, and the lily for France. They are part of the cultural identity and heritage. While the existence of significant aesthetic value for certain types of biodiversity is clear, it is hard to quantify.

The first two sources are instrumental values—they may vanish if we find synthetic substitutes for natural resources in these roles. They value the environment as a means rather than an end. The third source differs in that it recognizes an intrinsic, noninstrumental value in biodiversity.[3]

Some of these sources of value can, in principle, be quantified. For example, consider the role played by bacteria in a watershed area in purifying water that is collected in a reservoir. Adequate bacterial action may obviate the need for chemical water purification. If the bacteria are destroyed—for instance, via pesticides running into the soil—then this action is lost and chemical purification plants are needed. For a large city (such as New York city), these could involve capital costs in excess of $1 billion and then significant operational expenses. In such cases, looking at the provision that would have to be made in the absence of a certain species can provide a way of valuing the actions of that species. There are however, certainly dimensions of the value of biodiversity that are unquantifiable, such as the intrinsic aesthetic and cultural values.

## Efficient Allocation and Biodiversity

Four economic issues are central to an understanding of the extent to which market forces can realize the value of biodiversity. These are property rights, public goods,[4] time horizons, and uncertainty and irreversibility. Each of these represents a dimension of resource allocation problems in which the unaided market is weak, and which is also central to the management of biodiversity. In fact, biodiversity as an economic commodity strikes directly at the market's weakest links. Strengthening these adequately is not impossible, but it is a major challenge for the design of economic institutions

*Property Rights.* One can only buy or sell goods for which property rights are well-defined, that is, goods that the seller truly owns and has the right to transfer to others. This is the domain of law and economics, and is an issue that is discussed extensively by other contributors to this volume. The problem, of course, is that many biological resources are not owned in the conventional sense, so that they are not readily tradable. Even when the biological resources are owned, their diversity is not a tradable commodity. Therefore, their owner cannot appropriate their market value, assuming that the market would, in principle, give them a significant value.

This is a common problem with many natural resources, which are often *fugacious*, that is, prone to move around. The "Conally Hot Oil Act," a landmark piece of U.S. property rights legislation, was intended to establish property rights in oil reserves that could move from one property to another. So there are precedents for the successful establishment of property rights under apparently unpromising conditions. Chichilnisky and Stone (this volume) develop this point further.

*Public Goods.* Biodiversity is a public good, that is, a good with the property that if it is provided for one, then it is provided for all. Classic examples of public goods are law and order as well as defense. A basic proposition of welfare economics is that market economies will, if left to themselves, underprovide public goods relative to an efficient level. Markets work best for private goods (such as food, housing, and consumer durables), not public goods. They work for the private good bread, but not for the public good genetic variation in wheat types; they work for the private good beef, but not for the public good genetic variation in cattle types.

In fact, biodiversity is an interesting and special example of a public good. Most public goods are centrally provided, as the adjective in their title and the previous examples indicate. Biodiversity, however, is different. Our initial stocks of biodiversity, like initial stocks of extractive resources such as oil and coal, are part of the initial conditions of life on earth. We can only subtract from them. We can provide less biodiversity, but we cannot provide more. There is a ratchet effect here, an irreversibility. The process of subtraction from our initial stocks of biodiversity, which sets the amount that we have, is not the result of any central decisionmaking, as is the provision of the classic examples of public goods, such as defense or law and order. The total biodiversity remaining in the world is the result of millions of independent and decentralized decisions on what to grow, where to grow it, how to grow it, what land to clear, what fuels to burn, and so forth. Biodiversity is therefore a privately provided public good. Atmospheric composition, and hence climate, is also a privately provided public good (Chichilnisky 1994; Chichilnisky and Heal 1994; and Chichilnisky, Heal, and Starrett 1993), resulting from millions of independent decisions on how much fossil fuel to burn. This is a category of goods to which economists have not given much attention; environmental issues are forcing them to sharpen their tools.

*Depletion Permits.* One of the most attractive economic mechanisms for managing the use of the atmosphere, seen as a privately produced public good, is the use of *tradable emission permits*. A tradable emission permit regime is one in which a total acceptable level of emissions is set by a regulatory authority, which then issues a number of permits to emit that can be traded by their owners (for more details, see Chichilnisky and Heal 1994). By analogy, one can envisage tradable depletion permits as a way of managing the use of biodiversity. In fact, to the extent that tradable emission permits can be earned by the preservation or development of $CO_2$ sinks such as forests, the two will automatically be related.

Recent work (Chichilnisky and Heal 1994; Chichilnisky, Heal, and Starrett 1993) suggests that the efficient operation of these markets is particularly sensitive to the distribution of property rights. This is in contrast to the standard results in the welfare economics of competitive markets for private goods, which assert the orthogonality of efficiency and distribution. Two key results on the welfare properties of competitive markets for private goods are the first and second theorems of welfare economics, which state respectively that any equilibrium of a competitive market is efficient,[5] and that given any efficient pattern of resource allocation, this can be achieved as the equilibrium of competitive markets for private goods given some suitable distribution of property rights among the people in the economy. In other words, the mapping from distributions of property rights to equilibria is onto efficient patterns of resource allocation. This sensitivity of efficiency to distribution is also in contrast to the so-called Coase Theorem, which asserts that assigning property rights where there initially were none will lead to efficiency independently of to whom the rights are assigned.

The relationship between efficiency and distribution that emerges in the work of Chichilnisky, Heal, and Starrett (1993) in the context of permits to emit $CO_2$ is as follows. Suppose that a regulatory body has decided that it is appropriate for the world as a whole to emit 6 billion tons, and that this emission will be shared between 100 countries. It then has to decide how to allot the rights to the 6 billion tons of emission between the 100 countries. Several such schemes have been suggested, covering a spectrum that runs from allocation proportional to historical emissions to allocation proportional to population. Clearly, the former favors the industrial countries and the latter the developing ones. The conventional wisdom in economics, based on the first and second welfare theorems outlined above, has always been that from the perspective of attaining efficiency, it does not matter what allocation rule is chosen—proportional to preexisting emissions, to population, or any other rule. Chichilnisky, Heal, and Starrett conclude that this is not true: that in fact, only a small number of ways of allocating the emission rights between countries will lead to efficient outcomes. There is a presumption that these are ways that, in some loose sense, favor the developing countries.

How does this relate to the preservation of biodiversity? There has been no systematic analysis of this issue yet, so that what follows is a very preliminary set of conjectures. It abstracts altogether from the international political agreement that would be needed to implement it. Suppose, for the sake of argument, that we wish to preserve precisely the existing level of biodiversity. This is, of course, not obvious, and has to be justified by reference to the total social costs and benefits of such a decision. Given this target, we next ascertain the relationships between populations and minimum thresholds needed for survival, and categorize populations by their relationships to these thresholds—much as is done in determining which species are endangered. In fact, what has to be done at this stage is to forecast the relationship between population and threshold that will emerge over the near future, perhaps one or two decades, in the absence of corrective intervention, that is, on the basis of what is normally called the business-as-usual scenario. So far, then, there is nothing new.

Next, we take the species for which there is a danger of the population approaching the threshold, and decide on acceptable reductions in these populations. These reductions will be the total "depletion quotas" for these populations. In practice, for certain fisheries, this is already the standard method used in deciding acceptable catches for each year. Finally, the depletion quotas set in this way are allocated amongst possible users, who may trade them in an open market. The depletion quotas could be for forest clearing, planting grasslands, or harvesting species of animals or fish. As discussed above in connection with the results of Chichilnisky, Heal, and Starrett, the distribution of these quotas would have implications for efficiency. The key aspect of such a tradable depletion scheme, from an economic perspective, would be the establishment of a clear economic incentive *not* to deplete. It may seem paradoxical that giving tradable rights to deplete gives an incentive not to deplete, but the point is that unused depletion rights, resulting from a country depleting less than its allocation, have a market value and can be sold for hard currency on world markets. This is precisely how quotas for gaseous emissions operate: they establish a cost to emission, because if not used they can be sold. There are many practical details that would have to be resolved prior to the introduction of such a scheme; equivalent details for $CO_2$ emissions are reviewed in Chichilnisky and Heal (1994). One of these details would be the relationship between depletion quotas and $CO_2$ emission quotas, given that the emission of $CO_2$ may, in the long run, be one of the main threats to biodiversity.

*Valuing the Very Long Run: Biodiversity and Sustainability.* Biodiversity matters most in the very long run—over 50 to 100 years, at a minimum. In other words, the costs of a loss of biodiversity would be felt in full only over a period of at least this length. Most economic evaluation procedures are intrinsically biased against investments that pay off only over a long period. This is a consequence of our habit of ranking projects according to their present discounted value. Discounting at any positive discount rate (that is, weighing benefits in a way that decreases with their futurity) discriminates against future-oriented projects, but is nevertheless absolutely standard practice. Conventional economic criteria make it almost impossible to justify any investment in preventing climate change, preserving biodiversity, preventing leakage of nuclear waste, and so on, as all of these have time horizons that stretch over 100 years or more. The key point here is that even if we could value biodiversity accurately, then to the extent that its value is realized more than a quarter of a century hence,[6] conventional approaches will systematically undervalue it.

Many authors have expressed reservations about the balance that discounting strikes between present and future. Cline (1992) and Broome (1992) have argued for the use of a zero discount rate in the context of global warming; Ramsey (1928) and Harrod (1948) were scathing about the ethical dimensions of discounting in a more general context, commenting respectively that discounting "is ethically indefensible and arises merely *from the weakness of the imagination*" and that it is a *"polite expression for rapacity and the conquest of reason by passion"* (emphasis added; see also Heal 1993[7]). It may be fair to say that discounted utilitarianism dominates our approach more for the lack of convincing alternatives than because of the conviction that it inspires. It has proven particularly controversial with noneconomists concerned with environmental valuations.

Alternatives are being developed. An interesting approach far more consistent with an appreciation of the value of biodiversity is that of Chichilnisky (1993a), who formalizes the concept of sustainable growth in economically operational terms. I suggest that the essence of sustainability lies in two axioms (Heal 1995a): a symmetric treatment of the present and the long-term future, which places a positive value on the very long run, together with explicit recognition of the intrinsic value of environmental assets.

The first of these points is captured in a definition of sustainability proposed by

Chichilnisky (1993a). The second point relates to the way in which we value environmental assets: in their own right, rather than instrumentally, for their capacity to provide services to humanity.

In fact, the proper economic valuation of biodiversity is inseparable from a proper development of the concept of sustainable economic development. A test of the appropriateness of an economic approach to sustainability is the extent to which it enables one to argue satisfactorily about the value of biodiversity. Chichilnisky's ideas are developed further in Heal (1995a), and Beltratti, Chichilnisky, and Heal (1994a; 1994b). The key point is that it is possible to make operational a concept of valuation that systematically places more importance on the very long run than discounting does; Chichilnisky's approach does this by valuing an income stream as the present discounted value of that stream, plus an amount that depends on the very long-run characteristics of the stream. These very long-run characteristics, almost by definition, contribute nothing to the present discounted value of the stream of benefits. In effect, a pattern of payoffs wins additional points for continuing *ad infinitum*. If one compares two investments that have the same present value of payoffs according to the conventional approach, with one having a finite life and the other having a payoff that continues indefinitely, then according to Chichilnisky's criterion, the latter would be more valuable. Clearly, such an approach reorients the economic playing field in a way more favorable to biodiversity. So developing a "sustainable" approach to cost-benefit analysis is a necessary prerequisite to realizing the value of biodiversity. A corollary of success in this area would be a revised, "green" approach to the construction of national income accounts.[8]

*Value Uncertainty and Option Values.* There is great uncertainty about the exact value of certain types of biodiversity. This means that the concept of *option value* is relevant.[9] The issue here is as follows: If you preserve a resource whose value is uncertain, then you may be able in the future to get better information about its value, and so make a better-informed decision about its long-run disposition. If the information you get then suggests that its value is low, at that point, of course, you still have the option of destroying it. But if the information obtained in the future indicates that the resource is of great value, permanent preservation is still a policy option. Preservation now gives you the right, but not the obligation, to keep the resource should this seem appropriate when better information is available in the future. If you destroy the resource now, and then find it to be valuable in the future, you no longer have the option of preserving it, however worthwhile that may then seem. Hartwick (1994) provides an interesting application of related ideas to biodiversity.

This fundamental asymmetry between destroying and preserving gives rise to what is called an *option value*, a value that is related to the value of a financial option. A financial option gives you the right, but not the obligation, to buy or sell a security. There is a large literature in finance on how to value options, and a literature in accounting on how to account for them. A key point is that failing to recognize the option value in valuing irreplaceable assets, of which biodiversity is surely the prime example, will systematically lead to their being undervalued and, hence, underpreserved. It is not at all clear that markets do, in fact, recognize this option value in the case of biodiversity. The reason is one that we have touched on before: no one firm or individual owns or can appropriate the benefits from this option. It is society's option, a public good. The social value of this option must, in principle, be added as a benefit in any analysis of the costs and benefits of biodiversity preservation, although this value is hard to compute unless all of the previous problems are resolved.

### Conclusion

Can we rely on the market to preserve biodiversity? The answer is clearly "no." There are too many aspects of biodiversity that strike precisely at the weak points of the market. These relate to the public good aspects of biodiversity, to the time horizon one needs to appreciate the value of biodiversity, and to the uncertainties associated with its importance. Understanding how it is that biodiversity defeats the market's efforts to attain efficiency is a necessary first step in deciding how to alter the usual market approaches better to deal with biodiversity. Possible directions of such modifications have been indicated.

If we ask a more limited question—can we use the market to provide some incentives for the preservation of biodiversity?—the answer is positive. There are some aspects of biodiversity that are marketable and are being marketed. But they leave a lot out. For the time being, at least, we have to recognize biodiversity as a socially valuable resource whose value cannot even approximately be quantified, and that therefore, cannot be managed entirely by the market. We will have to invoke nonmarket measures, such as the Endangered Species Act or the Convention on the International Trade in Endangered Species (CITES) agreement on trade in endangered species. However, markets may come to play a more substantial role than they can at present via the development of a greater understanding of the socioeconomic role of biodiversity, the economics of privately produced public goods, and a framework for sustainable cost-benefit analysis. But we will never rely fully on the market to manage biodiversity. Intrinsic, aesthetic, and cultural values in other areas of society have always been protected by legislation, and that will continue to be true in the area of biodiversity. The balance between market and legal solutions may change, but certainly not the need for the two.

## Acknowledgments

I am grateful to Graciela Chichilnisky, Jeffrey McNeely, and Tom Lovejoy for their valuable comments. Support from National Science Foundation grants 91–0460 and 93–09610 is also acknowledged.

## Notes

1   For details of this and similar deals, see Chichilnisky (1994).
2   A discussion of some of these philosophical issues can be found in Kneese and Schultze (1985). A fascinating recent paper by Ng (1995) is also recommended.
3   Explicit recognition of the intrinsic value seems an essential element of the concept of sustainability. Formally, this means that utility is derived not just from a flow of consumption that can be produced from the environment, used either as a consumable good or an input to production, but also from the existence of a stock, so that instantaneous utility at each point in time can be expressed as a function of $u(c_t s_t)$, where $c$ is a flow of consumption at time $t$ and $s_t$ is an environmental stock at that date, as for example, in Krautkraemer (1995) and in Beltratti, Chichilnisky, and Heal (1994b).
4   A note for the cognoscenti: one might expect externalities to be included in this list. But public goods are a particular form of externalities, which are implicitly included in this category.
5   The concept of efficiency is Pareto efficiency. A pattern of resource allocation is Pareto efficient if there is no other feasible allocation at which everyone is no worse off and someone is better off. The first and second theorems of welfare economics make certain assumptions in addition to the absence of public goods (see Dasgupta and Heal 1979).
6   This is surely a great extent.
7   I have argued that a zero-consumption discount rate can be consistent with a positive utility discount rate in the context of environmental projects.
8   These issues are discussed at length in Heal (1995a).
9   For a more detailed development, see Chichilnisky and Heal (1993), and Beltratti, Chichilnisky, and Heal (1994a).

## References

Beltratti, A., G. Chichilnisky, and G.M. Heal. 1994a. "The green golden rule." *Economics Letters* 49:175.
———. 1994b. Sustainable growth and the green golden rule. In *Towards a Theory of Sustainable Development,* edited by I. Goldin and A. Winters. Paris: Organization for Economic Cooperation and Development.
Broome, J. 1992. *Counting the Cost of Global Warming.* London: White Horse Press.
Chichilnisky, G. 1993a. *What Is Sustainable Development?* Working paper, Stanford Institute for Theoretical Economics.
———. 1993b. *Property Rights and the Pharmaceutical Industry: A Case Study of Merck and INBIO.* New York: Columbia Business School.
———. 1994. The abatement of carbon emission in industrial and developing countries. In *The Economics of Climate Change,* edited by T. Jones. Paris: Organization for Economic Cooperation and Development.
Chichilnisky, G., and G.M. Heal. 1993. Global environmental risks. *Journal of Economic Perspective* 7:65.
———. 1994. Who should abate carbon emissions: An international perspective. *Economics Letters* 44:443.

Chichilnisky, G., G.M. Heal, and D.A. Starrett. 1993. *International Emission Permits: Equity and Efficiency*. New York: Columbia University, Department of Economics.

Cline, W.R. 1992. *The Economics of Global Warming*. Washington, D.C.: Institute for International Economics.

Dasgupta, P.S., and G.M. Heal. 1979. *Economic Theory and Exhaustible Resources*. New York: Cambridge University Press.

Harrod, R. 1948. *Towards a Dynamic Economy*. London: Macmillan Press.

Hartwick, J. 1994. *Decline in Biodiversity and Risk-adjusted N.N.P.* Working paper, Department of Economics, Queens University.

Heal, G.M. 1993. The optimal use of exhaustible resources. In vol. 3, *Handbook of Natural Resource and Energy Economics*, edited by A. Kneese and J. Sweeney. Amsterdam: North Holland.

———. 1995a. Interpreting sustainability. Forthcoming in *Social Sciences and the Environment: Proceedings of the Canadian Association of Social Sciences*, edited by L. Quesnel. Ottawa: University of Ottawa Press.

———. 1995b. *Lectures on Sustainability*. Book manuscript.

Kneese, A.V., and W.D. Schultze. 1985. Ethics and environmental economics. In vol. 1, *Handbook of Natural Resource and Energy Economics*, edited by A.V. Kneese and J. Sweeney. Amsterdam: North Holland.

Krautkraemer, J.A. 1995. Incentives, development, and population: A growth theoretic perspective. In *The Economics and Ecology of Biodiversity Decline: The Forces Driving Global Change*. Cambridge: Cambridge University Press.

Ng, Y. 1995. Towards welfare biology: Evolutionary economics of animal consciousness and suffering. *Biology and Philosophy* 10:255.

Ramsey, F. 1928. A mathematical theory of saving. *Economic Journal* 38:543–59.

# The Commercialization of Indigenous Genetic Resources as Conservation and Development Policy

by R. David Simpson, Roger A. Sedjo, and John W. Reid

This chapter questions the contribution of biological prospecting to conservation of tropical forests, suggesting that both economic development and the conservation of biodiversity might be achieved more efficiently with alternative investments. It provides a brief introduction to genetic prospecting activities, discusses the valuation of genetic resources and the form of contracts for transfer of rights of access, describes the challenges of vertical integration in the commercialization of genetic resources, and explores whether investments that might not be wise from the perspective of a single firm or government might be effective as part of an international strategy for conserving biodiversity. While property rights historically emerge when the benefits of their definition exceed the costs of their enforcement, property rights in indigenous genetic resources may be based more on high hopes than firm evidence. Regardless of the probability with which the discovery of a commercially useful compound may be made, the value of the marginal species is small if the set of organisms that may be sampled is large.

\*

There has been considerable recent interest in the potential of genetic prospecting for motivating the conservation of endangered ecosystems rich in biological diversity. Genetic prospecting is the search for naturally occurring compounds with commercial value in agricultural, industrial, or particularly, pharmaceutical applications. Suggestions that genetic prospecting constitutes an important part of a strategy for conserving threatened ecosystems, especially tropical rain forests, have become almost commonplace in popular, ecological, and increasingly, economic literatures (see, for example, Eisner 1989; Findeison and Laird 1991; Roberts 1992; Wilson 1992; Pearce and Puroshothamon 1992; and even our own earlier work, Sedjo 1992; Simpson and Sedjo 1993).

Regrettably, however, genetic prospecting may not help much in the struggle to preserve habitats rich in biological diversity. Potential values to be generated from pharmaceutical research are unlikely to generate significant funds for conservation. Economic value is determined by the value of the "marginal" species. What a pharmaceutical researcher is willing to pay for access to one collection of species is determined by how much it increases the researcher's chance of discovering novel compounds over and above the chance afforded by all the rest of the species that could have been chosen. There are simply so many untested species in the world that the value of any one collection—or, by extension, the value of any one hectare of endangered habitat supporting the collection—is unlikely to be great. While the total value of biodiversity as a storehouse of information for pharmaceutical research is immense, saving a few more species to test is unlikely to be of real importance.

In this chapter, we provide a brief and heuristic overview of valuation issues (for more formal arguments, see Simpson, Sedjo, and Reid 1996), but devote most of our attention to the implications of the valuation considerations we have raised for strategies for the commercialization of indigenous genetic resources. Commercialization is being accomplished by a combination of two mechanisms. The first involves contracts between source-country providers and pharmaceutical research organizations for access to indigenous genetic resources. The second is vertical integration into research: acquisition by developing countries of equipment and expertise to conduct their own processing and research operations.

A failure to appreciate that pharmaceutical researchers may be willing to pay very little for access to genetic resources may lead to unreasonable claims for the potential of contracts and vertical integration as conservation strategies. There is no indication that existing contracts are inconsistent with the view that the marginal value of the resources to which rights of access are granted is negligible. With respect to vertical integration, some prominent authors from the natural sciences and other fields (for example, Eisner 1989; Reid et al. 1993) suggest that significant earnings can be generated by enhancing the ability of source countries to engage in pharmaceutical research. Yet it is doubtful that substantial investments in pharmaceutical research capacity would be wise as a strategy for either economic development or the conservation of biodiversity. Both objectives might be achieved more efficiently with alternative investments. Moreover, the dedication of scarce conservation funds to unwise projects for the commercialization of genetic resources may be counterproductive if the funds are taken away from more worthwhile projects.

Six sections follow this introduction. The first section provides a brief introduction to genetic prospecting activities. Next, the value of genetic resources is examined, followed by the form of contracts for transfer of rights of access in the third section. The fourth section discusses vertical integration in the commercialization of genetic resources, and the fifth extends the discussion of vertical integration, asking whether investments that might not be wise from the perspective of a single firm or government might be effective as part of an international strategy for conserving biodiversity. A sixth section concludes.

As we have already indicated, we are pessimistic that genetic prospecting can provide important incentives for the conservation of endangered ecosystems and the biological diversity they shelter. We want to be careful to emphasize that we are not saying that no such incentives exist. There are a host of other reasons—ecological, aesthetic, and ethical—for which we ought to be concerned about the decline of biodiversity. Our findings do not argue that the need to conserve biodiversity is lower, so much as that the international community would be better advised to employ more direct mechanisms for conservation.

## The Nature of Indigenous Genetic Resources

Indigenous genetic resources are the genetically coded instructions for the production of specialized chemicals. Plants and animals use a variety of chemical mechanisms to communicate, improve reproductive success, capture prey, escape predators, and ward off infections. These capabilities may be of great value if they can be adapted for pharmaceutical products. Countless generations of the survival of the fittest may result in more imaginative chemical solutions than can be dreamed up by synthetic chemists attempting "rational design" of new products.

Pharmaceutical products are generally modeled after, rather than taken directly from, natural products. The natural product will provide a "lead," which may then be modified to enhance its effectiveness or reduce its toxicity. Even in instances in which stocks of the natural source material are required for commercial production, the source is more likely to be an *ex situ* plantation than the original habitat. Thus, indigenous genetic resources are nonrival goods (Sedjo 1992): one person's possession of a breeding stock or a chemical "blueprint" does not necessarily preclude another's.

If incentives for the production, or as in this case, the preservation, of nonrival goods are to exist, property rights must be established in them. That is, legal protection must be offered to prevent nonpayers from enjoying the benefits of possessing the nonrival good. Conventional wisdom has it that property rights emerge when the benefits of their definition come to exceed the costs of their enforcement (see Demsetz 1967; Barzel 1989; and, in this context specifically, Sedjo 1992). The emergence of property rights in indigenous genetic resources may be based more on high hopes than firm evidence, however.

Countries attempting to commercialize their genetic resources are taking steps to prevent their unauthorized appropriation. It has been reported from Costa Rica that "Unlicensed biological research involving export of samples or information is . . . becoming regarded as equivalent to 'industrial espionage'" (Caldecott and Alikodra 1992). National legislation is being drafted in a number of countries to prevent such thefts.

There is also a growing sentiment that international law should reflect countries' sovereign rights over genetic resources. The Convention on Biological Diversity offered for signature at the recent United Nations Conference on Economics and Development (UNCED) meeting in Rio de Janeiro codifies rights in indigenous genetic resources. Its terms recognize the

> sovereign rights of States over their natural resources, [and that] the authority to determine access to genetic resources rests with the national governments and is subject to national legislation. . . . access, where granted, shall be on mutually agreed terms and . . . subject to prior informed consent of the Contracting Party providing such resources. (appendix, this volume)

The combined effects of these terms and existing patent law might provide powerful protection for property rights in indigenous genetic resources. Medicines derived from natural products must be protected by patents if their creators are to earn sizable rewards for their development. Patent applications generally require a complete description of the genesis of the product to be protected. Thus, misrepresenting the source of materials leading to a patentable medicine might result in the loss of patent protection.

The successful commercialization of indigenous genetic resources may require transactions between the relatively technology-rich and gene-poor North and the nations of the genetically rich but technologically less-developed South. Either genetic resources may be transferred from the countries in which they are found to foreign research organizations with the capacity to develop them, or countries rich in genetic resources may acquire the technological capacity to develop them commercially. Complex contracts are evolving to facilitate transfers. At the same time, a number of countries are attempting to expand their domestic research capabilities.

The process of natural product pharmaceutical research is expensive, time consuming, and sometimes deceptively difficult. The first step involves the collection and classification of specimens. It is important that this be done systematically. In many cases, subsequent collection of promising samples must be performed in order to conduct further tests. Initial collections, then, must be done by trained biologists who can describe the organism, its location, its stage of development, and other characteristics in sufficient detail as to permit replication of finds.

Collection and classification are generally followed by drying, grinding, and extraction operations. These activities may be deceptively difficult. Dehydration to the standards required for pharmaceutical research necessitates specialized equipment. Grinding equipment must be kept meticulously clean to avoid contamination from one sample to another. Extraction—the process of removing active compounds by dissolving them in a solvent such as ethanol—also requires precise handling and special equipment. Pharmaceutical researchers emphasize that all the operations we have described must be done with extreme care; training and appropriate equipment are required.

After samples have been extracted, a number of assays are typically performed to discover potential uses. Small amounts of material are tested to determine if, for example, they contain compounds that bind to a certain receptor in a cell or react with certain chemicals. After assaying, there follow the extremely research-intensive tasks of isolating active compounds, "optimizing" them by chemical manipulations to enhance potency and/or diminish undesirable side effects, and *in vivo* and clinical testing. Finally, if all the tests have been successfully completed, regulatory approval is secured, production is planned, and commercial marketing begins. It is not unusual for a period of 10 or more years to pass between the identification of promising leads and the first sales of an approved drug. DiMasi et al. (1991) estimate the costs of new drug development at $231 million.

Groups in many countries have now begun to commercialize their indigenous genetic resources. Reid et al. (1993, 8–13) summarize the work of 21 research organizations active in genetic prospecting on all the populated continents. Several agreements have been entered into between source-country suppliers of samples and foreign pharmaceutical research organizations under which the latter are granted access to and use of genetic resources. The National Cancer Institute (NCI) in the United States has entered into contracts with organizations in Madagascar, Tanzania, Zimbabwe, and the Philippines for the provision of samples. Biotics Limited, a British firm, has signed agreements with organizations in Ghana and Malaysia. Perhaps the most innovative existing arrangement is that between Merck and Company, the world's largest pharmaceutical firm, and Costa Rica's Instituto Nacional de Biodiversidad (INBio). Under this contract, Merck has made an upfront payment of over $1 million, in addition to promising royalty payments in the event of commercially valuable discoveries.

In addition to contracting with foreign organizations, some countries rich in biodiversity have begun to develop their own processing and testing capacity. INBio has established relatively advanced facilities. Rather than simply selling access to Costa Rica's indigenous genetic resources, INBio performs a number of taxonomic and sample preparation operations as well. Biotics is in the process of setting up BioEx Associates, an association of extraction facilities in developing source countries. The Massachusetts Institute of Technology has recently initiated a $5 million research and technology transfer project in the Brazilian Amazon. Other organizations are contemplating the establishment of similar facilities.

## The Value of Indigenous Genetic Resources

It is difficult to make direct inferences concerning the value of indigenous genetic resources *in situ*. Markets for transactions in rights to access to indigenous genetic resources are just beginning to emerge. While payments of between $50 and $200 per kilogram for natural samples have been reported (Laird 1993, 99–130), the interpretation of fixed payments for samples as a measure of the value of untested resources is suspect. Sample collection is typically a much more difficult process than it may appear to the uninitiated. Payments made for samples may reflect compensation for collection and processing labor and taxonomic expertise rather than rents for the materials themselves, and thus greatly overstate the value of untested resources.

This point may be illustrated by INBio's agreement with Merck. Of the fixed, up-front payment of over $1 million, only $100,000 was directed to conservation activities; the remainder was designated to defray the costs of collection, classification, and processing activities (Sittenfeld 1992). It is not clear that even this $100,000 was a payment for access to rare genetic resources, as opposed to another form of payment for prospecting costs. That is, it is unclear whether any of the compensation INBio has received to date is for its genetic resources, as opposed to processing activities that it could have performed on samples of any origin.

The Merck-INBio arrangement illustrates another complication, however. Compensation for access to samples is often not made in the form of simple cash transactions, but in some combination of cash up front and royalty payments later. While the royalty provisions of the Merck-INBio arrangement have not been made public, it is clear that substantial payments could be made in the event of a major discovery (Sittenfeld and Gámez 1993, 69–98). Contingent payments are also called for in the Biotics, and, with less specificity, NCI contracts. Inasmuch as the terms of these provisions are generally secret, and both the probability of discovery and the payoff in the event that a valuable discovery is made are unknown, little can be inferred about the value of untested resources from public information concerning these contracts.

For these reasons, most existing attempts to estimate the value of indigenous genetic resources have been based on inferences from indicators other than observed transactions. Farnsworth and Soejarto (1985), Principe (1989), McAllister (1991), Harvard Business School (1992), Pearce and Puroshothamon (1992), and Aylward (1993)[1] have multiplied estimates of the value of successful commercial pharmaceutical products times the probability with which untested samples are likely to yield commercial products. These studies yield a wide range of estimates, from as much as $23.7 million per untested species *in situ* (Principe 1989) to as little as $44 (Aylward 1993).

While the more carefully conducted of these studies can be extremely helpful in advancing our understanding of the values to be realized in pharmaceutical research, they are all marked by a logical flaw: they fail to allow for the possibility of redundancy in discoveries—that is, that more than one wild organism may contain a cure for the same condition. In a recent paper (Simpson, Sedjo, and Reid 1996), we have constructed a model in which we demonstrate that the value of the "marginal species" is likely to be negligible under even the most optimistic expectations concerning the probability with which a species tested at random may contain a commercially useful compound.

The intuition driving the model can be described fairly simply. Suppose that one is searching for a chemical compound for a certain purpose. The marginal value of genetic information for medicinal purposes is measured by its contribution to the improvement of available health care. For example, the value of a new cancer treatment is determined by its capacity to improve remission rates, reduce side effects, lower costs, and so forth. A new drug that may be effective but is identical or inferior to an existing treatment is of little value. While the discovery of a novel compound with pharmaceutical potential may not often prove completely superfluous, it is often the case that one product will largely duplicate another, or that discovery of one effective compound will reduce the urgency, or even eliminate the need, to continue research on others.

There are other ways in which indigenous genetic resources may be redundant. The same organism may be found over a wide range. If all representatives of a species produce a particular compound, those individuals in excess of the number needed to maintain a viable population are redundant.

There are numerous instances in which identical drugs, or drugs with similar clinical properties, have been isolated from different species (Farnsworth 1988). It may also be the case that there are a host of other sources of common compounds that remain undiscovered because current sources are adequate. To give a recent example, the discovery of the anticancer drug taxol in the Pacific yew of western North America has set pharmaceutical researchers looking for similar compounds in its old-world relatives (Chase 1991). Given the numerous examples of parallel morphological development in the evolution literature, it should not be surprising to find that different organisms that have evolved in similar ecological niches have developed similar chemicals.

Finally, there is a dimension of what might be labeled clinical, or medicinal, redundancy. Very different compounds, perhaps even drugs working through different mechanisms, may be effective in treating the same set of symptoms. Moreover, while the inventiveness of nature in developing useful compounds is much extolled as a factor in the increased demand for natural products for pharmacological research (Findeison and Laird 1991), it is possible that synthesis from nonorganic sources would yield compounds chemically similar to natural products.

In essence, regardless of the probability with which the discovery of a commercially useful compound may be made, if the set of organisms that may be sampled is large, the value of the marginal species is quite small. At any given time, researchers will be searching for compounds effective in particular applications. If the probability that a species chosen at random will yield an effective compound is high, the probability that two or more species will be found to do so is also high. To the extent that additional species from which to sample are likely to be redundant, their marginal value will be low. Conversely, if potentially valuable compounds are so rare as to make their discovery in two or more species in a large aggregation highly unlikely, the probability of their discovery in any species will be unlikely. Hence, the value of the marginal species will again be low because the probability of it containing any useful compounds will be low.

Estimates of value can be extended to marginal hectares of endangered habitat by using models from the ecological literature on species-area relationships (for example, MacArthur and Wilson 1967; Myers 1988, 1990). If the value of the marginal species for use in pharmaceutical research is not great, the implied value of the marginal hectare is likely to be negligible (Simpson, Sedjo, and Reid 1996).

### Values, Risk Sharing, and Contracting Practices

A somewhat surprising feature of many existing contracts for access to indigenous genetic resources is their reliance on contingent payments. One might ask why transactions are structured in this way. Why not have cash sales instead? There are a number of reasons why sellers might prefer to receive fixed payments. Risk aversion is one. Genetic prospecting is an extremely uncertain proposition, even when large numbers of samples are involved. The citizens of poor, tropical, developing countries might well favor certain returns over the unlikely prospect of receiving high royalties. Even if decisionmakers are not risk averse in the technical sense of concave utility functions, they might prefer assured and immediate returns. If sellers' subjective discount rates are greater than those of would-be purchasers (as is likely to be the case if their countries face borrowing constraints in international credit markets), they would favor up-front payments to royalties. Similarly, if sellers face expenses of monitoring or auditing, they would again prefer to receive payments up front, avoiding the need to incur subsequent expenses.[2]

Of course, buyers of access to indigenous genetic resources may not be risk neutral themselves. Even large, well-diversified firms may choose to avoid risks in their research portfolios. Division managers may be unwilling to commit even modest expenditures to the purchase of access to indigenous genetic resources of uncertain value if it is their reputation that will suffer if the research is unsuccessful. Inasmuch as not all purchasers of genetic resources are large, well-diversified firms, there may be further reasons for them to avoid risks. Thus, contracts may call for royalty payments in order to shield both parties.

Another explanation for the drafting of contracts in which contingent payments are specified is simply force of habit. Contracts for oil, mineral, and timber concessions frequently call for risk sharing by specifying royalties (see, for example, McPherson and Palmer 1984; Broadman and Dunkerley 1985; Leffler and Rucker 1991; see also Downes et al. 1993, 277, who suggest that such agreements may provide guidance in structuring contracts for indigenous genetic resources). Contracts specifying royalties are also common in the dealings of major pharmaceutical companies with independent laboratories (Rostoker 1983). It may simply be the case that both pharmaceutical companies and organizations in source countries model the contracts they prepare after other agreements that they perceive to be in a similar category.

This begs the question of why agreements for the sale of access to indigenous genetic resources might be perceived to be similar to those of other resources. While we should not underestimate the importance of bilateral risk aversion, reasoning by analogy, and inertia in explaining the form of contracts, two features come to mind that may be important in the context of genetic, as well as more conventional, resources.

The first may arise as a consequence of source-country acquisition of research capacity. When source countries initially learn the value of their resources, they may not be able to credibly communicate this information to would-be buyers directly. Inasmuch as indigenous genetic resources are nonrival goods, to announce that a certain species exhibits promising characteristics might be to invite its unauthorized appropriation. Of course, unsubstantiated announcements that a source country wishes to sell access to an unspecified resource of supposedly great value may not be credible. Thus, the source country may be required to "put its money where its mouth is," in effect maintaining an equity position in the resources offered in order to signal confidence in their quality. It has been suggested that such considerations may explain the prevalence of risk-sharing contracts in the pharmaceutical industry in general (see Caves, Crookell, and Killing 1983; Gallini and Wright 1990; Simpson and Sedjo 1993).

There is a second reason why existing contracts may call for risk sharing by sellers that we might reasonably presume would be, or behave as if they were, risk averse: sellers may believe their products to be worth more than do buyers. Some researchers at pharmaceutical companies have claimed that sellers of access to indigenous genetic resources tend to overvalue them. Other researchers have said that they have refused to make cash payments when requested, but have entered into royalty arrangements instead. If this does explain some of the risk-sharing behavior observed in existing contracts—perhaps both for indigenous genetic resources and other commodities—sellers of access to genetic resources may be in for disappointment if buyers' expectations prove to have been more realistic.

More optimistic expectations concerning the value of resources may induce a seller to offer a riskier contract than they might if the buyer and seller were in agreement as to the value. The basic idea is simple. A contract calling for royalties in preference to up-front payments requires that the seller pay a large sum to the buyer only in the event that the resources do, in fact, "pan out." As the buyer infers this to be an unlikely event, his expected payments under the royalty contract are low, whereas the seller, with more optimistic expectations, anticipates higher payments.

We are, perhaps, attempting to be too clever in supposing that any rational and consistent explanation can be given for the structure of existing contracts. In addition to purportedly analogous experience, differences in subjective beliefs, and asymmetries of information, contracts may also be drawn up so as to comport with difficult-to-define notions of fairness. Given the monumental uncertainties involved, it may be unrealistic to suppose that any reliable formula could be derived for determining what share of earnings should be paid as royalties. It may also be the case that pharmaceutical companies do not devote a great deal of thought to the royalty figures; a couple of percent of profits may be perceived to be a small consideration in the event of the discovery of a blockbuster drug.

These factors notwithstanding, we believe the argument that indigenous genetic resources are unlikely to be very valuable to pharmaceutical research is compelling. To suppose that great surges of royalty income from pharmaceutical companies would be forthcoming seems unlikely. The most parsimonious explanation for the royalty provisions written into existing contracts may well be that buyers do not believe themselves to be parting with much by making such agreements.

## Vertical Integration in the Commercialization of Genetic Resources

Important questions revolve around the issues of how best to facilitate exchanges of access to indigenous genetic resources or the technology with which to evaluate them. In how many stages of the research and development process should the source countries engage? To what degree should arm's-length transactions be relied on? These decisions may be based on efficiencies, perceptions of value, aversion to risk, or strategic considerations.

Many tropical developing countries simply do not have the technological wherewithal to perform tasks beyond the fairly rudimentary stages of development. Even the requirements of systematic collection may be beyond the domestic capacities of some countries. Of course, given time and investment, any country could acquire the ability to perform any or all of the tasks of new drug development from natural materials. Several nations rich in genetic resources hope to improve their capacities.

Moreover, many commentators are advising gene-rich developing countries to undertake substantial investment in pharmaceutical research capacity. Eisner (1989) has suggested that source countries undertake tasks of new drug development up to and including assaying. Reid et al. (1993) suggest that source countries can capture substantial returns from their genetic resources, even if willingness to pay for access to organisms *in situ* is slight.

Two extremely important questions arise in the consideration of the wisdom of investment in source-country pharmaceutical research capacity. The first is whether such investments are financially wise; that is, will they "pay off" by returning profits commensurate with the expenses involved. The second is whether, even if they are not privately profitable, they may substantially enhance conservation incentives.

The expected profitability of investments in source-country pharmaceutical research capabilities must be examined. True profits are earned as a consequence of access to scarce resources. If potential purchasers of access to genetic resources regard them as scarce commodities, this will be reflected in the price they command. Performance of further processing steps in the source country will only increase net earnings if the source country has unique capabilities in these operations. Organizations that do not enjoy cost advantages in performing such operations will not add to their earnings by undertaking them.

In a world in which major corporations are frequently accused of transferring production to the low-wage *maquiladoras* of the developing world, the absence of drug development laboratories in most of the developing tropics suggests that they have no comparative advantage in this field. Of course, some synergies may be created by proximity. The tasks of collection and classification may be best accomplished by workers familiar with local terrain and species. This will certainly be true in cases in which indigenous ethnobiological knowledge is being tapped. That such synergies would extend very far up the chain of development processes seems highly doubtful, however. There would be no apparent advantage, for example, to performing chemical isolation or synthesis activities in source countries.

While it may be the case that investment in pharmaceutical research capacity would generate some increase in revenues gross of the fixed costs of acquisition, it would be remarkable if there were a substantial prospect of earning real profits from making such an investment. Given the quantity of genetic resources available for the picking, why have established pharmaceutical research organizations not already snapped up these excess returns?

In short, we doubt that there are any compelling reasons to make substantial investments in source-country pharmaceutical research facilities. It is difficult to infer the reasons such investments are being made; they may well not be rational, or at least, not in keeping with the types of motivations generally considered by economists. Part of the difficulty may lie in an incomplete appreciation of the notion of *value added* (see, for example, Reid et al. 1993, 34). The fact that products at a more advanced stage of development command greater (expected) compensation has been seen by some as evidence of greater profitability, although this observation is most likely consistent only with a different division of the costs of development between seller and buyer with no net gain to the former.

Similarly incomplete reasoning may underlie arguments for investments in source-country pharmaceutical research capacity as a "development" strategy. It is sometimes argued that investment in scientific capacity is desirable in and of itself; a better-educated, more capable workforce must be conducive to economic growth. The fact that such investments—especially if coming from foreign sources—may have ripple effects on the economy in general is not an argument for their optimality. The question should be, rather, would investments in other sectors of the economy have more desirable net effects? To the extent that investments in other sectors are likely to be more profitable, the answer is probably "yes."

Of course, investments financed by international aid will generally be made in those sectors of the economy in which private funds are not forthcoming. This does not mean that such investment should not be subject to a cost-benefit analysis, however. International aid should be targeted to those areas in which social returns are greatest, but market failures prevent private provision. Investments in education—especially for females—sanitation, and basic medical care are more likely to be effective development strategies than investments in very specialized and speculative enterprises.

Finally, a preference for the development of domestic research capacity may be traced, in part, to a fear of being exploited in transactions for raw materials (see, for example, the emotional arguments cited by Neto 1993). To the extent that would-be sellers of access to genetic resources tend to be small, inexperienced, and unsophisticated, and the buyers the opposite (Downes et al. 1993), there may be some legitimate concern regarding the exploitation of the former by the latter. That these concerns would be allayed by pursuing more extensive vertical integration is far from clear, however. If sellers of access to genetic resources fear large, sophisticated buyers, might not buyers of advanced research technology fear large, sophisticated sellers? More to the point, it may be difficult for an unsophisticated seller to distinguish between a conspiracy among would-be buyers to pay less than the resources are worth and a situation in which the bids of competitive buyers are merely informing the seller of the resources' objective worth. If the latter describes the case, large investments in capacity-building may prove to be extremely unwise.

It is not even clear that acquiring information regarding value of resources will be useful to the seller. If the value of untested resources is negligible, willingness to pay for them *a priori* should also be low. While there are some suggestions in the sales-mechanism-design literature that sellers can increase their expected returns by providing information to would-be purchasers (see, for example, Milgrom and Weber 1982; but also see McAfee and Reny 1992), if the value of untested resources on the margin is essentially zero, informing potential buyers of this cannot be useful.

Going beyond providing general information to informing would-be buyers of the identity of "winners" and "losers" among the species to which access is offered may raise the problems discussed above: those who claim to be selling "winners" may need to take costly actions to "put their money where their mouth is." With respect to these considerations, sellers may prefer not to have specific information regarding the resources they command. The better their information, the more risky may be the contract into which they enter. In the limit of complete vertical integration, the seller must take on all the risks of product development.

To conclude this section, there seems to be no good reason for a seller of access to indigenous genetic resources to invest in advanced capabilities for their processing. In some instances, however, the source of such investments is not the seller, but rather international conservation or development organizations. We ask next whether expenditures for the development of source-country pharmaceutical research capacity would be elements of an effective conservation strategy, even if they are not wise financial investments.

## Source-Country Capacity Building
## as a Conservation Strategy

Much of the investment in source-country research capacity building is coming from foreign funders. INBio has received substantial funding from the MacArthur Foundation. In the United States, the National Institutes of Health, the National Science Foundation, and the Agency for International Development have recently initiated a program to sponsor pharmaceutical research in the developing tropics. Biotics received initial funding from the European Community, and is seeking external funding from other sources, for its BioEx Associates program.

We should address several questions in considering the efficacy of investment in source-country research capacity as a conservation strategy. First, what are the effects on the source country's incentives to preserve endangered habitats? Second, might similar incentives be created more directly? Third, do such investments, even if less efficient than direct measures might be in theory, have advantages in practice to the extent that they provide "self-supporting" incentives to individuals in source countries? Finally, might considerations of political economy, social psychology, or other factors make it easier to raise funds for "roundabout" conservation programs; is investment in source-country pharmaceutical research capacity building easier to "sell" to donors than more direct measures?

It is important to be clear about the nature of incentives for conservation. In practice, the conservation of biodiversity entails the prevention of land conversion. If investment in research capacity is to have any effect on direct incentives for conservation, then it must increase the willingness to pay for the marginal hectare.

We might think of investments in source-country pharmaceutical research capacity as investments in assets complementary to the indigenous genetic resources themselves (for an analysis of these considerations, see Aylward and Barbier 1993). Just as investment in machinery raises productivity of labor and, hence, raises the wage rate, investment in research capacity may increase the marginal value of genetic resources.

Such payments may have little effect on source-country conservation incentives.

First, the willingness to pay for samples might have to increase by several orders of magnitude in order to become a serious consideration in conservation decisions. It is unlikely that even drastic decreases in sample processing costs could make such a difference. It also bears mentioning that increases in sample processing efficiency that make it more likely that desired compounds are detected, may actually decrease the willingness to pay for samples. If it becomes more likely that desired compounds will be detected in any given sample, different species become better substitutes for each other, and the willingness to pay for any one will decline.

A second reason for skepticism concerning the incentives created by investments in source-country research capacity is that foreign organizations are already willing to purchase samples from developing tropical countries. Investments in research capacity in those countries might not create demand so much as displace it. Effects on the willingness to pay for the marginal species would continue to be negligible.

A final consideration is that capacity-building investment is unlikely to be an effective strategy for promoting conservation where the need is greatest. We would argue—and this would seem to be the consensus among all who have studied the issue (see, for instance, Reid et al. 1993)—that Costa Rica's success in promoting the commercialization of its genetic resources is due as much to its stable institutions as its biological endowment. Building Costa Rica's capacity further makes it more likely that those interested in genetic prospecting will find all that they need there, and make prospecting in less-developed and stable regions still less attractive. In essence, Costa Rica is earning positive economic returns because it does offer an economically scarce asset: political stability in a tropical developing country. While the development of more stable and predictable political institutions in other nations would be a desirable thing for many reasons, it would most likely not increase the returns that they could earn from the commercialization of their indigenous genetic resources.

Investments in source-country research capacity might generate extra funds for those receiving them; while the value of the marginal species may not be substantially increased, the value of a country's total endowment of biodiversity may be increased enough so as to create some hope of substantial revenues. Thus, an organization that does score a "hit" may generate large sums of money. If, as is often the case, there is a legal requirement to devote a share of revenues from pharmaceutical discoveries to conservation activities—such as under the Merck-INBio agreement (Sittenfeld 1992) and Biotics's contracts (Thomas 1993)—major contributions to conservation efforts might be forthcoming. In a country with large stocks of genetic resources and substantial investments in research facilities, they might even be expected.[3]

Even if investments might reasonably be expected to result in some substantial receipt of revenues in the future, however, would these revenues be likely to exceed the investment expenditures that preceded them? The above arguments suggest that they would not. It should, then, be the case that a greater (and more certain) total contribution to the preservation of biodiversity could be realized by donating funds directly to conservation activities.

The history of direct conservation payments may raise some doubt as to their efficacy, however. The developing tropics contain many *paper parks*, areas designated for protection that receive little actual oversight to ensure that they are preserved. Experience with the establishment of and foreign support for nominally protected areas in developing countries has often been unsatisfactory (Southgate and Clark 1992; Deacon and Murphy 1994).

To suppose that financing conservation from the revenues of genetic prospecting operations will necessarily be more effective than outright grants from foreign donors assumes that those who receive revenues will be more diligent or honest conservators than those who receive grants. In the final analysis, it seems unreasonable to suppose that the people actually charged with preserving the resource would be any different in one scheme as opposed to the other. It is not clear why profits from pharmaceutical research should be any more likely to induce promised performance in the preservation of endangered ecosystems than would direct payments. While advocates of expanded genetic prospecting activities in tropical developing countries emphasize the need to structure contracts so as to require conservation expenditures, there seems to be little reason to suppose that performance of conservation requirements under these arrangements will prove to be any better than under alternative, more direct mechanisms. To the extent that less money might be available, and prospects for the availability of funds much less certain, performance might prove to be substantially worse.

The final argument for the adoption of source-country research capacity building as a conservation strategy is that it may capture the popular imagination in a way that more straightforward approaches do not. This is a difficult proposition to refute or indeed judge. It is probably not unreasonable to suppose, however, that potential funders find projects that purport to kill several birds with the same stone more attractive than straightforward approaches directed to a single issue. "Debt-for-nature swaps," for example, may have attracted more funding for habitat conservation than would have requests simply to pay for land set-asides. The simpler mechanism would have been more direct and probably equally effective, however (Deacon and Murphy 1994). The disadvantage of the direct approach was that it did not call attention to the solution of two crises—developing country debt and environmental degradation—simultaneously.

Similarly, a program that promises to simultaneously find important new drugs, promote sustainable development in poor tropical countries, and conserve endangered biological diversity may have greater appeal to the donor community than would other alternatives with less far-reaching goals. Regrettably, our review of the situation suggests that such programs may have negligible impacts with respect to the first and second goals, and may be inferior to more direct alternatives with respect to the third. While funders may be attracted to programs purporting to offer simultaneous solutions to several problems, more meaningful progress might be made by soliciting funds to address more modest goals.

## Conclusion

The incentives created for the conservation of biodiversity for genetic prospecting are unlikely to be substantial. Efforts to augment these incentives by investing in source-country pharmaceutical research and development facilities may well prove counterproductive with respect to both conservation and development objectives. It should be emphasized, however, that the value of biodiversity for pharmaceutical prospecting is but one of a multitude of possible arguments for the conservation of endangered habitats. Our point is not that the conservation of biodiversity is unimportant. On the contrary, there may be important ecological, ethical, and aesthetic reasons to suppose that it is of great importance. Rather than suggesting that there are no good reasons to support the conservation of biological diversity, it crucial that money and effort be devoted to defining and protecting the other values.

Given this priority, it is also important that funds not be diverted to ill-conceived projects. If the promise of biodiversity in pharmaceutical research were great, we should expect to see effective demand for conservation from private sources. The fact that, by and large, we do not, supports our conclusion that the marginal value attached to biodiversity by pharmaceutical researchers is low. Investments in pharmaceutical research capacity in areas where it is unlikely to be profitable would not only be financially unwise, but also likely less effective than would alternative conservation expenditures.

## Acknowledgments

We thank Jeremy Bulow, Walter Reid, participants at the Conference on Market Approaches to Environmental Protection (Stanford University, December 1993) and the Conference on Biological Diversity: Exploring the Complexities (Tucson, Arizona, March 1994), and especially Jeffrey McNeely for helpful comments on earlier drafts. All opinions, however, are those of the authors.

## Notes

1  An excellent summary of all these studies may be found in Aylward (1993).
2  Some arrangements have provisions for such monitoring and auditing. INBio, under its contract

with Merck, has the right to audit Merck's reported earnings on products derived from INBIO leads. Merck is also required to submit regular progress reports on materials it is testing. Another clause requires that Merck either develop products expeditiously or forfeit rights to the materials.

3   Note that even if—or perhaps, especially if—high revenues might be expected from the investigation of the pharmaceutical potential of a number of species, this does not imply that the marginal species is of any great value. To the extent that redundancy is a possibility, our basic point remains valid.

## References

Aylward, B.A. 1993. A case study of pharmaceutical prospecting. In vols. 1 and 2, *The Economic Value of Species Information and Its Role in Biodiversity Conservation: Case Studies of Costa Rica's National Biodiversity Institute and Pharmaceutical Prospecting*, edited by B.A. Aylward, J. Echeverria, L. Fendt, and E.B. Barbier. Report to the Swedish International Development Authority.

Aylward, B.A., and E. Barbier. 1993. Capturing the pharmaceutical value of biodiversity in a developing country. *Environmental and Resource Economics* 8:157–81.

Barzel, Y. 1989. *The Economic Analysis of Property Rights*. Cambridge: Cambridge University Press.

Broadman, H.G., and J. Dunkerley. 1985. The drilling gap in non-OPEC developing countries: The role of contractual and fiscal arrangements. *Natural Resources Journal* 25:415–28.

Caldecott, J., and H. Alikodra. 1992. *Biodiversity Management in Costa Rica*. Report submitted to the Ministry of State for Population and Environment, Indonesia.

Caves, R., H. Crookell, and J.P. Killing. 1983. The imperfect market for technology licensing. *Oxford Bulletin of Economics and Statistics* 45:249–67.

Chase, M. 1991. A new cancer drug may extend lives — at cost of rare trees. *Wall Street Journal* 9 April, A1.

Deacon, R.T., and P. Murphy. 1994. The structure of an environmental transaction: The debt-for-nature swap. Working paper, University of California, Santa Barbara.

Demsetz, H. 1967. Toward a theory of property rights. *American Economic Review* 57:347–59.

DiMasi, J.A., R.W. Hansen, H.G. Grabowski, and L. Lasagna. 1991. Cost of innovation in the pharmaceutical industry. *Journal of Health Economics* 10:107–42.

Downes, D., S.A. Laird, C. Klein, and B.K. Carney. 1993. Biodiversity prospecting contract. In *Biodiversity Prospecting: Using Genetic Resources for Sustainable Development*, edited by W.V. Reid et al. Washington, D.C.: World Resources Institute.

Farnsworth, N.R. 1988. Screening plants for new medicines. In *Biodiversity*, edited by E.O. Wilson. Washington D.C.: National Academy Press.

Farnsworth, N.R., and D. Soejarto. 1985. Potential consequences of plant extinction in the United States on the current and future availability of prescription drugs. *Economic Botany* 39:231–40.

Findeison, C., and S. Laird. 1991. *Natural Products Research and the Potential Role of the Pharmaceutical Industry in Tropical Forest Conservation*. Prepared by the Periwinkle Project of the Rainforest Alliance.

Gallini, N.D., and B.D. Wright. 1990. Technology transfer under asymmetric information. *RAND Journal of Economics* 21:147–60.

Harvard Business School. 1992. *The INBIO/Merck Agreement: Pioneers in Sustainable Development*. NI–593–015.

Laird, S.A. 1993. Contracts for biodiversity prospecting. In *Biodiversity Prospecting: Using Genetic Resources for Sustainable Development*, edited by W.V. Reid et al. Washington, D.C.: World Resources Institute.

Leffler, K.B., and R.R. Rucker. 1991. Transactions costs and the efficient organization of production: A study of timber-harvesting contracts. *Journal of Political Economy* 99:1061–87.

MacArthur, R., and E.O. Wilson. 1967. *The Theory of Island Biogeography*. Princeton, N.J.: Princeton University Press.

McAfee, R.P., and P.J. Reny. 1992. Correlated information and mechanism design. *Econometrica* 60:395–422.

McAllister, D.E. 1991. Estimating the pharmaceutical value of forests, Canadian and tropical. *Canadian Biodiversity* 1:16–25.

McPherson, C.P., and K. Palmer. 1984. New approaches to profit sharing in developing countries. *Oil and Gas Journal* 25:119–28.

Milgrom, P.A., and R. Weber. 1982. A theory of auctions and competitive bidding. *Econometrica* 50:1089–122.

Myers, N. 1988. Threatened biotas: "Hot spots" in tropical forests. *Environmentalist* 8:187–208.

———. 1990. The biodiversity challenge: Expanded hot-spots analysis. *Environmentalist* 10:243–56.

Neto, R.B. 1993. MIT's Amazon outpost. *Nature* 365:101.

Pearce, D., and S. Puroshothamon. 1992. Preserving biological diversity: The economic value of pharmaceutical plants. Discussion paper 92–27. London: Center for Social and Economic Research on the Global Environment.

Principe, P. 1989. *The Economic Value of Biodiversity among Medicinal Plants*. Paris: Organization for Economic Cooperation and Development.

Reid, W.V., S.A. Laird, C.A. Meyer, R. Gámez, A. Sittenfeld, D.H. Janzen, M.A. Gollin, and C. Juma, eds. 1993. *Biodiversity Prospecting: Using Genetic Resources for Sustainable Development*. Washington, D.C.: World Resources Institute.

Roberts, L. 1992. Chemical prospecting: Hope for vanishing ecosystems? *Science* 256:1142–43.

Rostoker, M. 1983. PTC research report: A survey of corporate licensing. *IDEA: The Journal of Law and Technology* 24:59–82.

Sedjo, R.A. 1992. Property rights, genetic resources, and biotechnological change. *Journal of Law and Economics* 35:199–215.

Simpson, R.D. 1992. Transactional Arrangements and the Commercialization of Biodiversity. ENR Discussion paper 92–11. Washington, D.C.: Resources for the future.

Simpson, R.D., and R.A. Sedjo. 1993. Adverse selection, risk aversion, and costly auditing: Implications for contract form and vertical integration. Resources for the Future discussion paper 93–08.

Simpson, R.D., R.A. Sedjo, and J.W. Reid. 1996. Valuing biodiversity for use in pharmaceutical research. *Journal of Political Economy* 104:1.

Sittenfeld, A. 1992. Tropical plant conservation and development projects: The case of the Costa Rican National Institute of Biodiversity (INBio). Presented at the Rainforest Alliance Tropical Forest Medical Resources and the Conservation of Biodiversity symposium, New York.

Sittenfeld, A., and R. Gámez. 1993. Biodiversity prospecting by INBio. In *Biodiversity Prospecting: Using Genetic Resources for Sustainable Development*, edited by W.V. Reid et al. Washington, D.C.: World Resources Institute.

Southgate, D., and H.L. Clark. 1992. Can conservation projects save biodiversity in Latin America? Discussion paper 92–38. London: Center for Social and Economic Research on the Global Environment.

Thomas, R. 1993. Personal communication, 16 March.

Wilson, E.O. 1992. *The Diversity of Life*. Cambridge, Mass.: Belknap Press of Harvard University Press.

# IV Institutional Responses

# Scales, Polycentricity, and Incentives:
# Designing Complexity to Govern Complexity

by Elinor Ostrom

Biological processes occur at all scales—small, medium, and large. Consequently, governance arrangements to regulate biological complexity also need to be organized at multiple scales and linked together effectively. The importance of nested institutional arrangements with quasi-autonomous units operating at quite small up to extremely large scales is stressed in this chapter. Design principles derived from a close study of long-surviving, self-governing institutions are presented, as are lessons derived for the future. A research program that will monitor both forest biodiversity and institutional diversity and performance is also described.

*

A concern for biodiversity is a concern for the importance of sustaining complex systems at multiple spatial and temporal scales. Much of the literature on biodiversity stresses the global nature of the gene pool and the consequent need for international institutional arrangements to articulate worldwide concerns for the preservation of biodiversity for future generations. Many biological processes, however, occur at extremely small scales that vary dramatically in climate, elevation, structure, and importance from one niche to the next. An overemphasis on the need for large-scale institutional arrangements can lead to the destruction or discouragement of institutional arrangements at smaller to medium scales. It is at these smaller scales that local knowledge about specific complex interactions and concerns about natural capital can be applied in daily life.

Central to this chapter is the idea that if the nature of the systems we have an interest in governing (regulating) are complex, it is essential to think seriously about the complexity in the governance systems that are proposed. Without a deep concern for creating complex, nested systems of governance, the very processes of trying to regulate behavior so as to preserve biodiversity will produce the tragic and unintended consequence of destroying the complexity we are trying to enhance.

W. Ross Ashby, an eminent biologist of an earlier era, wrote a book entitled *Design for a Brain: The Origin of Adaptive Behavior* (1960), in which he developed the "Law of Requisite Variety." Basically, the law of requisite variety can be roughly stated as: Any regulative system needs as much variety in the actions that it can take as exists in the system it is regulating. Or, translated into the discourse within this book:

> Any governance system that is designed to regulate complex biological systems must have as much variety in the actions that it can take as there exists in the systems being regulated.

This is a tall order. But it is one to which we need to pay serious attention. Otherwise, we will continue the trend to stress the importance of simple, large-scale, centralized governance units that do not, and cannot, have the variety of response capabilities (and the incentives to use them) that complex, polycentric, multilayered governance systems can have (see V. Ostrom 1991, 1995). Naive theories of institutions equate power and capability to regulate events with simple systems that are organized in a clear hierarchy of superior and subordinate relationships. Substantial recent research on forest institutions has challenged the presumption that centralized agencies achieve better regulation of forest resources than do more complex, polycentric institutions (see, for example, Agrawal 1994, 267–82; Ascher 1995).

Among the institutions that humans utilize for generating highly desirable future goods are open, competitive markets. Open, competitive market arrangements for producing private goods—those that are easy to exclude from noncontributors, and whose consumption is subtractive or rivalrous—have many advantages. One of them is that they create incentives for innovation and entrepreneurship. Those who are alert to the opportunities present at many different scales can enter tiny market niches to create benefits for others—and thus, for themselves. The environments in which markets work well are normally complex, polycentric, multitiered systems of individuals relating to one another in such a manner as to greatly enhance overall productivity. Efforts to achieve the same level of productivity by central direction have repeatedly failed, and they did so dramatically.

But, as Geoffrey Heal (this volume) carefully outlines, markets do not perform effectively in relation to public goods, where exclusion is very difficult, and thus very costly, to achieve. If one were to rely entirely on market institutions for enhancing biodiversity, this benefit would be dramatically underprovided. As Heal and many others point out, it is difficult, but not impossible, to develop property rights to some aspects of biodiversity. In some instances, however, we may not wish to allow any one entity to own key aspects of complex biological systems. Full ownership implies the complete power to control access to and use of a resource; the resource can be held for private use, or alienated or destroyed (Schlager and Ostrom 1993). Full owners have the power to destroy a resource as well as to use it for innovative, productive purposes.

Laura Jackson (this volume) provides us with a deep insight into what can happen when one large corporation becomes the full owner of large tracts of land that are developed for commercial purposes and then used or abandoned, given the opportunities to maximize profits through shifting operations to locations where the accounting costs of production may be lower and, thus, profits higher. Large agricultural corporations owned by distant shareholders managed out of central offices are not likely to enhance the biodiversity that exists in a particular ecological niche. Rather, incentives tend to press toward monoculture and the use of commercially available inputs rather than complex mixes of agricultural products that rely on intensive cultivation techniques. A similar argument is presented by McNeely (this volume).

Jackson contrasts the level of biodiversity that exists in Iowa today under an agricultural system dominated by large, commercial farms with the level of biodiversity that existed in 1910, when the agricultural system was dominated by smaller family farms. The amount of land devoted to agriculture is the same in both time periods. The key difference is in the complexity of the socioeconomic system. Biological diversity was higher in an era when large numbers of different cultivators had long-term stakes in the land they farmed. Jackson is working toward the restoration of complex, multitiered, ecological systems. She emphasizes that this type of restoration is not going backward—but rather, toward a better future. This chapter stresses the importance of restoring (or creating) complex, multitiered, sociopolitical systems. This, also, is definitely not going backward—instead, it is an essential step toward creating a better future.

## Looking Backward as a Foundation for Looking Forward

To go forward into a future that preserves high levels of biodiversity, however, may require serious attention to institutional arrangements where those directly involved have successfully managed complex resource systems over long periods of time. Many long-sustained, self-governing resource systems have been studied in depth by perceptive scholars such as Robert Netting, Thráinn Eggertsson, Gary Libecap, Daniel Bromley, Margaret McKean, Fikret Berkes, David Feeny, and others. The resources involved vary from irrigation systems to inshore fisheries, mountain grazing lands, and forests. The most notable similarity among them is the sheer perseverance of these resource systems and institutions. The institutions can be considered robust in that the rules have been devised and modified over time according to a set of collective-choice and constitutional-choice rules (Shepsle 1989). In other words, these systems have been sustainable over long periods of time. Most of the environments studied are complex, uncertain, and interdependent ones, where individuals continuously face substantial incentives to behave opportunistically. In *Governing the Commons* (E. Ostrom 1990), I addressed the puzzle of how individuals using these systems sustained them over long periods of time.

The specific rules-in-use differ markedly from one case to the next. Given the great variation, then, the sustainability of these resources and their institutions cannot be explained by the presence or absence of any particular rule or configuration of rules. Rather, part of the explanation that can be offered for the sustainability of these systems is based on the fact that the particular rules do differ. By differing, the particular rules take into account specific attributes of the related physical systems, cultural views of the world, and the economic and political relationships that exist in the setting. Without different rules, appropriators could not take advantage of the positive features of a local resource or avoid potential pitfalls that could occur in one setting but not others.

A set of seven design principles appears to characterize most of the robust institutions. An eighth principle characterizes the larger, more complex cases. A *design principle* is defined as a general organizing conception used either consciously or unconsciously by those constituting and reconstituting a continuing association of individuals. Let us discuss each of these design principles (drawn from E. Ostrom 1990).

*Clearly Defined Boundaries:* Individuals or households have the right to withdraw resource units from common property, and the boundaries of the resource itself are clearly defined.

Defining the boundaries of the resource, and of those authorized to use it, can be thought of as a "first step" in organizing for collective action. So long as the boundaries of the resource and/or the individuals who can use the resource remain uncertain, no one knows what they are managing or for whom. Without defining the boundaries of the resource and closing it to "outsiders," local appropriators face the risk that any benefits they produce by their efforts will be reaped by others who do not contribute to these efforts. At the least, those who invest in the resource may not receive as high a return as they expect. At the worst, the actions of others could destroy the resource itself. Thus, for any appropriators to have a minimal interest in coordinating patterns of appropriation and provision, some set of appropriators have to be able to exclude others from access and appropriation rights. If there are substantial numbers of potential appropriators and the demand for the resource units are high, the destructive potential of all users freely withdrawing from a resource could push the discount rate used by appropriators toward 100 percent. The higher the discount rate, the closer the situation is to that of a one-shot dilemma, where the dominant strategy of all participants is to overuse the resource.

*Congruence between Appropriation and Provision Rules and Local Conditions:* Appropriation rules restricting time, place, technology, and/or quantity of resource units are related to local conditions and to provision rules requiring labor, materials, and/or money.

Unless the number of individuals authorized to use a resource is so small that their harvesting patterns do not adversely affect one another, at least some rules related to how, when, and how much different products can be harvested are usually designed by those using the resource. Well-tailored appropriation and provision rules help to account for the perseverance of the resources themselves. Uniform rules established for an entire nation or large region of a nation can rarely take into account the specific attributes of a resource that are used in designing rules-in-use for a particular location.

In long-surviving irrigation systems, for example, subtly different rules are used in each for assessing water fees used to pay for maintenance activities and guards, but in all instances, those who receive the highest proportion of the water also pay approximately the highest proportion of the fees. No single set of rules defined for all irrigation systems in a region would satisfy the particular problems in managing each of these broadly similar, but distinctly different, systems.

*Collective-Choice Arrangements:* Most individuals affected by operational rules can participate in modifying them.

Resource institutions that use this principle are better able to tailor their rules to local circumstances, since the individuals who directly interact with one another and with the physical world can modify the rules over time. Appropriators who design resource institutions that are characterized by the first three principles— clearly defined boundaries, good-fitting rules, and appropriator participation in collective choice—should be able to devise a good set of rules if they keep the costs of changing rules relatively low.

The presence of good rules, however, does not account for appropriators following them. Nor is the fact that the appropriators themselves designed and initially agreed to the operational rules an adequate explanation for centuries of compliance by individuals who were not originally involved in the initial agreement. It is not even an adequate explanation for the continued commitment of those who were part of the initial agreement. Agreeing to follow rules *ex ante* is an easy "commitment" to make. Actually following rules *ex post,* when strong temptations are present, is the significant accomplishment.

The problem of gaining compliance to rules—no matter what their origin—is frequently assumed away by analysts positing all-knowing and all-powerful external authorities that enforce agreements. In many long-enduring resources, no external authority has sufficient presence to play any role in the day-to-day enforcement of the rules-in-use. Thus, external enforcement cannot be used to explain high levels of compliance. In all of the long-enduring cases, active investments in monitoring and sanctioning activities are quite apparent. These lead us to consider the fourth and fifth design principles.

*Monitoring:* Monitors, who actively audit resource conditions and appropriator behavior, are accountable to the appropriators and/or are the appropriators themselves; and

*Graduated Sanctions:* Appropriators who violate operational rules are likely to receive graduated sanctions (depending on the seriousness and context of the offense) from other appropriators, from officials accountable to these appropriators, or from both.

In long-enduring institutions, monitoring and sanctioning are undertaken primarily by the participants themselves. The initial sanctions used in these systems are also surprisingly low. Even though it is frequently presumed that participants will not take the time and effort to monitor and sanction each other's performance, substantial evidence has been presented that they do both in these settings.

To explain the investment in monitoring and sanctioning activities that occurs in these robust, self-governing, resource institutions, the term *quasi-voluntary compliance* used by Margaret Levi (1988) is useful. She uses it to describe taxpayer behavior in regimes where most everyone pays taxes. Paying taxes is voluntary in the sense that individuals choose to comply in many situations where they are not being directly coerced. On the other hand, it is "quasi-voluntary because the non-compliant are subject to coercion—if they are caught" (52). Levi stresses the contingent nature of a commitment to comply with rules that is possible in a repeated setting. Strategic actors are willing to comply with a set of rules, Levi argues, when:

1.  they perceive that the collective objective is achieved; and
2.  they perceive that others also comply.

In Levi's theory, enforcement is normally provided by an external ruler, even though her theory does not preclude other enforcers.

To explain commitment in many of the cases of sustainable community-governed resources, external enforcement is largely irrelevant. External enforcers may not travel to a remote village other than in extremely unusual circumstances. Resource appropriators create their own internal enforcement to deter those who are tempted to break the rules, thereby assuring quasi-voluntary compliers that others also comply. The Chisasibi Cree, for example, have devised a complex set of entry and authority rules related to the coastal and islaurine fish stocks of James Bay, as well as the beaver stock located in their defined hunting territory. Fikret Berkes describes why these resource systems, and the rules used to regulate them, have survived and prospered for so long:

> Effective social mechanisms ensure adherence to rules which exist by virtue of mutual consent within the community. People who violate these rules suf-

fer not only a loss of favor from the animals (important in the Cree ideology of hunting) but also social disgrace (1987, 87).

The costs of monitoring are kept relatively low in many long-enduring resources as a result of the rules-in-use. Rotation rules used in irrigation systems and in some inshore fisheries place the two actors most concerned with cheating in direct contact with one another. The irrigator who nears the end of a rotation turn would like to extend the time of his or her turn (and thus, the amount of water obtained). The next irrigator in the rotation system waits nearby for him or her to finish, and would even like to start early. The presence of the first irrigator deters the second from an early start, and the presence of the second irrigator deters the first from a late ending. Monitoring is a by-product of their own strong motivations to use their water rotation turn to the fullest extent. The fishing site rotation system used in Alanya (Berkes 1992, 161–82) has the same characteristic: cheaters are observed at low cost by those who most want to deter another cheater at that particular time and location. Many of the ways that work teams are organized in the Swiss and Japanese mountain commons also have the result that monitoring is a natural by-product of using the commons.

The costs and benefits of monitoring a set of rules are not independent of the particular set of rules adopted. Nor are they uniform in all resource settings. When appropriators design at least some of their own rules, they can learn from experience to craft enforceable rather than unenforceable rules. This means paying attention to the costs of monitoring and enforcing as well as the benefits that those who monitor and enforce the rules obtain. A frequently unrecognized "private" benefit of monitoring in settings where information is costly is obtaining the information necessary to adopt a contingent strategy. If an appropriator who monitors finds someone who has violated a rule, the benefits of this discovery are shared by all using the resource, as well as providing the discoverer a signal about compliance rates. If the monitor does *not* find a violator, it has previously been presumed that private costs are involved without any benefit to the individual or group. If information is not freely available about compliance rates, then an individual who monitors obtains valuable information from monitoring.

By monitoring the behavior of others, the appropriator-monitor learns about the level of quasi-voluntary compliance in the resource. If no one is discovered breaking rules, the appropriator-monitor learns that others comply and no one is being taken for a sucker. It is then safe for the appropriator-monitor to continue to follow a strategy of quasi-voluntary compliance. If the appropriator-monitor discovers rule infractions, it is possible to learn about the particular circumstances surrounding the infraction, to participate in deciding the appropriate level of sanctioning, and then to decide about continued compliance or not. If an appropriator-monitor finds an offender, who normally follows rules but happens to face a severe problem, the experience confirms what everyone already knows. There will always be times and places where those who are basically committed to following a set of rules succumb to strong temptations to break them.

A real threat to the continuance of quasi-voluntary compliance can occur, however, if an appropriator-monitor discovers individuals who repeatedly break the rules. If this occurs, one would expect the appropriator-monitor to escalate the sanctions imposed in an effort to halt future rule breaking by such offenders and any others who might start to follow suit. In any case, the appropriator-monitor has up-to-date information about compliance and sanctioning behavior on which to make future decisions about personal compliance.

Let us also look at the situation through the eyes of someone who breaks the rules and is discovered by a local guard (who will eventually tell everyone) or another appropriator (who also is likely to tell everyone). Being apprehended by a local monitor when the temptation to break the rules becomes too great has three results: it stops the infraction from continuing and may return contraband harvest to others; it conveys information to the offender that someone else in a similar situation is likely to be caught, thus increasing confidence in the level of quasi-voluntary compliance; and it imposes punishment in the form of a fine plus a loss of one's reputation for reliability.

The fourth and fifth design principles—monitoring and graduated sanctions—thus take their place as part of the configuration of principles that work together to enable appropriators to constitute and reconstitute robust resource institutions. To summarize to this point: When resource appropriators design their own operational rules (design principle three) to be enforced by individuals who are local appropriators or accountable to them (design principle four), using graduated sanctions (design principle five) that define who has rights to withdraw from the resource (design principle one) and that effectively restrict appropriation activities given local conditions (design principle two), the commitment and monitoring problems are solved in an interrelated manner. Individuals who think a set of rules will be effective in producing higher joint benefits and that monitoring (including their own) will protect them against being a sucker, are willing to make a contingent self-commitment of the following type: I commit myself to follow the set of rules we have devised in all instances except dire emergencies, if the rest of those affected make a similar commitment and act accordingly. Once appropriators have made contingent self-commitments, they are then motivated to monitor other people's behavior, at least from time to time, in order to assure themselves that others are following the rules most of the time. Contingent self-commitments and mutual monitoring reinforce one another, especially in resources where rules tend to reduce monitoring costs.

*Conflict-Resolution Mechanisms:* Appropriators and their officials have rapid access to low-cost, local arenas to resolve conflict among appropriators or between appropriators and officials.

In field settings, applying rules always involves discretion and can frequently lead to conflict. Even such a simple rule as "Each irrigator must send one individual for one day to help clean the irrigation canals before the rainy season begins" can be interpreted quite differently by different individuals. Who is or is not an "individual" according to this rule? Does sending a child below age 10 or an adult above 70 to do heavy physical work meet this rule? Is working for four or six hours a "day" of work? Does cleaning the canal immediately next to one's own farm qualify for this community obligation? For individuals who are seeking to slide past or subvert rules, there are always ways that they can "interpret" the rule so that they can argue they meet it. Even individuals who intend to follow the spirit of a rule can make errors. What happens if someone forgets about the labor day and does not show up? Or, what happens if the only able-bodied worker is sick, or unavoidably in another location?

If individuals are going to follow rules over a long period of time, some mechanism for discussing and resolving what is or is not a rule infraction is quite necessary to the continuance of rule conformance itself (see Crawford and Ostrom 1995). If some individuals are allowed a free ride by sending less-valuable workers to a required labor day, others will consider themselves to be suckers if they send their strongest workers, who could be used to produce private goods rather than communal benefits. Over time, only children and old people will be sent to do work that requires strong adults, and the system breaks down. If individuals who make an honest mistake or face personal problems that prevent them from following a rule cannot find mechanisms to make up their lack of performance in an acceptable way, rules can be viewed as unfair and conformance rates decline.

While the presence of conflict-resolution mechanisms does not guarantee that appropriators are able to maintain enduring institutions, it is difficult to imagine how any complex system of rules could be maintained over time without such mechanisms. In the cases described above, these mechanisms are sometimes quite informal and those who are selected as leaders are also the basic resolvers of conflict.

*Minimal Recognition of Rights to Organize:* The rights of appropriators to devise their own institutions are not challenged by external governmental authorities.

Appropriators frequently devise their own rules without creating formal, governmental jurisdictions for this purpose. In many inshore fisheries, for example, local fishers devise extensive rules defining who can use a fishing ground and what kind of equipment can be used. So long as external governmental officials give at least minimal recognition to the legitimacy of such rules, the fishers may be able to enforce the rules themselves. But if external governmental officials presume that only they can make rules, then it is difficult for local appropriators to sustain a rule-governed resource over the long run. At any point when someone wishes to break the rules created by the fishers, they can go to the external government and get local rules overturned.

Audun Sandberg (1993a; 1993b) provides an insightful analysis of what happens when the individuals using common-pool resources for many centuries do not have recognized authority to create their own rules. The formal rules for the northern Norwegian commons were first written as law in the eleventh century and have remained unchanged until 1993, thus representing "more than 1,000 years of unbroken traditions of oral and codified Common Law" (Sandberg 1993b, 14). The rules, however, outlined only generalized rights and did not recognize any local governance responsibilities. Since most commons, especially the northern ones, came to be conceptualized as the King's Commons, it was easy to conceptualize that the king was the only lawgiver with authority to change laws over time.

Through a long process, which started with the Protestant Reformation and accelerated around 1750, this eventually led to the conception that all forests and mountains in northern Norway that were not private property, and which would in other countries be considered a commons, were considered state property (19). The further effort of the state to then ration access to forests, grazing areas, fisheries, and other common-pool resources to those engaged in full-time specialized employment had the unintended effect of being disruptive to the mixed economic way of life of many northerners who were part-time farmers, part-time fishers, part-time foresters, and part-time herders. Converting this sustainable way of life into a modern system—including heavy reliance on transfer payments to specialized farming, fishing, and reindeer ranching—was probably not fully expected by anyone. Now, however, the economic and social base has been weakened substantially enough that simply assigning local authority to make rules related to the use of common-pool resources would probably not be a sufficient way out of a major dilemma. Maurizio Merlo (1989) provides a detailed and sorry tale of a similar effort in Italy to destroy local communal institutions, only to find that unanticipated consequences have led many to try to reestablish authoritative decisionmaking in local communities.

*Nested Enterprises:* Appropriation, provision, monitoring, enforcement, conflict resolution, and governance activities are organized in multiple layers of nested enterprises.

In larger systems, it is quite difficult to devise rules that are well-matched to all aspects of the provision and appropriation of that system at one level of organization. The rules appropriate for allocating water among three major branches of an irrigation system, for example, may not be appropriate for allocating water among farmers along a single distributary channel. Consequently, among long-enduring, self-governed resources, smaller-scale organizations tend to be nested in ever-larger organizations. It is not at all unusual to find a larger, farmer-governed irrigation system, for example, with five layers of organization each with its own distinct set of rules (Yoder 1994). In the Swiss Alps, day-to-day operational decisions have been made by local communities while larger governance units have had the responsibility to monitor local performance by doing periodic, careful, site visits on a rotational basis (Glaser 1987; Netting 1981). In the Italian Alps, the smallest units during the Middle Ages were "Family Communities," which were groups of families who collectively owned and governed land the size of a hamlet. "Then came the Village Communities as such which were in turn grouped in Federations which could be of the first degree (generally comprising one single valley) or second degree (comprising a group of valleys)" (Merlo 1989, 8).

## Looking Forward to High Levels of Biodiversity

How can these design principles be used as the basis for future policymaking that enhances the survival of biological resources beyond those that have caught the attention of electronic and print media? What one learns from looking backward is that there are several reasons why local users may more effectively manage smaller-scale resources than national agencies. One reason is the immense diversity of local environmental conditions that exist within most countries. The variation in rainfall, soil types, elevation, scale of resource systems, and plant and animal ecologies is immense, even in small countries. Some resources are located near urban populations or major highway systems; others are remote. Given environmental variety, rule systems that effectively regulate access, use, and the allocation of benefits and costs in one setting are not likely to work well in radically different environmental conditions. Efforts to pass national legislation establishing a uniform and detailed set of rules for an entire country are likely to fail in many of the locations most at risk. Users managing their resources locally may be a more effective way of dealing with immense diversity from site to site.

A second reason for the potential advantage of local organization in coping with problems of biodiversity losses is that the benefits local users may obtain from careful husbanding of their resources are potentially greater, when future flows of benefits are appropriately taken into account. At the same time, the costs of monitoring and sanctioning rule infractions at a local level are relatively low. These advantages occur, however, only when local users have sufficient assurance that they will actually receive the long-term benefits of their own investments. Extensive literature on the imposition of legal statutes by larger governmental units on the conservation of forest resources in India has consistently shown the adverse consequences of these efforts (see, in particular, Blaikie, Harriss, and Pain 1992, 274–64; Gadgil and Guha 1992; Fox 1993; Jodha 1992).

While there is agreement that the potential for effective organization at a local level to manage some of the smaller- to medium-sized biological resources exists in all countries, local participants do not uniformly expend the effort needed to organize and manage these resources, however, even when given formal authority. Some potential organizations never form at all. Some do not survive more than a few months. Others are dominated by local elite who divert communal resources to achieve their own goals at the expense of others (Arora 1994). In some cases, a local resource may be almost completely destroyed before local remedial actions are taken (Blomquist 1992), and then these actions may be too late. Still others do not possess adequate scientific knowledge to complement their own indigenous knowledge. Making investment decisions related to assets that mature over a long time horizon (25 to 75 years for many tree species) is a sophisticated task, whether undertaken by barely literate farmers or Wall Street investors. In highly volatile worlds, some organize themselves more effectively and make better decisions than others.

Thus, the romantic view that anything local is better than anything organized at a national or global scale is not a useful foundation for a long-term effort to improve understanding of what factors enhance or detract from the capabilities of any institutional arrangement to govern and manage biological resources wisely. Any organization or group faces a puzzling set of problems when it tries to govern and manage complex multispecies (including *Homo sapiens*), multiproduct resource systems whose benefits mature at varying rates. Further, any organization or group will come up against a variety of environmental challenges, from too much or too little rainfall, to drastic changes in factory prices, population density, or pollution levels.

The chapters that follow in this section (Reid, Gupta, Toenniessen, Anaya and Crider, and Sittenfeld and Lovejoy) propose new institutional arrangements (or the enhancement of earlier arrangements) that attempt to rely on both large- and small-scale institutions. An important aspect of all of these proposals is the effort to enable institutions of multiple scales to more effectively blend local, indigenous knowledge with scientific knowledge (Gadgil, Berkes, and Folke 1993). Key to the successful design of such institutions is their multiple scales and the generation of information that allows participants operating at many scales to learn from experience. The complexity of the environments involved is simply more than any single corporate entity can absorb and manage.

## The Role of Collaborative Research

Knowledge of and understanding about how institutional arrangements facilitated the survival of complex resource systems over the centuries are essential inputs into future policymaking to preserve biological diversity. It is also important to design studies that monitor a large sample of current institutions over time to ascertain which mix of institutional incentives tend to increase the learning capacity of participants so as to enhance biodiversity rather than reduce it. Essential knowledge can be gained from a carefully designed, systematic study of how many different kinds of institutional arrangements—including nascent groups, indigenous communal organizations, formal local governments, nongovernmental organizations, specialized forest and park agencies, and national ministries—cope with diverse types of forest resources. Much is to be learned from both successes and failures. Using multiple performance measures, one would expect to find some governance and management systems that are positive in regard to some evaluative criteria (such as the maintenance of species richness), but not necessarily in regard to others (such as gender representation, financial accountability, adaptability over time, or transparency of decisionmaking processes).

The International Forestry Resources and Institutions (IFRI) is a research program initiated by colleagues at the Workshop in Political Theory and Policy Analysis in response to a request by Marilyn Hoskins of the Forests, Trees, and People Programme of the Food and Agriculture Organization (FAO) of the United Nations (the final section of this chapter draws on E. Ostrom and Wertime 1994; copies of this research strategy can be obtained from the authors). In our effort to develop an integrated mode of assessing plant biodiversity, forest biomass, and other important measures of forest conditions with careful measures of socioeconomic and institutional variables, we have consulted researchers in all areas of the world and in many diverse disciplines. The research is being carried out by a network of research centers located in each region of the world. Prior theoretical and empirical studies provide an initial set of hypotheses about general factors that we expect to find associated with the more successful forest governance and management systems (see E. Ostrom, Gardner, and Walker 1994; Keohane and Ostrom 1995; McKean 1992; Moorehead 1994). A basic assumption is that investments in governance arrangements are costly, and the individuals (forest users and government officials) who are expected to make that investment must evaluate their long-term expected benefits to exceed their long-term expected costs. The initial set of working hypotheses will then be revised, added to, and refined over time. These hypotheses are strongly related to the design principles described above.

Our initial working hypotheses are that more effective organization to cope with the long-term, sustainable management of forest resources will occur where:

- local forest users participate in, and have continuing authority to design, the institutions that govern the use of a forest system;
- the individuals most affected by the rules that govern the day-to-day use of a forest system are included in the group that can modify these rules;
- the institutions that govern a forest system minimize opportunities for free riding, rent seeking, asymmetric information, and corruption through effective procedures for monitoring the behavior of forest users and officials. *Free-riding*

behavior occurs when individuals do not contribute to the provision and/or production of a joint benefit in the hopes that others will bear the cost of participating and that the free riders will receive the benefits without paying the costs. *Rent seeking* occurs when individuals obtain entitlements that enable them to receive returns that exceed what they would receive in an open, competitive environment. *Asymmetric information* occurs when some individuals obtain information of strategic value that is not available to others. *Corruption* occurs when individuals in official positions receive personal side payments in return for the exercise of their discretion;

- forest users who violate rules governing the day-to-day uses of a forest system are likely to receive graduated sanctions from other users, from officials accountable to these users, or both;
- rapid access is available to low-cost arenas to resolve conflicts between users, or between users and their officials;
- monitoring, sanctioning, conflict resolution, and governance activities are organized in multiple layers of nested enterprises; and
- the institutions that govern a forest system have been stable for a long period, and are known by and understood by forest users.

The variables in these hypotheses are all operationalized using multiple indicators in the IFRI research instruments. Further, we have included other variables noted in the literature as being of importance in explaining processes of deforestation and biodiversity loss—particularly those related to increases in population size and density, as well as the intrusion of external market forces on local economies.

In the design of this study, we have also been concerned with how national and regional governments can enhance or detract from the capabilities of local entities by the kind of information they provide; by the assurances that they extend to guarantee autonomy over the long run; by the provision of low-cost conflict resolution mechanisms; and by policies that allow localities to develop and keep financial resources that can be used to make local improvements. Detailed information about why some national policies tend to encourage successful self-organization, and others discourage it, will be provided. These results will help to reduce knowledge gaps about policy impacts and, thus, facilitate the development of more effective policies.

The IFRI research program is designed to examine relationships among the physical, biological, and cultural worlds in a particular location and the *de facto* rules that are used locally to determine access to and use of a forest. During data collection, researchers will use 10 instruments called coding forms (see table 1). Examination of the physical world includes the structure of forests and the species within. There are two coding forms that include rigorous forest mensuration methods in order to generate reliable and unbiased estimates of forest density,

TABLE 1    Data Collection Forms and Information Collected

| IFRI Form | Information Collected |
| --- | --- |
| Site overview form | Site overview map, local wage rated, local units of measurement, exchange rates, recent policy changes, and interview information |
| Forest form | Size, ownership, internal differentiation, products harvested, uses of products, master species list, changes in forest area, and appraisal of forest condition |
| Forest plot form | Tree, shrub, and sapling size, density, and species type within 1-, 3-, and 10-meter circles for a random sample of plots in each forest, and general indications regarding forest condition |
| Settlement form | Sociodemographic information, relation to markets and administrative centers, and geographic information about the settlement |
| User group form | Size, socioeconomic status, and attributes of specific forests user groups |
| Forest user group relationship form | Products harvested by user groups from specific forests and their uses |
| Forest products form | Details on three most important forest products (as defined by the user group), temporal harvesting patterns, alternative sources and substitutes, harvesting tools and techniques, and harvesting rules |
| Forest association form | Institutional information about forest association (if one exists at the site), including association's activities, rules, structure, membership, and record keeping |
| Governance form | Information about organizations that make rules regarding a forest(s), but do not use the forest itself, including structure, personnel, resource mobilization, and record keeping |
| Organizational inventory and interorganizational arrangements form | Information about all organizations (harvesting or not) that relate to a forest, including harvest and governance activities |

species diversity, and consumptive disturbances. The examination of cultural worlds includes gaining knowledge about the patterns of socioeconomic and cultural homogeneity, the number of individuals and groups involved, and diverse worldviews. Research conducted using a uniform set of variables as well as the best methods available for gaining reliable estimates of qualitative and quantitative data will enable scholars to analyze how different institutions work in the context of a large number of ecological, cultural, and political-economic settings. Diverse models of which variables, and which variable interactions, affect behavior and outcomes will be posed, tested, and modified so that policies based on revised and tested models will have a higher probability of being successful than past efforts to reduce deforestation and stop biodiversity losses.

Important steps have been taken in the last decade to increase the rigor and quantity of information known about forest cover and biodiversity losses in different parts of the world. In 1993, for example, the most "authoritative global tropical deforestation survey to be produced in more than a decade" (Aldhous 1993) was released by the FAO (1993). The FAO report attempts to document the extent of deforestation in tropical countries in an accurate fashion, but repeatedly stresses the problems that the project staff faced in obtaining reliable information for the task. After examining the current state of information about forest conditions in tropical countries, the project found that:

- there is considerable variation among regions with respect to completeness and quality of the information;
- there is considerable variation in the timeliness of the information—the data is about 10 years old, on average, which could be a potential source of bias in the assessment of change;
- only a few countries have reliable estimates of actual plantations, harvests, and utilization, although such estimates are essential for national forestry planning and policymaking;
- no country has carried out a national forest inventory containing information that can be used to generate reliable estimates of the total woody biomass volume and change; and
- it is unlikely that the state and change information on forest cover and biomass could be made available on a statistically reliable basis at the regional or global level within the next 10 or 20 years unless a concerted effort is made to enhance the country capacity in forest inventory and monitoring (FAO 1993, 5–6).

The report concludes its findings concerning information gaps by noting that "forest resource assessments are among the most neglected aspects of forest resource management, conservation and development in the tropics" (6).

The IFRI research program will immediately provide key information about variations in forest conditions, and the incentives and behavior of forest users within countries participating in the IFRI network. The first set of studies from Bolivia, Ecuador, India, Nepal, and Uganda provide useful microlevel information to complement the macrolevel information obtained from the FAO and other sources (Agrawal 1996; Banana 1996; Becker and Gibson 1996; McKean 1996; Schweik 1996). This information is essential for policy analysis and to test theories addressing knowledge gaps. Focusing on a sample of forests located in diverse ecological regions and governed by different institutional arrangements greatly reduces the cost of monitoring, as contrasted to national forest inventories. Further, it provides information about the variation of results achieved by different kinds of institutional arrangements.

Both quantitative and qualitative data will be collected about institutional arrangements, the incentives of different participants, and their activities; careful forest mensuration techniques will be used to assess the consequences in terms of density, species diversity, and species distribution. This information will immediately be made available to forest users and government officials, and used in regularized policy reports written by analysts who have a long-term stake in the success of the policies adopted. The results of projects adopted in one location can be compared with the results of other types of institutional arrangements in similar ecological zones within the same macropolitical regime. The data will also be archived in an IFRI-designed, relational database so that changes in institutions, policies, activities, and outcomes can be monitored over time and across regions within one, or more than one, country. Data will be collected, owned, assessed, stored, and analyzed by each countries' researchers. The IFRI research program fosters in-country development of information rather than sole reliance on the purchase of secondary data from international organizations. The program also encourages the development of "state-of-the-art" research conducted by researchers who have permanent roots in a country, rather than bringing in expertise from the outside.

The IFRI research program will work with a growing group of in-country Collaborating Research Centers, which obtain funding from donors and their own institutions to build their capabilities to become a permanent assessment site. The first three IFRI centers are located in Bolivia, Nepal, and Uganda. A training program to initiate centers in Cameroon, Ecuador, Guatemala, India, and Sweden began in 1996.

## Conclusion

This chapter summarized key findings from earlier studies and the design of an ongoing research program that helps provide essential information on how to design and reform complex, nested, institutional arrangements that enhance the preservation of complex, natural resource systems and their biodiversity. We can learn from the past by studying those institutional arrangements that have survived long periods of time while successfully coping with the complex balancing of different aspects of local resources. We need to enhance the efforts of current innovators of institutional arrangements that build nested institutions at both large and small levels of society. It is also important to design ongoing monitoring programs that simultaneously analyze both the social and biological system performance. The last section of the chapter describes an ongoing international research program to monitor forest biodiversity and the institutional arrangements that enhance its preservation.

## Acknowledgments

Support for the research on which this chapter is based has been provided by the National Science Foundation (grant number SBR 9319835); the Forests, Trees, and People Programme at the Food and Agriculture Organization of the United Nations; and the Beijer International Institute of Ecological Economics at the Royal Swedish Academy of Science. Comments by Susan Hannah and editing by Patty Dalecki were greatly appreciated.

## References

Agrawal, A. 1994. Rules, rule making, and rule breaking: Examining the fit between rule systems and resource use. In *Rules, Games, and Common-Pool Resources*, edited by E. Ostrom, R. Gardner, and J. Walker. Ann Arbor: University of Michigan Press.

———. 1996. *Group Size and Successful Collective Action: A Case Study of Forest Management Institutions in the Indian Himalayas.* Working paper, Forests, Trees, and People Programme, phase 2, Food and Agriculture Organization, Rome.

Aldhous, P. 1993. Tropical deforestation: Not just a problem in Amazonia. *Science* 259: 1, 390.

Arora, D. 1994. From state regulation to people's participation: Case of forest management in India. *Economic and Political Weekly* (March): 691–98.

Ascher, W. 1995. *Communities and Sustainable Forestry in Developing Countries.* San Francisco: ICS Press.

Ashby, W.R. 1960. *Design for a Brain: The Origin of Adaptive Behavior.* New York: John Wiley.

Banana, A. 1996. Successful forestry management: The importance of security and tenure and rule enforcement in Ugandan forests. Working paper, Forests, Trees, and People Programme, phase 2, Food and Agriculture Organization, Rome.

Becker, D.C., and C. Gibson. 1996. The lack of institutional supply: Why a strong local community in western Ecuador fails to protect its forest. Working paper, Forests, Trees, and People Programme, phase 2, Food and Agriculture Organization, Rome.

Berkes, F. 1987. Common property resource management and Cree Indian fisheries in subarctic Canada. In *The Question of the Commons*, edited by B. McCay and J. Acheson. Tucson: University of Arizona Press.

———. 1992. Success and failure in marine coastal fisheries of Turkey. In *Making the Commons Work: Theory, Practice, and Policy*, edited by D.W. Bromley et al. San Francisco: ICS Press.

Blaikie, P., J. Harriss, and A. Pain. 1992. The management and use of common-property resources in Tamil Nadu, India. In *Making the Commons Work: Theory, Practice, and Policy*, edited by D.W. Bromley et al. San Francisco: ICS Press.

Blomquist, W. 1992. *Dividing the Waters: Governing Groundwater in Southern California.* San Francisco: ICS Press.

Crawford, S.E.S., and E. Ostrom. 1995. A grammar of institutions. *American Political Science Review* 89: 582–600.

FAO (Food and Agriculture Organization) of the United Nations. 1993. *Forest Resources Assessment 1990.* Rome: FAO.

Fox, J., ed. 1993. *Legal Frameworks for Forest Management in Asia: Case Studies of Community/State Relations.* Hawai'i: East-West Center.

Gadgil, M., F. Berkes, and C. Folke. 1993. Indigenous knowledge for biodiversity conservation. *Ambio* 22:244–49.

Gadgil, M., and R. Guha. 1992. *This Fissured Land: An Ecological History of India*. New Delhi: Oxford University Press.

Glaser, C. 1987. Common property regimes in Swiss alpine meadows. Paper presented at the Conference on Comparative Institutional Analysis, Inter-University Center of Postgraduate Studies, Dubrovnik.

Jodha, N.S. 1992. Rural common property resources: The missing dimension of development strategies. Discussion paper no. 169, World Bank, Washington, D.C.

Keohane, R.O., and E. Ostrom, eds. 1995. *Local Commons and Global Interdependence: Heterogeneity and Cooperation in Two Domains*. London: Sage.

Levi, M. 1988. *Of Rule and Revenue*. Berkeley: University of California Press.

McKean, M.A. 1992. Success on the commons: A comparative examination of institutions for common property resource management. *Journal of Theoretical Politics* 4:247–82.

———. 1996. Common Property: What is it, what is it good for, and what makes it work. Working paper, Forests, Trees, and People Programme, Phase 2, Food and Agriculture Organization, Rome.

Merlo, M. 1989. The experience of the village communities in the north-eastern Italian Alps. In *Collective Forest Land Tenure and Rural Development in Italy*, edited by M. Merlo et al. Rome: Food and Agriculture Organization of the United Nations.

Moorehead, R. 1994. *Policy and Research into Natural Resource Management in Dryland Africa: Some Concepts and Approaches*. London: International Institute for Environment and Development.

Netting, R.M. 1981. *Balancing on an Alp: Ecological Change and Continuity in a Swiss Mountain Community*. New York: Cambridge University Press.

Ostrom, E. 1990. *Governing the Commons: The Evolution of Institutions for Collective Action*. New York: Cambridge University Press.

Ostrom, E., R. Gardner, and J. Walker. 1994. *Rules, Games, and Common-Pool Resources*. Ann Arbor: University of Michigan Press.

Ostrom, E., and M.B. Wertime. 1994. IFRI Research Strategy. International Forestry Resources and Institutions research program and database, Workshop in Political Theory and Policy Analysis, Indiana University, Bloomington.

Ostrom, V. 1991. *The Meaning of American Federalism: Constituting a Self-Governing Society*. San Francisco: ICS Press.

———. 1995. *The Meaning of Democracy: The Vulnerability of Democracies*. Working paper, Workshop in Political Theory and Policy Analysis, Indiana University, Bloomington.

Sandberg, A. 1993a. The analytical importance of property rights to northern resources. Working paper, Workshop in Political Theory and Policy Analysis, Indiana University, Bloomington.

———. 1993b. Entrenchment of state property rights to northern forests, berries, and pastures. Presented at the Mini-Conference on Institutional Analysis and Development, held at the Workshop in Political Theory and Policy Analysis, Indiana University, Bloomington.

Schlager, E., and E. Ostrom. 1993. Property-rights regimes and coastal fisheries: An empirical analysis. In *The Political Economy of Customs and Culture: Informal Solutions to the Commons Problem*, edited by T.L. Anderson and R.T. Simmons. Lanham, Md.: Rowman and Littlefield.

Schweik, C. 1996. Social norms and human foraging: An investigation into the spatial distribution of *Shorea robusta* in Nepal. Working paper, Forests, Trees, and People Programme, phase 2, Food and Agriculture Organization, Rome.

Shepsle, K.A. 1989. Studying institutions: Some lessons from the rational choice approach. *Journal of Theoretical Politics* 1:131–49.

Yoder, R. 1994. *Locally Managed Irrigation Systems*. Columbo, Sri Lanka: International Irrigation Management Institute.

# Halting the Loss of Biodiversity:
# International Institutional Measures

by Walter V. Reid

Action taken now to slow the rate of biodiversity loss and maintain the equilibrium level of surviving diversity will enable their sustainable use by nations to meet social and economic needs. International action that helps nations to anticipate and prevent threats to biodiversity, and align the goals of biodiversity conservation with local and national development goals, will advance conservation and sustainable development. The four major types of international action that influence biodiversity conservation are: international resource management coordination; "bargains" among nations; international financial and trade agreements; and international science. The Convention on Biological Diversity (CBD) provides a significant opportunity for advancing these goals. Further international institutional evolution to support conservation should include steps to build capacity in developing countries; control genetic and biochemical transfers; extend rights under the CBD to local communities; abandon the concept of global incremental costs; and address biosafety concerns through international negotiation.

\*

Our generation has the singular misfortune of living during—and helping to create—a period of biological simplification. There are few places on earth where biodiversity is not retreating from the combined pressures of human population growth, excessive resource demands, exotic species introductions, and toxic chemical and pollutant emissions.

The United States alone has a list of more than 950 threatened and endangered species, and on average, 38 species have been added to the list each year over the past decade. Of the listed species, some 69 percent are declining in population size or, at best, holding their own, while only 10 percent are increasing (USFWS 1990). And habitats and ecosystems are just as threatened. Old growth in the Pacific Northwest has been reduced to just 13 percent of its original extent, and some 290,000 acres of wetlands continue to be lost each year (Dudley 1992; DOI 1994).

As great as these problems are, prospects for biodiversity in the United States are far better than in many other nations. Worldwide, an estimated 2 to 5 percent of species per decade will be committed to extinction in the coming years, and habitats and genetic resources are disappearing at a comparable rate (Reid 1992; WCMC 1992). With growing populations and increasing demands on resources, the magnitude of the problem is overwhelming. If the United States, with the resources to tackle the problem and the benefit of a substantial decline in rural population size over the past half century, is at best only slowing the rate of biodiversity loss (and possibly only slowing the rate *of increase* in the loss of biodiversity), how are financially strapped countries with burgeoning populations to meet this challenge? What would it take to "stop biodiversity loss"? The habitat destruction that has already taken place has committed the world to a string of species extinctions that will continue for centuries, even if an immediate end were put to cutting forests, filling wetlands, and damming rivers (Reid and Miller 1989). Yet it is unrealistic to think of halting habitat loss anytime in the near future. World population is growing by nearly 1 billion people each decade, and both the number of poor people and the income disparity between rich and poor is increasing (WRI, 1994). Not only is there no technological fix in sight, but technology may well exacerbate the problem: new crops genetically engineered to grow on marginal lands may help meet subsistence and commercial needs, but they will also increase agricultural use of land that is now *de facto* protected from development.

The magnitude of the challenge of biodiversity conservation can be better appreciated by comparing it with another immense challenge: addressing the threat of global climate change. The International Panel on Climate Change estimates that approximately a 70 percent reduction in $CO_2$ emissions below 1994 levels is needed to stabilize atmospheric concentrations at current levels (Houghton et al. 1994). As unattainable as this target seems, given the anticipated growth in populations and economies around the world, it is far more likely to be achieved than the goal of stopping the loss of biodiversity. Dramatic reductions in the use of fossil fuels are conceivable by taking steps to raise fuel prices and invest heavily in the development of renewable energy sources. By contrast, halting biodiversity loss requires sweeping social and economic changes (indeed, it is likely to require the stabilization of climate as but one of many necessary conditions). To achieve success, the trend in habitat loss must be reversed and a substantial fraction of degraded habitats must be restored—again in the face of rapid growth in human populations and resource demands.

Ultimately, halting biodiversity loss requires social and economic progress, as well as technological change, that will stabilize human populations and their demands on natural resources (WRI 1992). But opportunities do exist to significantly increase the amount of the planet's biodiversity that survives the bottleneck that the world is entering. Reforms in national and international institutions are a case in point. Many of the benefits that biodiversity provides to humanity—such as clean air, clean water, genetic resources for agriculture, new pharmaceutical products, and aesthetic values—are public goods. Because these economic benefits of biodiversity are available to all, less incentive exists for any one individual or nation to conserve the resource than would make sense from the standpoint of society as a whole. This situation leads to the argument that biodiversity conservation ultimately requires a "subsidy" to correct for the market failures involved (Stone, this volume). It is worth noting that current public financing of biodiversity conservation only partially offsets subsidies supporting economically and environmentally damaging resource exploitation, such as excessive grazing or timber harvest that accelerate the loss of biodiversity.

Institutional reforms, however, can decrease the need for public financial support for conservation to the extent that they can align the goals of biodiversity conservation with local and national development goals. In particular, international institutions provide opportunities to establish property rights, along with regulatory and enforcement mechanisms, to help internalize the cost of biodiversity conservation within the market. Reforms of other international institutions can help reduce pressures on biodiversity, and additional international measures can enhance both the technologies for conservation and expand the economic benefits provided by biodiversity. This chapter presents an overview of the various international institutions influencing biodiversity conservation and examines the potential of the CBD (see appendix, this volume) for addressing key biodiversity conservation needs.

## International Action for Biodiversity

Both the rate of loss of biodiversity and the equilibrium level of diversity that survives will be influenced through actions taken now to protect species, genetic resources, and critical habitats, while using them sustainably to meet social and economic needs. International action can assist in the conservation and sustainable use of biodiversity to the extent that it helps nations to anticipate and prevent threats to biodiversity, and to align the goals of biodiversity conservation with local and national development goals. There are four major types of international action that influence biodiversity conservation:

1. *International Resource Management Coordination.* A long history of treaties and agreements seek to promote coordinated and complementary actions among countries that share resource management goals or concerns. For example, Mexico's efforts to maintain its shorebird populations would fail if the United States and Canada did not protect habitats for the birds during migration. Similarly, efforts in the United States to protect sea turtle populations would fail if other countries did not ban the import of turtle shells. Thus, nations have reached such agreements as the Convention on the Conservation of Migratory Species of Wild Animals (CCMSWA 1979), the Convention on International Trade in Endangered

Species of Wild Fauna and Flora (United Nations 1973), the Convention on the Conservation of Antarctic Marine Living Resources (U.S. Department of State 1980), the International Undertaking on Plant Genetic Resources (United Nations 1983), and numerous regional and species-specific councils and agreements addressing transboundary resources or migratory species, such as the United Nations Environment Programmes's Regional Seas Program.

2. *"Bargains" among Nations.* Another set of international mechanisms seeks not only the coordination of action, but also the establishment of reciprocity among nations. In a sense, these treaties or agreements strike "bargains" among countries with different needs and resource pressures. For example, the Convention on Wetlands of International Importance (United Nations 1971) obliges countries to take steps to conserve their wetland resources, and provides funding from relatively wealthy countries to support conservation activities in poorer nations. The pilot phase of the Global Environmental Facility provided financial support for conservation in developing countries, and in return, these countries took steps to protect their biodiversity—actions that resulted in both national and global benefits. The CBD establishes obligations for parties to undertake conservation activities, and ensures that wealthy countries will support conservation with financial and technical resources in return for maintaining their access to these resources. Additionally, a number of bilateral development assistance programs—including that of the United States—provide financial resources to help countries in their conservation efforts.

3. *International Financial and Trade Agreements.* Although not typically thought of in connection with biodiversity, international agreements that influence economic development and trade policies can have important conservation consequences—both positive and negative. Such agreements include structural adjustment programs, regional and global trade agreements, and commodity agreements, such as the International Tropical Timber Agreement, or ITTA (United Nations 1994). It is already standard for such agreements to address economic issues related to resource use, but increasingly the argument is being made that they should also address conservation concerns. Because of their major influence on patterns of development, these agreements may offer a far more valuable opportunity to promote the conservation and sustainable use of biodiversity than more narrowly focused conservation agreements.

For example, with appropriate modifications, commodity agreements such as ITTA could be made into instruments for promoting the sustainable use of resources (Repetto 1993). For decades, Southern producers have sought to improve their commodity terms of trade through commodity agreements designed to "stabilize" or reduce supply—efforts that have been resisted by consuming countries such as the United States. However, producers could improve export prices and revenues if they jointly agreed to adopt more sustainable production practices and internalize the full environmental costs of production in their pricing structure. Northern countries could not complain about agreements among producers to reduce logging to sustainable levels or protect biodiversity, and Southern producers could pass on the costs of these measures to Northern importers.

*4. International Science.* Finally, much of the basis for both the concern over the loss of biodiversity and knowledge of what steps need to be taken to protect biodiversity stems from the international scientific community, and the results of research in systematics, ecology, and conservation biology. Although science is not typically considered to be an international "measure," international scientific cooperation provides a model for the type of cooperation needed in such areas as technology and capacity building.

## The Convention on Biological Diversity

Although economic and trade agreements provide significant opportunities to embed concerns for sustainability into development plans, and other treaties could help coordinate resource management among nations, all attention is now focused on the CBD. This rapidly evolving mechanism provides by far the greatest opportunity for progress on international measures for conservation.

The CBD was one of the notable achievements of the United Nations Conference on Environment and Development—the Earth Summit—in Rio de Janeiro in June 1992. The 157 countries signing the convention believed that it struck a balance among the interests of nations differing widely in development needs, biodiversity endowments, and environmental threats. The CBD entered into force on December 29, 1993, and by December 1997, 171 countries had ratified the agreement.

The CBD is a *framework* agreement, meaning that it establishes an international legal structure for a coordinated response to the loss of biological diversity. It institutes general obligations for the parties to the CBD rather than legally binding targets for specific actions. Even as a framework agreement, its impact will be significant. The CBD increases the economic incentives for conservation—a key element of the long-term strategy to slow the loss of biodiversity. It also establishes short-term obligations to meet pressing conservation needs. Signatories have already begun to develop the required biodiversity action plans detailing how they will meet obligations to identify, monitor, and conserve biodiversity, and describing their priorities for training, education, and capacity building. Further, the CBD establishes a multilateral mechanism to provide financial support for these conservation needs in developing countries and sets up a scientific panel to help governments guide limited resources to the highest priorities.

The CBD marks a basic change in the international status of genetic resources. Prior to the CBD, these resources were considered to be the "heritage of humankind." This principle is a central tenet of the International Undertaking on Plant Genetic Resources and the Food and Agriculture Commission on Plant Genetic Resources—both established in the 1980s and involving some 135 countries (United Nations 1983). Under this regime, countries that violated the common heritage principle and restricted access to unimproved germ plasm (such as Ethiopia, which placed an embargo on the export of its coffee germ plasm in 1977) were soundly criticized (Fowler and Mooney 1990).

Though the intent of this open access regime was to ensure the widespread availability of genetic resources for agriculture and industry, it had one major flaw. By allowing free exchange of genetic resources, the source country received no direct benefit from their use. As a result, commercial use of these resources provided no additional economic incentive for conservation.

The CBD corrects this policy failure by establishing that states have sovereign rights over their own genetic resources, although it affirms that the *conservation* of biodiversity is a "common concern of humankind." It thus enables market incentives to be used to complement the various multilateral mechanisms that might directly fund biodiversity conservation. Moreover, the CBD adopted the principle that a portion of the benefits stemming from the productive use of genetic resources should flow back to the nations that act to conserve and provide access to these resources.

Not surprisingly, the negotiators envisioned that the flow of benefits would include both financial resources (for instance, royalties) and technologies. Given the central role of new technologies in addressing many environmental concerns, the inclusion of technology transfer articles is now commonplace in international agreements and understandings, including, for example, the Montreal Protocol on Substances that Deplete the Ozone Layer (United Nations 1987); Agenda 21, adopted by the 1992 United Nations Conference on Environment and Development (United Nations 1992a); and the United Nations Framework Convention on Climate Change (United Nations 1992b). Such provisions aid biodiversity conservation in two ways. First, technologies like seed banks, tissue cultures, and data management systems help meet conservation objectives directly. Second, technologies enabling countries to make better use of biodiversity in agriculture and industry add even greater economic incentives for conservation.

The technology transfer envisioned under the CBD is not a handout to poor countries, but a recognition that simple cash payments for access to resources are rarely the most effective mechanism to support their conservation. Indeed, the first country to take action with reference to the CBD to ensure technology transfer was not even a developing country. After the United States National Cancer Institute extracted a chemical, showing promise against HIV, from an Australian shrub, western Australia aggressively pursued its right to benefit from the resource. Australian scientists will now be involved in the research, and Australia will receive a share of any potential commercial benefits.

Already, the CBD has created a fundamental change in the understanding of the rights and obligations of countries and private firms. It is now accepted that countries or firms that obtain genetic resources from other countries assume the obligation under the convention to compensate those countries, both through the provision of financial resources and the facilitation of technology transfer. Because developed countries successfully negotiated language in the agreement ensuring that compulsory licensing is not authorized to meet these objectives, the burden falls squarely on the developed countries to take the actions needed to facilitate this technology transfer. There is an array of policy tools, including tax incentives and cooperative research programs, which can meet the convention's objectives without infringing on property rights. Rather than continuing to stress what actions developing countries *can't* take to meet the convention's objectives, it is essential that developed countries quickly demonstrate what actions they *will* take.

## Next Steps

A scaffolding for international action to support the conservation and sustainable use of biodiversity now exists. But what are the greatest needs for further international institutional evolution? There are five that are worthy of note.

*1. Build capacity in developing countries.* International institutional development in the area of biodiversity conservation has far outpaced the growth of local and national expertise and institutions in many countries. Countries have new obligations under the CBD to survey and protect species, as well as a new awareness of the need to control the transboundary flow of genetic resources. Moreover, there is a clear need for legal and technical expertise related to intellectual property rights and technology transfer, and there are new opportunities to benefit from biodiversity as more information about it is obtained. Faced with these needs and opportunities, most countries—with the exception of a handful of industrialized ones—confront a major shortage of expertise and capacity.

So long as many developing countries lack the expertise needed to deal with these issues, not only will progress toward goals of conservation and sustainable use suffer, but so too will these countries be placed at a disadvantage in the further elaboration of international institutional mechanisms. Without a focus on capacity building, the growing scaffolding of international measures could erect a house of cards. Every protocol, every project, and every new agreement needs to be screened to ensure that first priority is given to building local expertise and capacity in such areas as resource management, technology transfer, biological inventory, data management, environmental policy research, legal expertise, conservation biology, participatory methods, and biotechnology.

*2. Regulate Genetic and Biochemical Transfers.* Although the economic value associated with the transfer of genetic and biochemical materials is often exaggerated, effective mechanisms for dealing with these transfers and ensuring equity in the distribution of the benefits should nevertheless be a priority (Reid et al. 1993). Equity in the distribution of the benefits from these resources will be the standard against which international measures are judged. Several problems confront any such mechanism. First, the pharmaceutical industry—which receives the most attention today in relation to the CBD—is increasingly involved in the transfer of chemical extracts rather than samples of plants or animals. Because these are not strictly "genetic resources," they are not governed by the requirement in Article 15 of the CBD for the prior informed consent of the donor country. On the other hand, Article 8 enables countries to establish policies for the regulation and management of their biological "resources"—thus, it is within the rights of countries to require collectors to obtain a permit before exporting chemical extracts.

Second, mechanisms that make sense in relation to the use of *chemicals* in the pharmaceutical industry may not apply to the use of *genetic resources* in agriculture and medicine. There is no particular difficulty associated with tracking the development of a drug from a plant or animal extract. Countries could readily establish access regulations requiring contracts that stipulate royalties, technology transfers, and conservation conditions whenever chemical extracts are exported. Though it is possible for companies to evade the terms of such contracts, most would not consider it worth the risk, given the relatively low royalty rates typical for such agreements.

On the other hand, gene transfers present major monitoring problems. A soil sample sold to a pharmaceutical company might hold hundreds or thousands of microorganisms, many of which could easily be cultured. The supplier would have no means of knowing whether a microorganism that produced a valuable chemical came from a particular sample, and would not know if the organisms were transferred to another user. Similarly, a supplier would have no way of knowing whether a particular gene from a plant sample provided to a seed bank or a seed company was eventually patented. Thus, in the case of agriculture, the only "winner" from a scheme that would require the return of royalties to countries of origin of genetic material is likely to be the legal profession. Modern crop varieties incorporate genes from dozens of countries. The cost associated with parceling out royalties among those countries and litigating over the sources of various genes could quickly exceed the profits from seed sales.

Third, the CBD grandfathers seed collections that were already in existence when the CBD came into force. Thus, we face the prospect of an extraordinarily burdensome, two-tiered system of gene regulation and monitoring that would not only keep track of the country of origin, but also whether the material was collected before or after 1994.

There is a possible solution to these problems. An international fund could be established and financed through a small tax on the seed, pharmaceutical, and biotechnology industries and used to benefit countries which are the source of the genetic material. In fact, such a mechanism already exists in the International Undertaking on Plant Genetic Resources, but since contributions are voluntary, companies do not contribute (United Nations 1983). Perhaps faced with the specter of a Byzantine regulatory system, companies may now look more kindly on such a voluntary fund.

*3. Extend Rights under the Convention to Local Communities.* The CBD is premised on the notion that if countries receive greater benefits from the use of their resources, they will have stronger incentives to conserve them. Clearly, this same principle applies to local communities. If the CBD is to achieve its goals, then its key elements—prior informed consent obtained by collectors and benefit sharing on mutually agreed terms—should apply to private landowners, local communities, and indigenous groups with territorial claims. A case in point may be taxol—a chemical derived from the Pacific yew in the northwestern United States. The local support in the Pacific Northwest for using public forest lands for purposes other than timber extraction would be greatly strengthened if some of the economic benefits from the sales of taxol were returned to the region where it was discovered (Reid, Barber, and La Vina 1995).

*4. Abandon the Concept of Global Incremental Costs.* Although there are benefits that accrue globally from the conservation of biodiversity that exceed national benefits, it does not follow that international financial support for conservation should focus only on the need for funds for these international benefits. The resources that nations are now allocating to biodiversity conservation already fall far short of long-term national interests for biodiversity conservation. It thus makes little sense for international funds to support only incremental costs, when financial support is still needed simply to help nations meet their national goals.[1]

In contrast, the definition of incremental costs in the Global Environmental Facility encourages countries not to approach biodiversity conservation as something that is in their self-interest, but rather, as something to be paid for with international funds. International resources should build the capacity and demand for biodiversity conservation, and should support local actions that are already seen to be of national value. The result may well be the preservation of biodiversity that would otherwise not have survived. But this should be an outcome, not a premise.

*5. Address the Entire Array of Biosafety Concerns through International Negotiation.*
At the second meeting of the Conference of Parties to the CBD in November 1995, a working group was established that is now negotiating a protocol on the international transfer of genetically engineered organisms. Since the negotiations are still underway, there isn't much to report. Yet there is substantial concern in developing countries about the risks of releasing genetically modified organisms. These countries do not have the regulatory mechanisms in place to monitor or enforce testing guidelines; scientists could take advantage of the lax rules in developing countries to test organisms that could not have been tested in the developed nations.

Ideally, however, such a protocol would address not only the transfer of genetically modified organisms, but the transfer of biodiversity more generally. Indeed, the more pressing biodiversity conservation problem is not genetically modified organisms, but common, garden-variety, genetically unmodified "weeds"; it is the introduction of exotic species. Already, the United States has some 4,500 introduced species—accounting for some 2 to 8 percent of such groups as insects, vertebrates, and mollusks—and more frequent and rapid international travel is resulting in more frequent introductions (U.S. Congress 1993). Not only are exotic species often serious threats to native biodiversity and extremely difficult to control, they can have substantial negative economic impacts. Between 1906 and 1991, 79 introduced species caused a documented $97 billion in losses in the United States (U.S. Congress 1993). It is estimated that a recently introduced species—the zebra mussel—will cause more than $3 billion in losses in the coming years (U.S. Congress 1993). The negotiation of a biosafety protocol should not be restricted to genetically modified organisms, but should address the broader issue of the introduction of exotic species, both their social and ecological impacts.

## Conclusion

Fundamental changes in national policies and the allocation of public resources are needed to significantly slow the loss of biodiversity. Investments in parks and protected areas can slow the loss today, but if significant changes in public policy are not forthcoming, then these investments will only be stopgap measures. The current framework of international measures is encouraging, because it is premised on the need to integrate the concern for biodiversity conservation with development needs. But international and national action is needed to make this framework effective. The greatest current obstacle to progress on achieving the goals of biodiversity conservation and sustainable use is the shortage of capacity in developing countries. In the haste to elaborate the international framework, we must ensure that we do not outpace the growth of this capacity.

## Notes

1 Under the CBD, the *incremental cost* is the added cost to developing countries of implementing the convention. Thus, if a country is obligated to prepare a biodiversity strategy that it would not have done otherwise, this could be considered an incremental cost of implementation (Glowka, Burhenne-Guilmin, and Synge 1994).

## References

CCMSWA. (Convention on the Conservation of Migratory Species of Wild Animals). 1979. *International Legal Materials* 19:11.

DOI (United States Department of the Interior). 1994. *The Impact of Federal Programs on Wetlands; A Report to Congress by the Secretary of the Interior.* Washington, D.C.: Department of the Interior.

Dudley, N. 1992. *Forests in Trouble: A Review of the Status of Temperate Forests Worldwide.* Gland, Switzerland: World Wide Fund for Nature.

Fowler, C., and P. Mooney. 1990. *Shattering: Food, Politics, and the Loss of Genetic Diversity.* Tucson: University of Arizona Press.

Glowka, L., F. Burhenne-Guilmin, and H. Synge. 1994. *A Guide to the Convention on Biological Diversity.* Environmental policy and law paper, no. 30. Gland, Switzerland: World Conservation Union.

Houghton, J.T., ed. 1994. *Climate Change, 1995.* Cambridge: Cambridge University Press.

Reid, W.V. 1992. How many species will there be? In *Tropical Forests and Species Extinction*, edited by T.C. Whitmore and J.A. Sayer. London: Chapman and Hall.

Reid, W., and K. Miller. 1989. *Keeping Options Alive: The Scientific Basis for Conserving Biodiversity.* Washington, D.C.: World Resources Institute.

Reid, W.V., C.V. Barber, and A. La Viña. 1995. Translating genetic resource rights into sustainable development: Gene cooperatives, the biotrade, and lessons from the Philippines. *Plant Genetic Resources Newsletter* 102:1–17.

Reid, W.V., S.A. Laird, C.A. Meyer, R. Gámez, A. Sittenfeld, D.H. Janzen, M.A. Gollin, and C. Juma, eds. 1993. *Biodiversity Prospecting: Using Genetic Resources for Sustainable Development.* Washington, D.C.: World Resources Institute.

Repetto, R. 1993. Trade and environmental policies: Achieving complementarities and avoiding conflicts. *Issues and Ideas.* Washington, D.C.: World Resources Institute.

United Nations. 1983. International undertaking on plant genetic resources. 23 November. UN Doc. C 83/REP.

———. 1987. Montreal protocol on substances that deplete the ozone layer. 1 January. *International Legal Materials* 26:1550.

———. 1992a. Agenda 21. 13 June. UN Doc. A/CONF. 151/26, vols. 1–3.

———. 1992b. United Nations framework convention on climate change. 21 March. *International Legal Materials* 31:849.

———. 1994. International tropical timber agreement. 26 January. *International Legal Materials* 33:1014.

United Nations. Treaty Series. 1971. Convention on wetlands of international importance. 21 December. *Treaties and International Agreements Registered or Filed or Reported with the Secretariat of the United Nations*, vol. 996, no. 245 (1976).

———. 1973. Convention on international trade in endangered species of wild fauna and flora. 3 March. *Treaties and International Agreements Registered or Filed or Reported with the Secretariat of the United Nations.* vol. 993, no. 243.

U.S. Congress. 1993. Office of Technology Assessment. *Harmful Non-Indigenous Species in the United States.* OTA-F-565.

U.S. Department of State. 1980. Convention on the conservation of Antarctic marine living resources. TIAS no. 10240. Treaties and Other International Acts.

USFWS (United States Fish and Wildlife Service). 1990. Report to Congress: Endangered and Threatened Species Recovery Program. Washington, D.C.

WCMC (World Conservation Monitoring Centre). 1992. *Global Biodiversity: Status of the Earth's Living Resources.* London: Chapman and Hall.

WRI (World Resources Institute), IUCN (World Conservation Union), and UNEP (United Nations Environment Programme). 1992. *Global Biodiversity Strategy: Guidelines for Actions to Save, Study, and Use Earth's Biotic Wealth Sustainably and Equitably.* Washington, D.C.: World Resources Institute.

WRI (World Resources Institute), UNEP (United Nations Environment Programme), UNDP (United Nations Development Programme). 1994. *World Resources, 1994–1995.* Washington, D.C.

# Rewarding Local Communities for Conserving Biodiversity: The Case of the Honey Bee

by Anil K. Gupta

Perhaps the economic class most critical in conserving biodiversity is the one made up of extremely poor people residing in areas of rich biodiversity. As traditional caretakers of abundant biological wealth, these people have developed a vast reservoir of knowledge concerning the sustainable uses of nature. Thus, in order to preserve the biological resources surrounding such local communities, we also need to preserve the knowledge about those resources. One way to preserve such knowledge is to grant intellectual property rights to the individual innovators within those communities or to the communities as a whole. Along with such protection, grassroots networks—such as the Honey Bee Network outlined here—provide a decentralized method of disseminating important knowledge to other indigenous communities. These two approaches—creating intellectual property rights and sharing helpful knowledge—also offer some hope in alleviating the poverty that currently degrades the most important resource in protecting our biological heritage: local people.

\*

A large number of local communities across the world have unhesitatingly shared their knowledge about local biodiversity and its different uses. Many continue to share despite an awareness that by withholding this knowledge they could receive pecuniary advantage from interested researchers, corporations, and gene collectors. In some cases, cultural and spiritual taboos exist against receiving compensation, arising from the fear that accepting payment will reduce the effectiveness of their knowledge. In other cases, the local party might accept a transfer payment, or an offering either to specific animals or nature in general. There are even instances where the scale of the offering is proportional to the capacity of the person helped and not the degree of the help itself. In all of these situations, it is clear that the compensation system favored by the richer nations of the world will not necessarily suffice in providing benefits to local communities for their efforts to save biodiversity.

## Impoverished Economics and Biodiversity

It is against this backdrop of ethical and ecological concerns that we must begin to discuss the issue of recognizing, respecting, and rewarding the contribution of local communities. The challenge becomes even more difficult given the abject poverty of many of these communities. Yet it is precisely this economic class, which tends to live in areas of both economic poverty and biodiverse riches, which is perhaps the most crucial to the preservation of biodiversity. Several factors have contributed to this linkage between greater biodiversity and poverty (Gupta 1991a; 1991b; 1993a, 31–56; 1993b, 1). One global initiative exploring this connection, the Society for Research and Initiatives for Sustainable Technologies and Institutions (SRISTI), takes note of the following factors:

1. The biodiversity is high in these areas primarily due to diversity in soil, climate, and other physical and social structures.
2. The poverty is high because markets are often unable to generate demand for diverse colors, tastes, shapes, and qualities of natural products. In short, mass production has devalued unique indigenous production, leaving these areas with a lack of capital for adequate investment.
3. The regions of high diversity also have poor public infrastructure (in tandem with weak private market forces) because the people have limited surplus to attract public servants, and they are less articulate and less organized in their efforts to create political pressure (except through insurgent movements).
4. The low demand for ecological and technological skills of people in these communities often leaves them characterized as "unskilled" labor. Once the knowledge system is devalued, a cultural and social decline follows. With the rupture of the tenuous link to nature, the ecological degradation spurred by various external resource extractors is aided and abetted by many poor people, for whom survival in the short term seems possible only through ecologically degrading strategies.

There are many characteristics of indigenous communities that contribute to their economic strife. The preferred household crops generally lead to a low surplus and prices that are particularly susceptible to market and environmental fluctuations. In addition, the livestock breeds, though well-adapted to the environment, suffer huge losses due to drought and disease, which also causes wide fluctuations in price. The price instability and market imperfections in the nonfarm sector further impair the capability of the households to adjust with the risks. This variability in income, in turn, has something of a domino effect.

Most of the typical households described above tend to run deficits in their budgets (Gupta 1981; 1983, 573; 1989, 115). Thus, dependence on other social groups and informal institutions, such as moneylenders or traders, is enormous. This dependence—in its most politically exploitative forms—tends to divide the community into different subgroups of mutually conflicting identities. The end result is that collective action, for economic purposes, is at times extremely difficult for people in these communities.

Four mechanisms, which arise due to the economic burdens on indigenous communities, cast an ominous shadow over the hopes of these communities to develop in a sustainable fashion:

1. In order to reduce their economic hardship, some of the communities end up resorting to resource-degrading strategies. This happens, despite their great affinity with nature, in order to meet their immediate livelihood requirements.
2. Because of the high variability in income over space and time, the communities are vulnerable due to their limited ability to purchase essential needs. This sometimes makes them an involuntary agent of resource degradation in the hands of outside commercial extractors. The indifference of the state and markets, as well as a lack of value-adding technological alternatives, further contribute to the high income variance and associated overexploitation.
3. In the context of the regions rich in biological diversity, with a majority of the people having a low average income with high variability, the official attempts to provide short-term relief, employment, and other means of subsistence in order to alleviate poverty have a debilitating effect. These measures, along with the continued economic stress on the communities, in many cases weaken the will of the people to struggle and innovate so as to survive in harmony with nature. Thus, not only are the resources no longer conserved, but more important, the knowledge around these resources is slowly destroyed.
4. The systematic pattern in the movement of people out of biodiversity-rich, economically poor regions creates certain anomalous situations. First, because of male emigration in the absence of local employment opportunities, the preponderance of households headed or managed by women makes the communities vulnerable. Second, in a world where traditional biological knowledge has little value, the only way markets can deal with these people is by classifying them as unskilled labor.[1] The result is that the valuable ecological knowledge carried by these people gets lost as they leave the local communities for bigger towns or metropolitan areas to overcome their poverty.

Contrary to conventional understanding, however, poverty does not prevent people from thinking. For them, the knowledge gained through experimentation and innovation is a matter of life and death, given the uncertainties of nature. For this reason, the term *resource poor farmer* is one of the most inappropriate and demeaning contributions from the West. If knowledge is a resource, and if some people are rich in this knowledge, why should they be called resource poor? At the same time, the market may not price peoples' knowledge properly in the current economic system. Nonetheless, they have great riches in their tacit knowledge base that has immense value to all of humankind.

Thus, in order to bring about a viable system of sustainable development, two events must take place. First, resource conservation must become more economically attractive than the economic benefits of overexploitation. And second, the knowledge about natural resources must be placed into a system that both preserves and economically values it appropriately by providing some sort of compensation designed to reward local communities for their contributions. The solution, therefore, must contain elements that eliminate poverty, preserve biological richness, and reward local communities for their indigenous knowledge, creativity, and innovation.

Much has been written about the conservation and valuation of natural resources with regard to indigenous communities, as well as their role in patterns of sustainable development and the preservation of biodiversity. The conservation and valuation of indigenous knowledge, however, has virtually escaped notice. Thus, the two topics that are the subject of the remainder of this article are the preservation and documentation of indigenous knowledge, and the economic valuation of such knowledge.

## The Honey Bee Network

When one thinks of conservation, instinctively the focus is on preserving natural resources. It is equally vital, however, to conserve the knowledge about the resources. Knowledge may be produced and reproduced through cultural, social, and even individual innovations. Some of these innovations have been carried forward from one generation to another, thereby becoming part of what is popularly called traditional wisdom (Verma and Singh 1969; Richards 1985; Gupta 1980, 92; Warren 1988). The question remains: How do we discover these innovations, build on them, generate experimentation, and help the transition of experimentation into enterprise through support of markets as well as self-design institutions?

Organizations of creative people—whether in the form of networks, informal cooperatives, or even loose associations—can generate a more sustainable society. The spirit of excellence, critical peer group appraisal, competitiveness, and entrepreneurship—so vital for self-reliant development—may emerge best within networks of local "experts," innovators, and experimenters.

Thus, in order to stem knowledge and resource erosion, the Honey Bee Network (hereinafter Honey Bee) was launched six years ago.[2] A global voluntary initiative, its purpose is to network people actively engaged in eco-restoration and the reconstruction of knowledge about precious ecological, technological, and institutional systems. Honey Bee has been documenting the ecological institutions that indigenous peoples have developed, in an effort to preserve and manage such knowledge as the property of local individuals and communities. The knowledge is then freely passed to the various indigenous communities whose existence depends on it.

More specifically, this network aims to identify innovators who have tried to break out of existing technological and institutional constraints through their own imagination and effort. What is remarkable about these innovations is the fact that most require quite low external inputs, are extremely eco-friendly, and improve productivity at a low cost. To date, the network has collected more than 1,400 innovative practices, predominantly from dry regions, proving that disadvantaged people may lack financial and economic resources, but remain rich in knowledge.

Honey Bee insists that two principles are followed without fail: whatever is learned from people must be shared with them in their language; and every innovation must be sourced to individuals/communities with the correct name and address to protect the intellectual property rights of the people.

The Honey Bee Network newsletter is brought out in seven languages in India (Hindi, Gujarati, Malayalam, Tamil, Kannada, Telugu, and English), as well as Zonkha in Bhutan. In fact, the network extends into 75 countries at present, with additional offers received from Nepal, Sri Lanka, Uganda, Paraguay, and Mali. Indeed, the creative people of one place should be able to communicate with similar people elsewhere to trigger mutual imagination and fertilize respective recipes for sustainable natural resource management.

As an example of the efficacy of the network, consider the topic of pesticides. The hazards of pesticide residues and associated side effects on the human, as well as entire ecological system, are well-known. The second issue of the Honey Bee Network newsletter documented 34 practices dealing with indigenous, low-external-input ways of plant protection. One ingenious practice involved cutting many 30-to-40-day-old sorghum or *Calotropis* plants and placing them in an irrigation channel in a successful effort to control termite attack in light, dry soils. Perhaps the hydrocyanide present in sorghum and similar toxic elements in *Calotropis* contribute toward this effect. This example confirms the hope that a large number of other plants of pesticidal importance exist in indigenous regions that might eventually provide sustainable alternatives to highly toxic chemical pesticides.

The Honey Bee database contains a large number of examples of the use of local materials to solve plant-protection problems. Farmers have found new uses of existing plant biodiversity to control the pest and disease problems in the crops. For instance, naffatia (*Ipomoeae fistulosa*) is a plant often used for fencing purposes. Animals don't eat it and there are not many other popularly known uses for this toxic plant. In some places, the branches have been dried and used for making baskets for storing seeds or grains. During 1973, when there was a steep oil price hike, many farmers started to look for substitutes for chemical pesticides. Thus, new inventions took place in the field of nonchemical pest control. Later, when many pests became resistant to pesticides, the farmers' search for alternatives became widespread. There are many tales as to how the use of one particular plant as a pesticide originated. According to one account, a farmworker's wife covered her husband's tea with the leaf of the naffatia. When the worker returned and drank the tea, he developed toxic symptoms, and survived only with great difficulty after seeing a doctor. But the idea was born to use the plant as a herbal pesticide. Subsequent research has found it to be quite effective against not only some pests but also against certain microbial and fungal cultures.

In another case, a tribal person named Bhogilal Rajwadia of the Bharuch district devised a unique method of pest control. He enlisted the help of 8 to 10 farmers or laborers who stood in a line, each placing the leaves of a creeper (*Combretum ovalifolium*) in a shoulder bag. The line then moved in the windward direction and caught blister beetle from the air, crushing these with the leaves already collected. The combined effect of insect and leaf extract seemed to produce some signals that repelled other insects. Such a heuristic approach to the combining of plant and insect extract does not exist in modern science. Similarly, there are large numbers of other plant extracts that have been developed by farmers that could help make crop cultivation in marginal regions more profitable.

Most countries do not have a fast-track approach for developing or registering herbal pesticides derived from plants. Perhaps one answer is a special fund to support formal research on farmers' innovations in public or private sector labs, so as to develop a whole range of sustainable and cost-effective technologies. These technologies may help transform agriculture not only in developing countries, but also in economically developed ones lacking traditional farmer knowledge and creativity. These innovations, therefore, may not only help transfer technologies between countries in the South, but from the South to North as well.

## The Role of Intellectual Property Rights

While knowledge can be preserved through networks such as Honey Bee, the issue of eliminating economic hardship in the indigenous communities must still be addressed. The solution lies in establishing intellectual property rights (IPRS) in favor of local communities and individuals in order to provide those rural areas with an economic return for their specialized knowledge.

The local knowledge systems would need several mechanisms of recognition in addition to IPRS, as illustrated through these four categories of reward: material-individual; material-collective; nonmaterial-individual; and nonmaterial-collective. One can appraise all four in the short and long term. The material-individual category includes royalties, license fees, or other individual compensation, such as patent rights in monetary form. Even here, given the historical legacy of deprivation, the intervention of too much money too quickly can disturb individual preferences and choices. Therefore, a combination of short-term monetary reward coupled with medium- to long-term bonds might provide the portfolio of assurances that a creative individual may need. The material-collective category deals with trust funds, venture capital funds, insurance funds, risk funds, and other forms of collective payments for insurances/assurances. The nonmaterial-individual category deals with recognition, honor, and social esteem that can motivate individual conservators of diversity. Finally, the nonmaterial-collective category deals with policy, as well as pedagogical and curricular, changes that incorporate local ecological knowledge and generate esteem for communities that conserve diversity; in turn, this approach may beneficially alter the mind-set of future leaders as they look at opportunities.

Regardless of which category, or blend of categories, of reward that the international community chooses, the scope for linking scientific searches and farmers' knowledge is enormous. Out of 114 plant-derived drugs, for example, more than 70 percent are used for the same purpose originally discovered by native peoples (Farnsworth 1988, 83–87). This proves that the basic research—conducted by native people—had already been accomplished in a majority of the cases. In these cases, modern science and technology only supplement the original efforts; it is native people who have done the creative work, linking cause and effect.

After the adoption of the 1992 Convention on Biological Diversity (appendix, this volume), sensitivity to the subject has certainly increased. There is a realization that biodiversity cannot be prospected or used without making the conserving communities and innovative individuals the stakeholders in any plan for adding value to the resource. Under Article 8(j), the convention requires the "approval and involvement of the holders of such knowledge, innovations and practices and encourage[s] the equitable sharing of benefits" with local communities. On a more general level, Article 15(5) mandates the prior informed consent of biologically rich countries before economically rich countries can extract samples. Similarly, the Food and Agriculture Organization of the United Nations (FAO) has articulated an undertaking on plant genetic resources, calling for an international gene fund in the name of "farmers' rights." The fund, administered by an international civil service, would distribute generated resources to various governments for conservation purposes. Yet neither the FAO nor the convention provides specific mechanisms for achieving the goal of compensating local communities. In fact, the FAO undertaking as it stands today is highly misleading. It celebrates the contribution of farmers, but provides for no direct incentives to those who conserve genetic diversity.

The mechanism that would best serve to channel an economic return to indigenous communities is the establishment of IPRs. Patents are available for inventions, whether products or processes, in all fields of technology, provided that they are new, involve an inventive step, and are capable of industrial application. Indigenous knowledge, which is not known to a biotechnology or drug company, or a company interested in producing herbal pesticides or veterinary drugs, is often patentable. The same plant should not be used for a commercial purpose without providing compensation to the stakeholding community.

The legal and institutional framework associated with IPRS has been discussed thoroughly elsewhere, and is not in need of elaboration with the exception of two points. First, the key to preserving the IPRS in a given piece of knowledge for an indigenous community would have to lie with the network. Under most systems of law, once an idea has been disclosed or published, a property right may no longer be established. Thus, prior to the publication of the knowledge in the newsletter, the network attempts to aid the community or individual in legally establishing a right to its knowledge. An international registration system of innovations by farmers and communities is necessary so that, in the absence of costly legal aid, communities do not get robbed of their creative outputs. The developed nations also should ensure that before the IPRS of any corporation are recognized, the patent applicant should declare that the knowledge and resource to be patented were collected "lawfully and rightfully." *Lawful* implies that if a country does not have a law requiring prior informed consent, then it is not illegal to take germ plasm from such a country. *Rightful*, however, implies that such conduct is not morally right.

Second, given the weak record of most governments in making the machinery of government accountable to local, disadvantaged communities, entrusting the task of routing compensation from national or international funds through the same machinery will be counterproductive. Whether nongovernmental organizations can help in the delivery of such funds depends to a great extent on their ethical position, as well as their accountability to local communities. This is one area in which the values of provider, receiver, and intermediary would inevitably require reconciliation.

## Conclusion

The conservation and preservation of diversity must be attacked on two fronts: the natural resources themselves must be conserved; and the indigenous knowledge about the natural resources, which is itself a localized resource, must be fostered and preserved. In the often overlooked concept of knowledge as a resource, the Honey Bee Network exists as an effort to mold markets of ideas and innovations in favor of sustainable development. With a concentration on high-risk environments, the key objectives of SRISTI are thus to strengthen the capacity of the innovators to: protect their IPRS; experiment to add value to their knowledge; evolve entrepreneurial ability to generate returns from this knowledge; and enrich their cultural and institutional basis for dealing with nature. The economic reward mechanism offered by IPRS and other material-collective means—such as trust, risk, venture capital, and guarantee funds—provides a way to channel monetary benefits into the local communities as compensation for the exploitation of their knowledge resources. This economic return would serve to address the widespread poverty currently devastating these communities. Through the use of networks and IPRS, both the conservation and valuation of indigenous knowledge can be accomplished. In the end, only by rewarding grassroots innovation can we hope to foster the most crucial force in preserving biodiversity: local people.

## Notes

1 Some suggest the inadvisability of attempting to stem the migration of people out of the less-developed regions, lest the supply of cheap labor for infrastructural projects in the cities becomes inadequate.
2 The Honey Bee Network is headquartered at SRISTI (Society for Research and Initiatives for Sustainable Technologies and Institutions, c/o Professor Anil K. Gupta, Indian Institute of Management, Ahmedabad 380015), an autonomous nongovernmental organization.

## References

Farnsworth, N.R. 1988. Screening plants for new medicines. In *Biodiversity*, edited by E.O. Wilson. Washington, D.C.: National Academy Press.
Gupta, A.K. 1980. *Communicating with Farmers: Cases in Agricultural Communication and Institutional Support Measures.* New Delhi: S.N.
———. 1981. A note on internal resource management in arid regions, small farmers-credit constraints: A paradigm. *Agricultural Systems* 7:157–61.
———. 1983. Impoverishment in drought prone regions: A view from within. Joint field study SDC/NABARD/IIM-A, CMA, IIM, Ahmedabad.
———. 1989. Managing ecological diversity, simultaneity, complexity, and change: An ecological perspective. Working paper no. 825, IIM, third survey on public administration. New Delhi: Indian Council of Social Science Research.
———. 1991a. *Why Does Poverty Persist in Regions of High Biodiversity? The Case for an Indigenous Property Right System.* Presented at the International Conference on Property Rights and Genetic Resources sponsored by the World Conservation Union and the United Nations Environment Program, Kenya.
———. 1991b. *Sustainability Through Biodiversity: Designing Crucible of Culture, Creativity, and Conscience.* Copenhagen: International Conference on Biodiversity and Conservation.
———. 1993a. Biotechnology and intellectual property rights: Protecting the interests of third world farmers and scientists. In *Commercialization of Biotechnologies for Agriculture and Aquaculture: Status and Constraints in India,* edited by U.K. Srivastava and S. Chandrasekhar. New Delhi: Oxford and IBH Publishing Co.
———. 1993b. Biotechnology and IPR: Third world issues for farmers and scientists. In *Biotechnology Monographs: Focus on Third World Issues.* Paris: Organization for Economic Cooperation and Development.

Richards, P. 1985. *Indigenous Agricultural Revolution*. London: Hutchison Press.

Verma, M.R., and Y.P. Singh. 1969. A plea for studies in traditional animal husbandry. *Allahabad Farmer* 43 (2):93–98.

Warren, D.M. 1988. *Linking Scientific and Indigenous Agriculture Systems in the Transformation of International Agriculture Research and Development: Some U.S. Perspectives*, edited by J. Lin Compton. Boulder, Colo.: Westview Press.

# International Research on Crop Plants:
# Strategies for Utilizing Biotechnology
# and Proprietary Products

by Gary H. Toenniessen

Developing countries need to increase their crop production to meet future demand for food. At the same time, they need to avoid significant further losses in biodiversity caused by converting additional land to agriculture. In order to do so, they must achieve substantial increases in sustainable crop yields on land already in production. The existence of an effective international agricultural research system and the development of biotechnology as a powerful new set of tools for genetic monitoring and manipulation make this a feasible task. New mechanisms should be established to facilitate the flow of technology from the corporate sector to the international system. This chapter reviews the challenges that lie ahead and describes several new institutional structures designed to help meet them.

*

To feed the world's current population of roughly 5.6 billion people, today's farmers cultivate nearly 1.5 billion hectares of cropland, or about 11 percent of the earth's land area (WRI 1994). If, as demographers project, the world's population increases to 10 billion or more before leveling off, it can be assumed that at least a doubling of food production will be necessary to meet future demand (Bongaarts 1994). If there were no further increases in crop yield per hectare, more than a doubling of the land area committed to crop production would also be necessary, with most land conversion occurring in tropical countries where the food is needed and where much of the earth's biodiversity exists. Alternatively and preferably, if the increased food could be produced on the same land that is already in production, there would be much less pressure to expand agriculture and much more biodiversity would be saved (see Horsch and Fraley, this volume).

Achieving the necessary crop yield increases in developing countries is a significant challenge, but not an unreasonable task. Projections made by Waggoner (1994) indicate that if all farmers lifted their yields to six tons per hectare, a level currently not uncommon for major cereals in industrialized countries, 10 billion people could be fed on slightly less land than that committed to crop production today. Admittedly, tropical agriculture does not currently have the technologies needed to achieve average yields at this level, and there is a real threat, already materializing in some locations, that farmers will irreparably damage the resource base as they seek to feed growing populations. Success will be dependent on the discovery of new knowledge and the development of new yield-enhancing and resource-conserving technologies for tropical agriculture, which when combined with the broader adaptation and application of existing technologies will allow for greater intensification of crop production on a sustainable basis.

Many of the institutional structures and financial support systems necessary to advance food science and technology in developing countries are already in place, and agriculture is on the verge of a substantial boost in its productive potential as a result of progress in biotechnology. Unfortunately, most advances in the application of biotechnology to agriculture now occur in the corporate sector and are often protected; their use is restricted through patenting and other forms of intellectual property rights (IPR), thus complicating the process of technology transfer. These new technologies, however, must be developed for tropical crops and made available for use in developing countries. This chapter reviews the nature of IPR constraints to technology transfer, and describes several new institutional structures designed to facilitate the application of biotechnology to developing-country agriculture as well as to help protect biodiversity and the natural resource base.

## IPRS and the International Agricultural Research System

IPRS are legal systems established by national governments and are intended to benefit the people of the nation granting the proprietary rights. The international dimensions of IPRS are derived from treaties and conventions in which each country affirms its own self-interest. Of late, industrialized countries have made IPRS an important component of international trade negotiations that promote national competitiveness as well as encouraging research and development.

IPR systems are intended to benefit society by rewarding and therefore encouraging invention, by promoting the commercialization and delivery of useful products resulting from invention, and by laying the foundations for future advancements in science and technology through disclosure requirements and research exemptions. Most often it is increased profits from the marketing of proprietary products that directly or indirectly generates financial rewards for IPR owners. In industrialized countries, IPR systems have been effective in stimulating increased research and development in areas that have, or are perceived to have, the potential for generating profits. This has clearly happened in the plant sciences over the past decade. The perceived profit potential of crop biotechnology has generated substantial increases in the funding for, and in the total research effort committed to,

TABLE 1    Crop-Focused International Agricultural Research Center (IARC)

| International Center | Program Focus | Headquarters Location |
|---|---|---|
| International Rice Research Institute (IRRI) | Rice | Philippines |
| Centro Internacional de Mejoramiento de Maize y Trigo (CIMMYT) | Maize, wheat, triticale | Mexico |
| Centro Internacional de Agricultura Tropical (CIAT) | Cassava, field beans, tropical forages, rice | Colombia |
| International Institute of Tropical Agriculture (IITA) | Cassava, yams, cowpeas, maize, soybeans, plantains | Nigeria |
| International Crops Research Institute for the Semi-Arid Tropics (ICRISAT) | Sorghum, millet, groundnuts, chickpeas, pigeon peas | India |
| Centro Internacional de la Papa (CIP) | Potatoes, sweet potatoes | Peru |
| International Center for Agricultural Research in Dry Areas (ICARDA) | Wheat, lentils, chickpeas, barley, pasture legumes | Syria |
| Asian Vegetable Research and Development Center (AVRDC) | Chinese cabbage, tomatoes, mung beans, soybeans, sweet potatoes | Taiwan |
| West Africa Rice Development Association (WARDA) | Rice | Ivory Coast |
| International Plant Genetic Resources Institute (IPGRI) | Germ plasm conservation | Italy |
| International Network for the Improvement of Banana and Plantain (INIBAP) | Bananas, plantains | France |
| Center for International Forestry Research (CIFOR) | Forestry ecosystem conservation | Indonesia |

the plant sciences and crop improvement. A few products are beginning to reach consumers; many others are in the research pipeline, and most should provide some benefit to society.

Research and development of foreseeable benefit to society, but with little or no profit potential, is usually of scant interest to the private sector. It is left to the public sector, where it may or may not get supported. Until recently, most research on plants, including crop plants—particularly crop plants grown in developing countries—fell into this category. Fortunately, an international system of publicly supported agricultural research institutions was established in the 1960s and 1970s to address many of the crop improvement needs of developing countries. Its record of accomplishment includes a dozen international centers (table 1) with

TABLE 2   IARC Germ Plasm Conservation and Distribution

| | Number of Accessions | |
| Center | Conserved (1991) | Distributed (Average of Five Years) |
| --- | --- | --- |
| CIMMYT | 106,300 | 13,421 |
| ICRISAT | 103,085 | 76,506 |
| ICARDA | 87,769 | 19,397 |
| IRRI | 86,660 | 10,663 |
| IITA | 39,858 | 14,874 |
| AVRDC | 35,502 | 11,866 |

*Source:* M. Iwanaga, 1993

worldwide or regional responsibility for particular crops, including the conservation of germ plasm resources. The international centers conduct strategic and applied research, and link the components of the system together, assuring that it functions as a backstop for national agricultural research efforts (figure 1). This system has the ability to take relevant scientific discoveries that occur in the ivory towers of academe, and through a series of technology transfers and collaborative research projects, incorporate this new knowledge and technology into improved seeds grown in fields throughout the developing world.

The improved cultivars and agronomic practices generated by this research system have benefited literally billions of people who daily consume the end product. The so-called green revolution in rice and wheat is no doubt the most dramatic result of this research, ending the periodic famines that had plagued Asia for centuries and enabling most of these countries to become food self-sufficient. High-yield cultivars produced by this system now account for nearly 70 percent of the area planted to these crops in Asia. Over the past 20 years, the proportion of the population affected by undernutrition declined from roughly 40 to 20 percent (ACC/SCN 1992). Steady progress has also resulted in improved production of maize, cassava, legumes, sorghum, millet, potatoes, and other crops. These international centers have taken responsibility for the collection, evaluation, preservation, distribution, and wide-scale utilization of the biodiversity existing in their mandated crops and related species (table 2). In aggregate, they store and manage an estimated 610,000 accessions for the benefit of the world community, representing much of the biodiversity existing within these important crop species.

Still, there is no doubt that agriculture also challenges biodiversity. Because agricultural ecosystems are inherently more uniform than natural ones, agricultural land can diminish biodiversity. The problem is that over the next half century we will need to feed twice as many people as today, and we do not know how to do so without agriculture. The best we can hope for is to increase food production still further in those areas where sustainable agricultural systems are available,

FIGURE 1  The International Agricultural Research System in the Era of Biotechnology

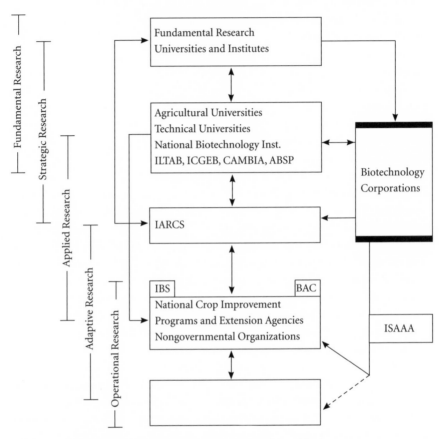

| IARCS | International Agricultural Research Centers (see table 1) |
| ILTAB | International Laboratory for Tropical Agricultural Biotechnology, La Jolla, California |
| ICGEB | International Center for Genetic Engineering and Biotechnology, New Delhi, India, and Trieste, Italy |
| CAMBIA | Center for the Application of Molecular Biology to International Agriculture, Canberra, Australia |
| ABSP | Agricultural Biotechnology for Sustainable Productivity project, East Lansing, Michigan |
| ISAAA | International Service for the Acquisition of Agri-biotech Applications, Ithaca, New York |
| IBS | Intermediary Biotechnology Service, The Hague, Netherlands |
| BAC | Biotechnology Advisory Commission, Stockholm Environment Institute, Stockholm, Sweden |

thereby reducing the need to extend agriculture over greater and greater areas, and into regions where it is nonsustainable. For example, rice paddy production is one of the most sustainable systems known to agriculture. Following the introduction of hybrid seed during the 1980s, rice production in China increased by roughly 32.5 million tons (22 percent) while the area planted to rice decreased by roughly 2.2 million hectares (6 percent), and there was an increase of mixed cropping with rice (FAO 1992). Further development and broader application of such yield-enhancing technologies on land already in production will enable other lands to remain as natural repositories of biodiversity, and as described by Gupta (this volume), to serve the needs of local communities.

Much of the success of the international agricultural research system has resulted from the free exchange of genetic diversity and information. For example, the gene for dwarfness used by Nobel Peace Prize–winner Norman Borlaug to produce high-yielding wheat was originally discovered in Japan. A Japanese cultivar containing this gene was brought to the United States, where breeders at Washington State University used it to produce semidwarf winter wheat. It was then sent to Mexico for Borlaug to use in breeding semidwarf spring wheat. The semidwarf wheat did well in Mexico; had its greatest impact on production in India, Pakistan, and Bangladesh; and is now widely grown throughout the world (Borlaug 1968). Waggoner (1994) calculates that in India alone, over the period 1961–1994, roughly 40 million hectares of land were spared from being cultivated by increased wheat yields.

Today, the modern rice and wheat varieties produced by the international centers are the product of over 30 years of accumulating useful genes from numerous countries into elite breeding lines. Unlike developments in many other areas of technology, the genetic improvement of plants is a derivative process in which each enhancement is based directly on preceding generations and the process of adding value requires access to the plant material itself. Much of the value in today's seeds was originally added over the centuries as farmers selected their best plants as the source of seed for the next planting. These landraces and other sources of biodiversity have traditionally been provided free of charge by developing countries to the international system. The international centers add value via breeding, and the elite lines they generate are widely distributed without charge — benefiting both developing and developed countries. As an example, the pedigree of the broadly utilized elite breeding line of rice, IR72, is based on 22 land races and 65 varieties or breeding lines from China, India, the Philippines, Indonesia, Thailand, Vietnam, Malaysia, and the United States (Chrispeels and Sadava 1994, 304–5).

In developing countries, public sector institutions are primarily responsible for introducing and disseminating improved varieties. In the case of self-pollinating cereals, such as rice and wheat, farmers can keep a portion of their harvest as improved seed for subsequent plantings, as well as to sell or trade with other farmers. Once improved varieties are introduced, such farmer-to-farmer distribution greatly speeds their dissemination to contiguous areas. In India, for example, about 60 percent of the farmers' total seed requirements are estimated to be met by this "interfarmer" trade (Alam 1994). Farmers throughout a region are thus provided with elite cultivars at the lowest possible price. Since there is little profit to be made in providing seed to these farmers, the private sector has not played a significant role in this overall process. The principal factor eventually limiting the spread of new rice and wheat varieties is not the lack of a distribution system, but rather the lack of new varieties well suited to the needs of farmers in marginal land areas.

In the case of cross-pollinating crops such as maize, adoption of improved varieties is also occurring, albeit at a slower pace. Seed of improved, open-pollinated varieties of cross-pollinating crops needs to be replaced every three to five generations due to outcrossing. This limits farmer-to-farmer distribution and makes hybrid seed more competitive despite having to be purchased for each planting. In some developing countries, a private seed industry has emerged that primarily markets hybrid seeds to commercial farmers. The seed industry receives benefits from the public sector in the form of well-adapted parental lines. The public breeding programs also benefit from having a more efficient overall seed multiplication and distribution system. In some countries, such as Kenya and Zimbabwe, this mutually beneficial relationship has resulted in wide-scale adoption of improved maize varieties by small-scale farmers (Byerlee and Lopez-Pereira 1994, 94–101). By itself, the private sector is never likely to be a major developer or supplier of improved seed to farmers who have only limited purchasing power.

The scientific advances responsible for the green revolution will not be sufficient to feed the projected world population, expected to stabilize at 10 billion people. The demographics demand that the international agricultural research system make significant further contributions toward improved nutrition and economic growth in developing countries. Moreover, experience to date suggests that no country can expect to achieve sustained fertility reduction without first having a major decline in infant and child mortality through improved nutrition and primary health care (Bongaarts 1994). The international agricultural research system can accomplish its goals if it receives continued strong public sector support, maintains access to traditional forms of crop biodiversity, and develops mechanisms for gaining access to new technology and new types of genetic materials, such as cloned and/or synthesized genes, all of which are increasingly being protected under expanding IPR systems.

The corporate sector dominates applied research in crop biotechnology. Even public sector plant scientists in industrialized countries are now encouraged by their home institutions and governments—and in some cases, their own self-interests—to seek IPR protection for their inventions and to license technology to the corporate sector. Most universities and government laboratories in the United States today have their own patent attorney or engage outside agencies to defend their IPR interests. Individual plant scientists may not even have control over who they can share their discoveries and breeding materials with. No longer can the international system expect to have ready and free access to the results of strategic and applied crop research conducted at land grant universities and other public sector research institutions, unless specific arrangements are made to gain such access. In the future, much of what the international system needs from industrialized countries will be proprietary property in these countries. Furthermore, the increased value of plant genetic resources is causing developing countries to rethink their policies concerning access to the natural biodiversity they control. International negotiations related to genetic resources are currently taking place in four main fora:

1. The Conference of the Contracting Parties to the Convention on Biological Diversity.
2. The Commission on Plant Genetic Resources of the United Nations Food and Agriculture Organization, and related discussions within the context of its fourth International Technical Conference on Plant Genetic Resources.
3. The World Trade Organization, and in particular the Forum on Trade Related Intellectual Property.
4. The Union for the Protection of New Varieties of Plants.

The results of these negotiations will affect the global policy environment for genetic resources. Governments around the world are responding by modifying their own policies and enacting new legislation. While many within the international agricultural research system are unhappy and/or uncomfortable with all these new proprietary arrangements, the days of unencumbered free exchange of plant genetic materials are no doubt over.

## New Institutional Structures

The task ahead is to develop new institutional mechanisms that will allow the international system to gain access to the results of the increased investments in crop research generated by the profit potential of biotechnology. While their effectiveness and impact remain unproven, the following efforts are aimed at this goal.

The Rockefeller Foundation's International Program on Rice Biotechnology is designed to bring the benefits of biotechnology to poor rice producers and consumers in developing countries. The research is expected to contribute to the production of improved seed and other materials used by farmers through a series of cooperative projects and transfers of technology. Over the past 10 years, the foundation has invested roughly $27 million in relevant research conducted at premier biological research institutions located primarily in industrialized countries, with another $40 million committed to rice biotechnology capacity building in rice-dependent countries. The results include molecular genetic maps and markers of rice and important rice pathogens, routine procedures for the genetic engineering of rice, a variety of gene constructs of potential use for rice improvement, assessments of possible environmental and socioeconomic impacts, and recommendations concerning the establishment of national biosafety procedures for testing transgenic rice. Through collaborations established within the program, rice plants containing potentially useful transgenes are now being tested in several developing countries.

The program allows grantees to pursue IPRs on their discoveries and improved materials in industrialized countries, but requires that they share materials and technologies with others in the program at zero royalty for use in developing countries. A variety of fellowships enable scientists from participating developing-country institutions to work in the premier laboratories, then transfer materials and technology to their countries. Other fellowships support scientists from industrialized countries that periodically work at and transfer technology to collaborating developing-country institutions. A "shuttle scientist" employed by the International Rice Research Institute was stationed at Cornell University for several months each year with technology-transfer responsibilities in addition to research responsibilities. At the same time that technology is transferred, patent applications may be filed in developed countries by grantees in the hope of generating income for further research. Overall, this program, and others sponsored by bilateral aid agencies, demonstrate that with appropriate donor support, public sector research institutions in industrial countries can continue to be the most important source of advanced technology for the international agricultural research system without seriously jeopardizing their potential for financial gain from such research.

While the rice biotechnology program to date has had limited need to access the results of corporate sector research, the Rockefeller Foundation's Health Sciences Division recently announced an innovative, albeit unproven, strategy that might be useful. The basic idea is to announce a prize, in this case $1 million, to be awarded to the first group producing a desired product through research that the corporate sector is uniquely well suited to pursue, but for which there is limited profit potential. It is anticipated that such a prize will serve as an inducement for considerably more than $1 million worth of corporate sector research, and that the desired product will be generated quickly, since the first to deliver wins the prize.

The International Service for the Acquisition of Agri-biotech Applications (ISAAA) is a newly established, not-for-profit international organization committed to the acquisition and transfer of proprietary agricultural biotechnologies from the industrial countries for the benefit of the developing world. ISAAA currently has offices in the United States, Europe, Japan, and Africa, and it plans to open offices in Southeast Asia and Latin America. Initiated with a $1 million start-up grant from the McKnight Foundation, it is now supported by several foundations, bilateral and multilateral donors, and corporations. ISAAA assists in identifying biotechnology needs and opportunities, evaluates the availability of proprietary technologies, serves as an "honest broker" matching needs with available technology, and when necessary, mobilizes the financial resources required to implement brokered proposals. It has already demonstrated that a surprisingly large amount of proprietary technology can be made available at limited cost, or as donations, provided a mutually acceptable use agreement can be reached.

To date, ISAAA's most advanced project involves the donation of cloned genes by the Monsanto Company to Mexico for control of virus diseases in potatoes, and includes training for Mexican scientists in the use of the technology. As a companion project, ISAAA assisted Mexico in developing biosafety and regulatory procedures for testing transgenic plants, something that should benefit Monsanto in the long run. Through the work of Mexican scientists, local potato varieties containing the Monsanto genes have been developed and are being tested in Mexico. ISAAA has several other technology transfers underway, and numerous others are in negotiation.

The International Laboratory for Tropical Agricultural Biotechnology (ILTAB) is a collaboration between the Scripps Research Institute and the French technical assistance organization, ORSTOM, with facilities located on the Scripps campus in La Jolla, California. At the Plant Division of Scripps, scientists are doing pioneering work to develop disease-resistant plants through genetic engineering. This technology is transferred by Scripps to five ORSTOM scientists and over a dozen research fellows from developing countries working next door at ILTAB. These ORSTOM-ILTAB scientists use the Scripps technology to produce new sources of disease resistance in tropical crops such as cassava, rice, sweet potatoes, and yams. The products of this research are now reaching crop improvement programs in the tropics. Once proven effective in field tests, the engineered genes will be incorporated in elite and local cultivars disseminated to farmers. Scripps, meanwhile, has identified several products as a result of this research that are believed to have profit potential in developed countries and are worth patenting.

The Center for the Application of Molecular Biology to International Agriculture (CAMBIA) located in Canberra, Australia, is another recently established research and technology transfer organization committed to the application of biotechnology to international agriculture, with a special interest in producing inexpensive biotechnology tools that can be employed in developing countries. The founder of CAMBIA, Richard Jefferson, is the inventor of "GUS," a reporter-gene system now widely used in plant molecular biology research. Jefferson has been issued patents on GUS in the United States and Europe, and intends to use income from the patents to support CAMBIA's research programs and the dissemination of products to developing countries. In this case, the income generated by IPR protection should directly benefit international agriculture.

The Intermediary Biotechnology Service (IBS) is a unit of the International Service for National Agricultural Research, located in The Hague. IBS provides national agricultural research agencies with information, advice, and assistance to help strengthen their agricultural biotechnology capacities and enable them to establish collaborative arrangements with international biotechnology programs.

The Agricultural Biotechnology for Sustainable Productivity (ABSP), headquartered at Michigan State University and funded by the United States Agency for International Development (USAID), is a unique bilateral project in that it supports research at, and technology transfers from, both public and private sector crop research institutions in the United States. It includes training of developing-country personnel in IPR and biosafety issues, as well as crop research.

Within the U.N. system, there are several relevant initiatives. One, established by its Industrial Development Organization, is the International Center for Genetic Engineering and Biotechnology (ICGEB), with headquarters and information-gathering and dissemination facilities in Trieste, Italy, and agricultural biotechnology research and training facilities in New Delhi, India. ICGEB has established collaborative arrangements that give it access to proprietary technologies for use in developing countries.

In another U.N. project, the Food and Agriculture Organization (FAO) conducts research on and facilitates transfer of plant biotechnologies through its Plant Production and Protection Division in Rome. Together with the International Atomic Energy Agency, it supports the Joint Division of Nuclear Techniques in Food and Agriculture, which is located in Vienna and works on mutation breeding and other plant biotechnologies. In addition, the International Fund for the Conservation and Utilization of Plant Genetic Resources was established by the FAO in 1987. The idea was to finance the fund through a tax on seed sales, and then use the proceeds to support public sector breeding and crop germ plasm conservation programs in developing countries. This was seen as a way of repaying farmers for their prior selection and improvement efforts, as well as a means of conserving biodiversity as a common resource for the future. While, to date, no contributions to the fund have been made, the concept is still being discussed within various fora.

The Biotechnology Advisory Commission of the Stockholm Environment Institute is an independent resource for impartial biosafety advice. It was established to help developing countries assess the possible environmental, health, and socioeconomic impacts of proposed biotechnology introductions. It does this on receiving an official request from an appropriate institution or agency in the country where the introduction is to occur.

## Conclusion

A highly effective international agricultural research system is now in place that has made, and can continue to make, important contributions toward meeting the crop improvement and food production needs of billions of poor people in developing countries. This research establishment must be sustained both financially and technologically for it to foster agricultural systems that are both sustainable and sufficiently productive to spare other lands for the preservation of biodiversity. With incentives provided by IPRs, private sector agricultural research will generate many of the future advances in crop improvement technology, but will not directly address the needs of a majority of the people in developing countries. Mechanisms need to be developed that will allow technology to flow from advanced laboratories, and the private sector, to the international agricultural research system while simultaneously maintaining economic incentives for further research. Several such mechanisms are currently being tested, and some, such as those initiated by ISAAA, have the potential to be more broadly replicated. All, however, require the continued existence of a strong, well-supported public sector agricultural research system committed to meeting the needs of those with limited purchasing power.

## References

ACC/SCN (Administrative Committee on Coordination/Subcommittee on Nutrition). 1992. *Second Report on the World Nutrition Situation: Global and Regional Results*. Geneva: World Health Organization.

Alam, G. 1994. Biotechnology and Sustainable Agriculture: Lessons from India. Technical paper no. 103, Organization for Economic Cooperation and Development, Paris.

Bongaarts, J. 1994. Population policy options in the developing world. *Science* 263:771–76.

Borlaug, N.E. 1968. *Wheat Breeding and Its Impact on Wild Food Supply*. Canberra: Australian Academy of Science.

Byerlee, D., and M.A. Lopez-Pereira. 1994. Technical change in maize production: A global perspective. Economics working paper, Centro Internacional de Mejoramiento de Maize y Trigo, Mexico.

Chrispeels, M.J., and D.E. Sadava. 1994. *Plants, Genes, and Agriculture*. Boston: Jones and Bartlett.

FAO (Food and Agriculture Organization). 1992. *1990 Selected Indicators of Food and Agricultural Development in the Asian-Pacific Region*. Rome: RAPA Publications.

Waggoner, P.E. 1994. *How Much Land Can Ten Billion People Spare for Nature?* Council for Agricultural Science and Technology, report no. 121, Ames, Iowa.

WRI (World Resources Institute). 1994. *World Resources, 1994–95: A Guide to the Global Environment*. Oxford: Oxford University Press.

# The Mayagna Indigenous Community of Nicaragua: Moving from Conflict to a Convergence of Interests

by S. James Anaya and S. Todd Crider

Tropical forests are the greatest remaining repository of global biodiversity, and a large segment of these forests are the ancestral lands of indigenous peoples, such as the Sumo Indians of Nicaragua's Atlantic coast region. This case study reviews a trilateral agreement negotiated between a Nicaragua timber company, the government of Nicaragua, and the Sumo Indian community, with the participation of an international nongovernmental organization, the World Wildlife Fund. The chapter examines the difficulties in meeting the challenge of sustainable development, in perfecting the property rights of the Sumo, and in negotiating and implementing a viable agreement.

*

Tropical forests are a vital source of biodiversity on the planet. The earth's forests, however, are being depleted at an alarming rate, due to unrestrained commercial harvesting of valuable timber and the stripping of forestlands for agricultural use (WRI 1994).[1] A large portion of the world's remaining tropical forests are on the ancestral lands of indigenous peoples, such as the Sumo Indians. On May 15, 1994, the Community of Awas Tingni, which is indigenous Sumo, or Mayagna, signed a trilateral agreement with a foreign-owned timber company and the government of Nicaragua for the large-scale harvesting of timber on lands claimed by the community. The agreement, negotiated under the watchful eye of a major international environmental organization, is an effort at a new model of forestry development that is economically beneficial, environmentally sound, and respectful of the rights of indigenous peoples. As this essay explains, however, the negotiated agreement did not come easily, nor is its faithful implementation secure. The Awas Tingni case is at the intersection of diverse interests and value categories that find a theoretical basis for convergence in the concept of sustainable development. This case is a lesson in both the potential applications and limitations of this theoretical convergence.

This essay begins with a brief overview of the theoretical underpinnings of sustainable development as a synthesis of diverse values and interests. Next, there is a description of the players involved in the Awas Tingni case, the events leading up to the signing of the trilateral framework agreement, and the first annual timber harvesting cycle. The final part evaluates the Awas Tingni case critically in order to make observations that may be useful in gauging the desirability or feasibility of subsequent partnerships between commercial logging interests and indigenous peoples, especially in developing countries where a high level of demand for capital investment is often not matched by a stable legal and political climate.

## The Modern Synthesis

Environmentalism, with its embrace of the concept of biodiversity, is a value and interest category that has primarily been promoted by Western progressive elites and nongovernmental organizations (NGOs). The environmentalist agenda has traditionally emphasized the preservation of the natural world with its diversity of animal and plant species, paying only secondary attention to the economic or material welfare of human beings. Indeed, the traditional environmental movement has typically cast human beings as the culprit, the antagonist to be defeated in the quest to preserve nature.

A second, and often competing, value and interest category is that of economic development, which has been promoted especially by governments of the less-developed world and fueled by conditions of poverty as well as by the flow of capital transnationally. The economic development agenda has as its goals the raising of material standards of living and the accumulation of material wealth. This agenda thrives on the nexus of common interests among impoverished sectors of humanity and profit-motivated industry, a commonality of interests in which the resources of the natural environment are to be exploited to serve the ends of human consumption.

The concept of sustainable development represents the contemporary synthesis of environmentalism and economic development.[2] It is reflected prominently in the concluding documents of the Rio Earth Summit of 1992, particularly the environmental program and policy statement known as Agenda 21 (United Nations 1992). Sustainable development refers to the enhancement of human well-being through methods that do not lead to the long-term decline of the natural environment and its biological diversity, thereby maintaining nature's potential to meet the needs and aspirations of present and future generations (WCED 1987). The concept thus embraces normative elements of both the environmentalist and economic development agendas, while relying on science and the existence of objective criteria to provide its methodology. However, sustainable development is a largely theoretical concept that is only beginning to be implemented in large-scale projects in the less-developed world.

Another value and interest category that arises in relation to the foregoing is that associated with the rights of indigenous peoples. The rubric of indigenous peoples is generally understood to include the Indians of the American continents and other culturally distinctive groups that are the descendants of the original inhabitants of lands now dominated by others (United Nations 1986). Especially in the developing world, many indigenous peoples live in forestlands and other resource-rich areas that developers have sought to exploit and environmentalists have sought to preserve. Large-scale industrial development programs have, in many cases, resulted in indigenous groups being dispossessed of their ancestral lands (Burger 1985). By the same token, environmentalists' efforts to preserve ecosystems frequently have ignored the legitimate interests of indigenous peoples (Wiggins 1993).

Through the United Nations and other international institutions, the world community has become increasingly concerned with promoting and protecting the rights of indigenous peoples. Indigenous people are now widely acknowledged to have rights to continue as distinct communities on their ancestral lands, to benefit from the natural resources of their lands, and generally to be in control of their own destinies. Such rights have been included in the International Labour Organization Convention, no. 169, on Indigenous and Tribal Peoples (ILO 1989), and arguably can be identified as part of customary international law (Anaya 1991).

Indigenous peoples' rights have begun to be incorporated in the philosophy of sustainable development (Breckenridge 1992) as manifested most notably by Chapter 26 of Agenda 21, which states in part:

> In view of the interrelationship between the natural environment and its sustainable development and the cultural, social, economic and physical well-being of indigenous people, national and international efforts to implement environmentally sound and sustainable development should recognize, accommodate, promote and strengthen the role of indigenous people and their communities. (United Nations 1992, para. 26.1)

Furthermore, Chapter 26 calls for: legal mechanisms to secure indigenous peoples' access to and control over their lands; recognition of their traditional dependence on renewable resources and ecosystems; and procedures by which indigenous peoples are able to be active and meaningful participants in decisions concerning the development process as it affects them (United Nations 1992, paras. 26.1–21.6).

Thus, within the conceptual framework of sustainable development, there is a theoretical joining of environmentalism, economic development, and indigenous peoples' rights. But how does the synthesis of these three value and interest categories play out in practice? The case of ongoing efforts at forestry development involving the Sumo Indian Community of Awas Tingni in Nicaragua illustrates the problems in this regard and offers some important lessons.

## The Case of Awas Tingni

The Players

This case involves the following players, each representing diverse yet interrelated interests.

The Community of Awas Tingni (or the community) is one of numerous ethnically Sumo, or Mayagna, indigenous communities in the isolated Atlantic coast region of Nicaragua.[3] It is comprised of approximately 150 families. Like other indigenous communities of the Atlantic coast, Awas Tingni is organized under a customary leadership structure that functions as the effective authority within the community and represents it to outside interests. On the basis of historical use and occupancy, the community lays claim, but has no formal title, to over 100,000 hectares of land. Community members are generally impoverished and look to the forest resources within their claimed lands as a potential source of income. At the same time, community members express a strong ethic of environmental stewardship toward their forest lands, which is common to many indigenous peoples.

Maderas y Derivados de Nicaragua, S.A. (MADENSA, or the company) is a relatively young Nicaraguan timber company established with foreign (Dominican Republic) capital. It operates a sawmill in the Atlantic coast region and is interested in large-scale development of the forest resources, including a part of those claimed by Awas Tingni. Company officials—active members of the left-leaning Sandinista Party, formerly in power in Nicaragua—profess a modern, environmentally sensitive approach to the timber industry and are active in international efforts to promote environmentally sound forestry development. The company has close, yet ambiguous, links to a private Nicaraguan forestry consultants group called Swietenia.

Ministerio de Recursos Naturales y del Ambiente (MARENA, or the ministry) is the Nicaraguan ministry for the environment and natural resources,[4] and the Servicio Forestal Nacional (SFN, or the forest service) is an agency within the ministry. As its name suggests, the ministry is responsible for environmental protection, the administration of state-owned forestlands, and regulating the use and development of the country's natural resources; the forest service is specifically charged with oversight of forest resources. The ministry and its forest service are disadvantaged by a lack of financial and human resources, and an unstable political environment affecting the Nicaraguan government generally. Both these elements plague government institutions in many developing countries, but they are particularly acute in Nicaragua in the aftermath of the civil war that gripped the country throughout the 1980s.

The principal ministry official involved in the Awas Tingni case—from its initial phases through the signing of the trilateral framework agreement—was the director of the forest service. In the apparent absence of rigid conflict of interest guidelines in Nicaragua, the forest service director was also at all relevant times, in his unofficial capacity, a member of the private forestry consultant group, Swietenia.

The World Wildlife Fund (WWF)—now also known as the Worldwide Fund for Nature—a well-known international environmental NGO, has been working with the natural resources ministry to promote sound forestry development policies. The WWF views responsible industry as a potential ally for the sustainable development of forests in less-developed countries. The WWF has had extensive prior dealings with officials within the ministry and its forest service.

The University of Iowa's Awas Tingni Resource Development Support Project (the Iowa project) is a project organized by the University of Iowa College of Law in coordination with the WWF, and funded by the WWF, for the purpose of providing legal and technical advisory assistance to the Awas Tingni community with regard to its forest resources. The Iowa project encompasses an advisory team that includes the authors of this chapter: Iowa Law Professor James Anaya as coordinator, and Todd Crider from the New York law firm of Simpson, Thacher and Bartlett (who is provided by that firm on a *pro bono* basis). The team also includes a Nicaraguan attorney, Maria Luisa Acosta, and a forestry expert with extensive experience in Nicaragua, Hans Akesson.

Initial Strategies Pursued by MADENSA

In late 1991 and early 1992, prior to any involvement by the WWF or the Iowa project, MADENSA identified the commercial value of tropical forestlands in the area of Awas Tingni. MADENSA approached the Awas Tingni community about a joint project to harvest timber in the area, and arranged for the community to petition the forest service for a harvesting permit.

In 1992, the forest service issued a permit to the community for the harvesting of 1,500 cubic meters of timber, making an exception to a moratorium against timber harvesting in the Atlantic coast region. The moratorium was imposed because of overharvesting that had occurred in the absence of a sufficiently developed regulatory apparatus. As justification for the exception, the forest service cited the economic necessities of the community. The permit was conditioned on receiving progress reports on the harvesting and restricting the harvest to a certain mix of species, requirements aimed at maintaining the forest as a viable resource.

In 1992, MADENSA carried out timber-harvesting operations under the permit, using members of the community as laborers. It would later become apparent that the procedures followed in this limited harvesting were environmentally unsound, and that the company's financial dealings with the community and its members were problematic at best.

While harvesting under the initial permit took place, MADENSA proceeded with initiatives for long-term harvesting in an area of approximately 43,000 hectares in the vicinity of Awas Tingni. MADENSA, through the forestry consultant group, Swietenia, developed a management plan for a 25- to 30-year harvesting cycle in the 43,000-hectare area. The management plan is a voluminous document based on principles of sustainable forestry.

While government forestry regulations were still being developed, the forest service entertained MADENSA's petitions for approval of its management plan. Simultaneously, MADENSA proposed to enter into a 25-year contractual relationship with the Awas Tingni community in order to gain exclusive rights to the forestlands included in the management plan. The apparent assumption on the part of all concerned was that the 43,000 hectares were entirely within the lands to which the community had a valid claim. Without the benefit of legal counsel, the community leaders signed a vaguely worded, 25-year agreement with few safeguards to the company's control of the community's lands. With the signed agreement in hand, MADENSA continued to press the forest service and its parent ministry for approval of the management plan.

The Assertion of the Community's Rights

The nature and course of the community's relations with the company and government changed with the introduction of legal and technical advisers to work for the community. In early 1993, the WWF became interested in the MADENSA project as part of its ongoing concern for environmentally sound forestry development in Nicaragua. The WWF determined that technical assistance and legal counsel should be made available to Awas Tingni so that the community could better make informed choices and protect its interests in dealings with the company. Both MADENSA and the forest service initially voiced support for this view. The WWF sought out available legal resources and eventually coordinated with the University of Iowa College of Law to organize the Iowa project.

During the summer of 1993, Iowa project personnel, including Anaya and the project's Nicaraguan attorney, Acosta, traveled to the remote community to meet with its members and leaders. The Iowa project personnel offered advisory assistance and legal representation in connection with the community's interests in the harvesting of timber on its ancestral lands. After several hours of discussion with community members gathered in open assembly, the community accepted the offer of assistance and ceremoniously designated the Iowa project attorneys as its representatives.

Awas Tingni leaders told their newly acquired attorneys of several complaints they had about dealings thus far with MADENSA, including the company's efforts to collect on a disputed several thousand dollar debt of dubious origin. (The company would later tell the Iowa project attorneys that labor costs had exceeded the value of the timber.) Community leaders further explained that they felt pressured into signing the agreement with MADENSA for large-scale harvesting pursuant to the company's management plan, and that they thought the agreement was only for a 5-year term rather than the 25-year term specified in the signed document. The community directed the Iowa project team to pursue its concerns with MADENSA, to have the prior agreement annulled, and to develop an alternative contract proposal for a relationship with the company along certain specified lines and with terms that would avoid past problems. (The community rejected entering into a 25-year relationship with the company, preferring a shorter-term trial period.)

When the Iowa project attorneys first approached MADENSA with the community's concerns, company officials reacted angrily, petitioning the WWF to cancel the Iowa project and acquire new attorneys for the community. MADENSA officials complained that the Iowa project attorneys were being obstructionist and not helping the company secure the agreement it desired with the community. The WWF responded that it would not withdraw its support from the Iowa project; the project attorneys were pursuing the community's interests, as they had been expected to do. For its part, the Iowa project maintained that, at this stage, the terms or termination of its legal representation of the community could only be affected or brought about by the community itself.

In the fall of 1993, acting in part on some of the concerns raised by the community through the Iowa project team, the WWF funded the forest service to conduct an evaluation of the timber harvesting that had taken place under the 1992 permit for 1,500 cubic meters. The evaluation showed that the required mix of species had not been cut; instead, primarily mahogany, the most valuable species, had been harvested such that the long-term value of the forests was reduced and forest regeneration would be economically impaired. Also, other kinds of environmental damage were found.

Forging a Change in the Climate for Negotiation

Through the end of 1993, while the Iowa project attorneys advocated on behalf of the community, MADENSA avoided any substantive dealings with them. Meanwhile, unbeknownst to the community, its attorneys, or the WWF, MADENSA proceeded to engage the natural resources ministry and its forest service in negotiating a 30-year concession contract for timber harvesting within the 43,000-hectare management area, under a new theory that a substantial portion of the area belonged to the Nicaraguan state, not the community. In December 1993, MADENSA and the ministry signed the 30-year concession contract for timber harvesting on "state lands" within the 43,000-hectare management area. Like the negotiations leading to it, the 30-year contract between MADENSA and the ministry was signed without the knowledge of the community, the WWF, or the Iowa project. Shortly thereafter, MADENSA proceeded with plans to begin harvesting operations in early 1994.

Even assuming the existence of state lands within the 43,000-hectare area, the boundary between state and indigenous communal lands was, and still is, very much in question. The particular area in which MADENSA was about to begin operations, however, was in a location all agreed was within Awas Tingni lands. Furthermore, the regulatory scheme remained in flux. While forestry regulations were approved in fall 1993 (Nicaragua 1993), substantial ambiguity remained; the separation of powers and functions between the national government, represented by the ministry, and the autonomous government for the northern Atlantic region, where Awas Tingni is located, was unclear. The autonomous government for the northern Atlantic region, and its counterpart to the south, were created by a constitutionally mandated statute enacted in 1987 (Nicaragua 1987; Nicaragua 1995, art. 5) in response to demands by the indigenous and other minority peoples of the Atlantic coast. Since the autonomous governments were installed in 1990, a power struggle has ensued between the central and regional governments over regulatory control and ownership of natural resources, fueled by an absence of clear guidelines in the organic law.

In late December 1993, MADENSA officials told Awas Tingni leaders that timber operations were about to begin pursuant to the large-scale forestry concession negotiated with the natural resources ministry. Alarmed, the community directed its attorneys to intervene. In January 1994, after learning for the first time of the 30-year concession, the Iowa project attorneys issued a letter to the ministry demanding that no forestry development go forward on the lands claimed by Awas Tingni without community consent to the terms. The Iowa project attorneys implicitly threatened to seek legal recourse through available domestic and international channels. The WWF also made known to the ministry and forest service its disapproval of the 30-year concession.

The director of the forest service attempted to justify the signing of the 30-year contract on the grounds that it would encourage the foreign investment offered through MADENSA, bring much-needed capital into the country, and provide work for community members. No explanation was forthcoming for having concealed from the Iowa project and the WWF through numerous discussions the existence of the contract or the negotiations leading to it, other than to question whether the Iowa project advisory team in fact knew or represented what the community desired.

Nonetheless, faced with potential legal action and public denunciation internationally, the natural resources ministry and its forest service agreed to delay MADENSA's harvesting until a suitable arrangement could be reached with the community. Eventually, the combination of the persistence of the Iowa project, pressure by the WWF, and the intervention of a mediator retained by the WWF resulted in the ministry and MADENSA recognizing the legitimate role of the Iowa project team. The mediator, from Plan de Acción Forestal Tropical para Centro América, a Central American NGO dedicated to environmentally sound forestry in that region, was helpful in defining the issues and imposing a calendar of meetings. A constant problem was simply creating a degree of accountability so that the company would be forced to negotiate seriously. With the introduction of a mediator and some committed government professionals into the process, the company was discouraged from pursuing an obstructionist strategy intended either to outlast the interest or resources of the Iowa project or to divide the community from the project advisers. As late as January 1994, and intermittently since then, the company attempted to drive a wedge between Awas Tingni and its advisers by persuading members of the community that, but for the intervention of the Iowa project, the foresting activity would have already commenced, generating jobs and revenues. While this claim was probably true, it neglected that any such forestry activity would have been under terms much less favorable to Awas Tingni and the long-term value of its forests than those advocated by its advisers on the basis of the community's own articulated goals. Nevertheless, the combination of factors, including the company's eventual recognition that the Iowa project team was genuinely interested in negotiating a mutually beneficial agreement, resulted in a steady increase in the professionalism exhibited by MADENSA in dealings with the project advisers.

Another breakthrough was the government's specific commitment to not prejudice the community's land claim. The Nicaraguan Constitution and the 1987 autonomy statute guarantee the rights of indigenous peoples to their traditionally held communal lands (Nicaragua 1987, art. 36; Nicaragua 1995, arts. 89, 180). However, the communal lands of most indigenous communities in Nicaragua have not been officially titled or demarcated, and as a result, Awas Tingni and other communities exist on officially designated state lands. The natural resources ministry agreed early in the process that Awas Tingni had a valid land claim; the problem was determining the extent of that claim. Eventually, the ministry agreed that, for the purposes of timber harvesting, the entire 43,000-hectare management area would be considered community land. Based upon the legal delimitation of those lands, the community's rights under any contract for harvesting in that area would be adjusted accordingly.

Tripartite Negotiation of a Framework Agreement
The process entered into a more standard contract negotiation once MADENSA and the natural resources ministry abandoned surreptitious dealings exclusive of Awas Tingni, accepted the role of the Iowa project team, and settled on a framework for recognizing the community's land rights. The community sought a contract with the following features: protection of the long-term value of the forest and the community's rights as a landowner; steady income for the community; the ability to get out of the contract in order to establish its own forestry operation; protection of the way of life of community members, which depends heavily on the multiple forest resources; and employment for individual members of the community

working through a cooperative. MADENSA, on the other hand, sought an ensured source of timber for the longest period possible with a minimum of interference.

A number of the community's objectives were in tension with the company's. Awas Tingni's assertion of its rights as a landowner and interests in preserving the long-term sustained value of the forests placed a practical limitation on the company's operations as well as its ability to function with a minimum of supervision. From the company's perspective, instead of a convenient source of cheap labor and timber, the community would be transformed into a directly engaged owner overseeing the development of its principal asset, the forests.

Additionally, the community's desire for a steady income flow from the timber harvesting posed indirect difficulty as a result of the unclear definition of land rights. In order for Awas Tingni's income to be ensured on a steady, annual basis, the harvesting had to be sustainable within community lands. The company's forestry management plan was for a specific 43,000-hectare land area, with annual yields determined on the basis of an inventory of the forest resources within that area. Initial annual harvesting under the plan as originally designed might result in overharvesting within the community's lands, if those lands were subsequently defined as only partly inclusive of the 43,000-hectare management area. Foresting within the defined community lands would then have to be adjusted downward, imperiling the stability of Awas Tingni's principal source of income and employment. The community's position, therefore, was that it had to limit the volume harvested, commence operations in the part of the 43,000-hectare management area farthest from the community, or have two logging camps, one near the community and another removed. The alternatives were opposed by the company, which insisted that two camps were too expensive, and it had intended to finance the construction of logging roads to the more distant portions of the management area by logging the closer areas.

The community's desire to leave open the possibility of getting out of a contractual relationship with MADENSA was the most controversial. Any exit strategy for the community clashed with the company's insistence on attaining a stable, long-term resource. In addition, ministry officials supported the company's position, arguing that foreign experts and international lending agencies, including the World Bank, had suggested long-term, large-scale concessions to encourage capital investment and provide business concerns with a vested interest in the sustainability of the resources. While this position has some logical basis, the Iowa project advisers felt that an approach based on long-term concessions depended on both a responsible business partner with an established track record and a reliable regulatory body to supervise the activity. Unfortunately, as a result of undercapitalization by the company on the one hand, and governmental instability on the other, both of these elements were missing. In addition, and most important, even had they been present, the community expressed a strong preference for retaining the possibility of managing their forests on their own. This would cause a shift from a management approach, based on making the company entirely responsible for the logging and forestry, to an approach that emphasized owner (Awas Tingni) oversight and delegated certain functions to the company.

By the time serious negotiations were underway, certain events had turned business incentives against the community's interests in important ways, weakening Awas Tingni's position. First, the moratorium on logging activity, in place when the company initiated contact with the community, had been lifted. Without a full implementation of the government's newly developed forestry regulation, the end of the moratorium allowed the company to purchase wood from numerous sources other than Awas Tingni at low cost and with minimum responsibility. Second, the ministry had signed the agreement granting the company a 30-year concession to undefined lands within the management area, lands presumed to be state lands adjacent to the community's. Even though the ministry took the position that it would not implement the concession until the extent of Awas Tingni's territorial rights were defined, the concession established the company's claim to the area and reduced the community's alternatives. As a result, for reasons largely attributable to regulatory action or inaction, the community's bargaining leverage was decreased significantly.

Awas Tingni's disadvantage, however, was at least partly offset by the WWF's role in the process. The WWF weighed in heavily in favor of an agreement that would include the community as an equal partner and provide a model for sustainable forestry in Nicaragua. To some extent, both MADENSA and the ministry perceived that their interests were best served by maintaining the WWF as an agent of cooperation rather than condemnation.

Terms of the Framework Agreement
After prolonged negotiations, compromises were made leading to at least partial solutions to each of the items of contention, and a framework agreement was signed. First, under the agreement, the community participates in the annual planning procedures implementing the management plan developed by the company and approved by the forest service. The level and areas of harvesting are determined annually within the framework of the management plan, which is designed to secure a sustainable yield of timber through selective harvesting of diverse species. Prices are also negotiated annually along with the level of discounts from the price of the standing timber to reflect the capital investments and reforestation activity required.

Second, as an additional party to the contract, the government, represented by the natural resources ministry, agrees to treat the community as owner of the entire management area until the land rights are officially defined, and to assist the community in securing official demarcation and recognition of its territory. Accordingly, Awas Tingni receives the stumpage value of the timber harvested, less applicable taxes, from anywhere in the management area. In order to ensure a steady revenue stream for the community, harvesting in the areas most proximate to Awas Tingni is limited to a specified volume estimated to be well within what could be sustained within subsequently demarcated community lands. Once territorial rights are defined, the annual volume to be harvested may be adjusted as long as the harvesting within the community's lands remains sustainable over time.

Third, the company agrees to enter into a separate labor agreement with the community's forest workers' cooperative and to give hiring preference to members of the community. Community members formed the cooperative, with the assistance of the Iowa project, as part of an effort to gain greater control over the development of its forest resources.

Fourth, the framework agreement is for a 5-year term. At the conclusion of this term, the community may opt out of a renewal. If the community opts out, the company will retain a right of first refusal on purchasing any timber from the management area that is sold to a third party. Required capital investments by the company are to be paid by deductions from the cost of the timber or they are amortizable within the 5-year term. No deductions are made for heavy equipment and other itemized costs that are borne by the company.

The framework agreement provides for a yearly work cycle as follows: preparation by the community and company of an operational plan specifying the volume and species to be cut, as well as the details about the year's plan of operations; submission of the plan to the forest service for approval; marking the trees to be cut; entering into a contract specifying terms and prices for that year's cut; and an evaluation after the completion of the logging season. This cycle provides for close monitoring of activity by both the forest service and the community in order to safeguard the long-term value of the forests. And once normalized, the implementation of the cycle should not be overly cumbersome.

At the signing ceremony at Awas Tingni on May 15, 1994, all the participants, including community leaders and a Nicaraguan vice minister, spoke to the assembled group about the importance of the event, describing it as a symbolic moment in Nicaraguan history. Even after discounting for rhetorical excesses (the vice minister said it was a "transcendent day in the history of Nicaragua"), it was the consensus of those gathered that the contract represented an important departure in Nicaragua both in terms of indigenous peoples' rights and environmental stewardship.

Implementation of the Framework Agreement
The extent to which the framework agreement can or will be faithfully implemented, however, is another question. The experience of setting in motion the first annual harvesting cycle was problematic within a still weak scheme of government regulation. Despite the enactment in late 1993 of an extensive regulatory apparatus

designed to support environmentally sound forestry in Nicaragua, forest service officials have been slow to conform entrenched practices to the newly enacted regulatory regime (Nicaragua 1993). Harvesting of the most valuable tropical wood at unsustainable levels has continued in many parts of the Atlantic coast region with the acquiescence and, apparently, even the permission of the forest service. MADENSA has been—and, to these writers' knowledge, continues to be—a regular purchaser of valuable hardwoods, mostly mahogany, harvested from lands outside of Awas Tingni at unsustainable levels. Under these conditions, the company's incentive for making the required capital investments for sustainable forestry, and for harvesting trees other than the most valuable species, as required by the agreement with Awas Tingni, is diminished significantly.

The framework agreement requires an annual operational plan that identifies the area of activity, the volume of each species to be cut, the procedures to be followed, and the reforestation required. Although each year's plan is to be submitted jointly by the community and MADENSA to the forest service for its approval, the company is responsible for preparing an initial draft. For the first year of harvesting, the company agreed to provide its initial draft by August 1994 in order to ensure approval by fall. Due to the rains during the summer and fall months, the forest harvesting season in eastern Nicaragua is roughly from December to early June. But with apparently plentiful alternative sources of cheap valuable timber, the company was clearly in no hurry to commence operations under the Awas Tingni agreement. At every step along the way toward setting in motion the first annual harvesting cycle, the Iowa project had to take the initiative or prod the company into action.

When the company finally presented an initial draft in September 1994, the annual plan provided for a harvest limited to far fewer species than the number of diverse species required by the framework agreement. The draft plan was also contrary to the principles of sustainability incorporated into the agreement. Diversified harvesting is an important component in the theory of sustainable forestry. The selective harvesting of only the most valuable hardwood, such as mahogany, frequently strips a forest of its value. Especially in developing countries, once forests lose their economic value, they are often converted to more economically viable nonforest uses. The sustainable course is for the mahogany and other valuable species to act as an incentive; in order to access the mahogany, a forestry company agrees to harvest other species as well and manage the forest as a sustainable resource. A diversified harvest allows the forest to support a broader and more permanent laboring class, which, in theory, creates several constituencies with interests in seeing it managed properly: the forestry company, the owner, and the workers, as well as the government and the broader international environmental community.

The company took the position that its commitment to diversified harvesting was to be fulfilled over time; moreover, its sawmill and international market conditions would not allow it to produce or sell the species omitted from its draft plan. The company's position revealed significant constraints arising from separate supply and demand limitations.

First, because MADENSA was undercapitalized, it was relying on the forestry operations to generate a flow of capital sufficient to finance its expansion and diversification. According to company representatives, investments intended for the installation of additional sawmill facilities and a plywood plant were delayed due to the political and economic instability of Nicaragua. As a result, the company did not have the resources to handle some of the species it had committed to harvest, thereby creating a supply-side constraint. Separate, specially adapted sawmill lines were required for the denser hardwood species, which are useful principally as railroad ties and marine pilings; other species are only useful as plywood, making a plywood plant a prerequisite for truly diversified harvesting.

A second constraint on the process was the alleged absence of demand for certain species. MADENSA insisted that it only had access to purchasers in the international markets for mahogany and cedar, and that even if the company harvested other species, it would be unable to sell them. While this demand-side limitation is largely attributable to the company's lack of investment (for example, if a plywood plant were constructed, there would be both domestic and regional markets for plywood), it also reveals a problem analogous to the supply and demand problem in the drug trade. The voracious markets of the developed world establish high levels of demand for certain species, while at the same time, international pressure on developing countries restricts and monitors the production of those species. As various NGOs, including the WWF, have recognized, in the effort to promote sustainable forestry, a change in the patterns of demand offers one of the most promising ways of improving the means of supply. This observation has led to initiatives by the WWF and others to establish a system to rate timber according to the soundness of the environmental procedures followed in its harvesting. To the extent that the market can be educated to pay a premium for timber harvested on a sustainable basis and discount timber not harvested according to specified guidelines, stronger market incentives can be implemented to assist the efforts to preserve forests as economically viable assets.

Unfortunately, Awas Tingni was faced with a much more immediate dilemma: it desired to protect the long-term viability of its forests, but it required income from the harvesting and related work for its underemployed population. Faced with this situation, the community's advisers were compelled to seek leverage to increase the diversification of the proposed harvesting and institute mechanisms to ensure that, to the extent viable species were overlooked one year, the company would be required to return and harvest them in subsequent seasons.

The result was a period of difficult negotiations, which culminated in an annual operational plan with a proposed harvest slightly more diversified than that of MADENSA's initial draft plan, and a commitment by the company to increase the density and diversity of each annual harvest progressively to certain minimum levels. In addition, the company agreed to return to areas intervened so as to harvest species as they become financially viable. The annual operational plan as agreed on by the community and MADENSA was proposed to the forest service in November 1994; it was approved in early 1995.

After the approval of the 1995 annual operational plan, the parties proceeded to negotiate the 1995 annual timber contract as well as the labor contract provided for in the framework agreement. Each of these proved much less difficult to negotiate, largely because their principal function was to incorporate the prices to be paid for the relevant year, and such prices were determined on the basis of the international benchmarks for the relevant species and the prevailing Nicaraguan compensation rates for forest workers. The international market prices for standing timber were discounted to reflect taxes payable by the company, the cost of transportation, and the infrastructure investments to be made. Due to the limited intervention planned for 1995, no discount was included for reforestation. The community's forestry engineer, a member of the Iowa project team, considered that the forest would best recover if left to its natural regeneration, since only two to four trees per hectare would be cut in 1995.

With regard to the labor agreement, the community and the company indicated a preference for an agreement treating Awas Tingni's labor cooperative as a subcontractor to the company. While this imposed greater responsibilities and liabilities on the members of the community, it was in keeping with their objective of maximizing their autonomy and preparation. Under the labor agreement, the cooperative agreed to provide services to the company in return for compensation at specified rates for tasks performed. In addition, under the framework agreement, the company is obligated to hire members of the community for other tasks related to the forestry operations in Awas Tingni, to the extent that there are skilled workers available.

In March 1995, that year's annual contract and labor agreement were signed by company and community representatives. On March 23, however, the company failed to make its first payment due to the community in the amount of approximately $13,500. The company blamed this failure on what it characterized as a *force majeure*; namely, the failure of the forest service to mark the trees to be harvested as provided by the 1995 annual operational plan. Apparently, the company's audit of the results of the forest service's markings showed too few trees had been marked and that the proportion of mahogany was inferior to the agreed-on amounts. The Iowa project attorneys, acting on behalf of the community, responded that this was a disagreement with the ministry to be resolved in the ordinary course, not one that impeded the commencement of operations or rising to the level of a *force majeure*. The forest service immediately agreed to conform the markings to the operational plan. By late April 1995, the company had made its initial payment to Awas Tingni and operations commenced.

## Evaluation

An assessment of this case gives rise to the following observations, which lead, in turn, to a series of recommendations or guidelines.

*Political, Legal, and Regulatory Instability.* A major difficulty that plagued this case was the lack of stability in two areas of fundamental importance: first and generally, the Nicaraguan political and legal systems, and second, the regulatory regime for natural resources. The overall lack of stability in Nicaragua illustrates the crucial role of a political and legal system that protects the expectations of parties through property rights and other mechanisms that yield reliable outcomes. Repeatedly, from the absence of definition of land rights, to the government's inability to implement forest regulations, to the undetermined relationship between the central and regional governments, to MADENSA's delaying of its capital investment plans on account of domestic uncertainty, a climate of instability jeopardized—or at least made more difficult—the efforts of everyone involved.

Particular difficulty was presented by the evolving nature of the regulatory regime for natural resources. While this is not unique—indeed, in the United States, the fortunes of both environmental and logging interests are subject to routine realignment every two to four years—in Nicaragua, the combination of ambiguity and a failure to implement regulations is especially frustrating. The lack of definition and implementation, among other things, creates at least the appearance, if not the reality, of regulatory decisions being influenced or driven by industry concerns. This problem is further complicated in a small country by the close linkages between the individuals involved, as discussed below.

*Unclear Definition of Territorial Rights.* In Nicaragua, a progressive constitution recognizing the rights of indigenous communities to traditional land areas has not been supported by specific implementing legislation or by the controversial task of defining the relative rights over the resource-rich lands in the eastern half of the country. The absence of any definition as to the land or territorial rights of indigenous groups, the national government, the regional government, and individual or corporate owners, results in an absence of responsibility in which land and property is frequently abused. In addition, as each group has an interest in maximizing its own reach, efforts at economic development are often thwarted by rival interest groups. For example, at various points, the regional government of the North Atlantic coast region interposed threats to the process by disputing the central government's role and contending for a greater share of the tax proceeds from forestry operations. Each time, these interventions would delay the process or, at the least, provide other parties with an excuse for delay.

As Awas Tingni asserted its interests in its ancestral lands and appurtenant forest resources in the face of MADENSA's development initiative, both the company and the natural resources ministry attempted to treat those lands as subject to state ownership and control. After initially assuming and sustaining the validity of the community's claim, once Awas Tingni began to assert its corresponding rights, the company began to question the extent of the community's claim and to ally itself with the central government.

Fortunately, these difficulties were largely overcome due to the persistence of the community and the eventual goodwill of the ministry and the regional government, both of which agreed in principle, in February 1994, to treat the lands in question as Awas Tingni property for the limited purposes of the framework agreement. As a result, the land tenure question mainly presented difficulties in terms of ensuring a sustainable plan for properties not yet defined.

The Iowa project continues to work with the community in an ongoing effort to define land rights. Even outside the context of resource development initiatives, the delimitation of indigenous lands tends to be complex, involving a mix of historical, ethnological, and legal issues. Indigenous land claims should be resolved on the basis of principled criteria, and it is difficult to accomplish this in the face of development initiatives that exacerbate conflicting political and economic interests. Ideally, land tenure questions should be settled before large-scale or long-term development projects are initiated; this is especially important where geographically substantial claims by indigenous communities are involved.

*Conflicts of Interest.* A related concern was the close relationship between the company and its forestry consultants, on the one hand, and certain ministry officials, on the other. While this situation is not surprising in a small country like Nicaragua, it is precisely this type of circumstance, in which the legitimacy of government agency decisions is called into question, that conflict of interest guidelines are intended to protect against. During the first several months—including, notably, the period in which the company surreptitiously negotiated a 30-year concession with the ministry for national lands—the ministry's efforts were spearheaded by the director of the forest service, who, in his presumed unofficial capacity, was also a member of the forestry consultant group Swietenia, which in turn worked for the company. From December 1993, other individuals less tied to industry concerns became primarily responsible for the ministry's positions. The result was a substantial improvement in the professionalism exhibited by the ministry and, consequently, in the rate of progress.

*Participation of Indigenous Peoples.* A threshold issue for this type of activity is the recognition by all parties of the status of the indigenous community in question as a full and coequal partner. While this may appear self-evident to enlightened individuals, it conflicts with the ingrained, paternalistic tendency of many government agencies and politically empowered elites to either act on behalf of indigenous groups or, at best, provide symbolic recognition of their interests. Accordingly, for both the company and the ministry, it was a difficult adjustment to recognize the right of Awas Tingni to utilize advisers who pursued the community's interests with vigor.

Even after the engagement of the Iowa project team, the apparent expectation of the ministry and the company was that the role of the community's advisers would be limited to providing an anthropological, interpretive interface with the community and making recommendations from an "objective" standpoint. During the first few months of the Iowa project's involvement, the company—and to a lesser extent, the ministry—characterized the external advisers as obstructionist. As late as March 1994, the company continued to question the project's credentials as well as its legitimacy as the community's representatives, exploiting their closer continuous physical proximity to Awas Tingni in an attempt to divide the community from its advisers.

*The Attorney-Client Relationship.* A related issue is the relationship between the community and the Iowa project advisers. The community did not actually initiate contact with its advisers, although it did enthusiastically accept their offer of assistance and approve of their retention. Anaya was sought out by the wwf, and he determined who comprised the advisory team. The Iowa project did go to great lengths to legitimize its role as an instrument of the community, and to ensure that its negotiation strategy was that preferred and determined by the community leadership. This process involved a time-consuming effort to build trust and maintain lines of communication between the community and the project.

A complicating factor was that the Iowa project was funded by the wwf, which positioned itself in the process as a neutral party interested primarily in promoting sound forestry practices and securing a fair process of negotiation. Attorneys advocating for indigent clients are often in the position of being funded by sources other than their clients as a necessary result of the clients' indigence. The Iowa project, despite its funding source, attempted to insulate itself from the environmental policy-driven decisionmaking of the wwf, subjecting itself entirely to the decisionmaking apparatus of the community. For the most part, nonetheless, the positions adopted by the community coincided with the wwf's environmental objectives.

The relationship between the Iowa project team and the wwf no doubt created ambiguity for MADENSA and the government. The negative reactions of the company and ministry officials, however, appeared to be motivated by a strong initial resistance against a shift in the balance of power vis-à-vis the community, a shift resulting from the presence of legal and technical advisers advocating the community's interests and backed by an influential international environmental organization.

It is now generally accepted internationally that indigenous peoples should be full and equal participants in the decisions leading to development projects that affect them. To this end, the decisionmaking capacities of an indigenous community implicated in a development project should be empowered with legal and other technical assistance that is answerable to the community alone. Just as government and industry are entitled to legal and technical counsel, so too are indigenous peoples. Industry should encourage and respect efforts by attorneys and other technicians to pursue the interests of their indigenous clients above all other interests involved. This means, among other things, that governments and industry should be prepared to negotiate in good faith and at arm's length with a community, and to respect the attorney-client relationship.

*The Capitalization of the Forestry Company.* To a significant measure, any long-term forestry venture between an indigenous community and a forestry company should be viewed as a commercial joint venture with all the traditional commercial concerns. Of particular importance in this regard is the reputation, financial integrity, and resources of the business partner. The sustainable nature of the Awas Tingni project over the long term is dependent on the capital investment of the company. Unless the additional production facilities are installed, the agreed-on diversification will be impossible.

Unfortunately, Nicaragua, because of its history and current situation, does not offer the necessary conditions to attract large-scale capital investment. The few companies brave enough to function in this environment, such as MADENSA, have a frustrating task of overcoming high capital costs along with legal and regulatory instability. As a result, from the community's perspective, there were few alternative suitors besides MADENSA. Furthermore, once the ministry signed its own agreement with the company, the community's ability to seek an alternative partner was further reduced.

*Government as a Partner, Not Competitor.* Though the role of government in the protection of the environment and the regulation of business remains controversial, it would appear that, minimally, the government should provide incentives— through its tax and regulatory policies—for beneficial behavior and disincentives for counterproductive behavior. In the context of the Awas Tingni project and Nicaraguan forestry, this translates into a policy that would encourage the development of forest resources on a sustainable basis with respect for indigenous rights. Though it was eventually helpful in bringing about the negotiation of a framework agreement, the natural resources ministry, and particularly its forest service, initially acted ambiguously. Moreover, it has since failed to take a number of steps helpful to both the community and the protection of forest resources.

In any country with forest resources, government must curtail and penalize unregulated harvesting by attacking not only the wildcat loggers, but also the companies that purchase timber from such sources. Relevant government agencies must also achieve a high level of proficiency in fieldwork and faithfully implement applicable regulations. Further, government should avoid making decisions that effectively compel indigenous communities to deal with specific companies, as occurred in this instance. Finally, in order to advance the process of sustainable development of forestry resources and the exercise of the rights of indigenous groups, government should clearly and justly define the governing land tenure and property regime.

In sum, the Awas Tingni case represents a new model of forestry development on land owned or claimed by indigenous peoples in developing countries. This model involves a partnership between industrial and indigenous sectors, and the application of normative elements arising within the global concern for environmental protection, indigenous peoples' rights, and the need for economic development, especially among impoverished populations. The Awas Tingni case is an important step in the evolution of this model, in understanding the myriad problems that are likely to be faced when applying the model, and in working toward solutions of these problems.

## Acknowledgments

This article is based on a project funded, in part, by the wwf, with monies granted by the U.S. Department of State. The opinions, findings, and conclusions or recommendations expressed here are those of the authors; they do not necessarily reflect those of the wwf or State Department.

## Notes

1   The Food and Agriculture Organization of the United Nations estimates that between 1980 and 1990, tropical forest areas shrank 15.4 million hectares (0.8 percent) per year on average (WRI 1994).
2   An important agent in the evolution of the concept of sustainable development was the World Commission on Environment and Development, which was established by the United Nations General Assembly in 1984 to study issues of environmental protection and development. It published its report and recommendations, featuring the theme of sustainable development, in the now widely circulated document, *Our Common Future* (WCED 1987), also called the "Brundtland Report" after the commission's chair, Norwegian prime minister Gro Harlem Brundtland. For a commentary on *Our Common Future*, see Caldwell 1990, 207.
3   The members of this indigenous ethnic group, which encompasses a number of distinct communities, traditionally call themselves *Mayagna*, although the term *Sumo*, which originated outside of the group, is more commonly used among outsiders.

The Atlantic coast region of Nicaragua is traditionally understood to include roughly the eastern third of the country. The geographically isolated region has a unique history and cultural milieu, being home to the Miskiito, Sumo, and Rama Indians as well as a substantial Black Creole population. For a demography and history of the Atlantic coast region, see Vilas (1990) 19–27.

4 MARENA was previously the Instituto Nicaragüense de Recursos Naturales, known by its Spanish acronym IRENA. This arm of government was reconstituted into MARENA by executive decree in early 1994, in part to add an institutional emphasis to environmental protection.

## References

Anaya, S. 1991. Indigenous peoples in international law. *Arizona Journal of International and Comparative Law* 8:1, 8–15, 27–28.

Breckenridge, L.P. 1992. Protection of biological and cultural diversity: Emerging recognition of local community rights and ecosystems under international environmental law. *Tennessee Law Review* 59:735, 750–60.

Burger, J. 1985. *Report from the Frontier.* London: Zed Books.

Caldwell, L.K. 1990. *International Environmental Policy: Emergence and Dimensions.* 2d ed. Durham, N.C.: Duke University Press.

ILO (International Labour Organization). 1989. Concerning indigenous and tribal peoples in independent countries. *International Legal Materials* 28:1382.

Nicaragua. 1987. Statute of autonomy for the Atlantic coast of Nicaragua, law no. 28. *La Gaceta, Diario Oficial* 238.

———. 1993. Executive decree, no. 45–93. In *La Gaceta, Diario Oficial* 197:3370.

———. 1995. Political constitution of the republic, arts. 5, 181.

United Nations. 1986. *Study of the Problem of Discrimination Against Indigenous Populations.* Geneva: United Nations.

———. 1992. Agenda 21. 13 June. UN Doc. A/CONF. 151, vols. 1, 2, and 3.

Vilas, C.M. 1990. *Del colonialismo a la autonomia: Modernizacion capitalista y revolucion social en la costa atlantica.* Managua, Nicaragua: Editorial Neuva Nicaragua.

WCED (World Commission on Environment and Development). 1987. *Our Common Future.* Oxford: Oxford University Press.

Wiggins, A. 1993. Indian rights and the environment. *Yale Journal of International Law* 18:349–50.

WRI (World Resources Institute). 1994. *World Resources, 1994–95.* Oxford: Oxford University Press.

# Biodiversity Prospecting Frameworks:
# The INBio Experience in Costa Rica

by Ana M. Sittenfeld and Annie Lovejoy

This chapter presents a conceptual framework for biodiversity prospecting (bio-prospecting) activities and discusses strategies that can facilitate widely beneficial bioprospecting agreements. Bioprospecting may enable tropical countries to integrate biodiversity into scientific, technological, and economic development that improves the quality of life for their citizens while ensuring that resources are protected and used sustainably. Recent experience in managing INBio's Bioprospecting Program demonstrates the multisectoral and multidisciplinary nature of this activity. It requires the creation of bioprospecting frameworks that facilitate cooperation among governments, intermediary institutions, private enterprise, and academia. Bioprospecting frameworks also need to incorporate local communities and entities, and the involvement of scientists, lawyers, managers, and economists from developing and developed countries. Finally, the framework agreements should require the merger macropolicy and other essential components, including biodiversity inventories, information management, technology access, business development, and strategic planning.

\*

Traditionally, tropical forests have been valued exclusively for their lumber and potential agricultural land, an assessment that has led to high rates of deforestation and species loss. While tropical forests do not comprise the whole of the earth's biological diversity, a high level of species diversity is concentrated in these areas, where misuse is leading toward irremediable circumstances. Because the loss of species is irreversible, the potential impact on evolutionary processes, the human condition, and the fabric of life itself is inestimable (NRC 1992). These dangers point to the need for creating biodiversity strategies that address the causes for this loss, and provide for conservation mechanisms and the sustainable use of natural resources (WRI, IUCN, and UNEP 1992). A conservation strategy arising from these growing concerns seeks to increase the value of biodiversity in intellectual, spiritual, and economic terms. Broad in scope, it can be broken down into the following three basic steps: saving, knowing, and using biodiversity.

Only by saving biological diversity can we come to understand its value to society. Once representative samples of biodiversity have been preserved, ongoing conservation can only be ensured through knowing what biodiversity exists, where it is located, its ecology and natural history, and its distribution and role in the ecosystem where it is found (NRC 1992). Studying biodiversity builds awareness of its value in all sectors, at all levels, and provides ways for the public to appreciate it through information access or educational outreach (WRI, IUCN, and UNEP 1992).

Knowledge and understanding of nature's diversity and complexity is also the key to using it sustainably, in a way that neither diminishes nor damages the resource and ensures that its existence is guaranteed for future generations. The diversity of nature itself is reflected in the manifold ways it can be made useful and, therefore, valuable to humankind. They range from intellectual, ecological, and spiritual uses to economic uses; from educational tools, watershed protection, and aesthetic interests to the discovery of new food sources. As one of these many approaches to making use of the natural world that surrounds us, bioprospecting may offer tropical countries a means to integrate biodiversity into scientific, technological, and economic development; improve the quality of life for their people; and ensure that the resource is protected and used sustainably. It involves the systematic search for and development of new sources of chemical compounds, genes, microorganisms and macroorganisms, and other of nature's economically valuable products (Eisner 1989; Joyce 1991; Sittenfeld and Villers 1993a).

Bioprospecting is nothing new; humanity has always used biodiversity. What is new is that current prospecting activities arising from new biodiversity conservation strategies are working to provide returns to source countries by generating resources for conservation purposes and promoting economic development. This chapter will focus on how bioprospecting activities might be constructed and orchestrated by taking a closer look at bioprospecting frameworks and the experience of Costa Rica's National Biodiversity Institute's (INBio) Bioprospecting Program in exploring and conserving biodiversity in collaboration with private enterprise.

Events following the Global Biodiversity Strategy meeting in Bogotá, Colombia, which took place in September 1988, and the October 1988 Congress on the Costa Rican National Conservation Strategy for Sustainable Development, provided the opportunity for the save, know, and use strategy to be implemented in Costa Rica. With two decades of conservation efforts completed and consolidated within an integrated National System of Conservation Areas, INBio was created in 1989. INBio's purpose was to participate in the second and third steps of the strategy; it operated on the philosophy that society is unlikely to continue paying the high maintenance costs of preservation, and to resist the political pressures that lead to resource mismanagement, unless biodiversity is shown to be economically and intellectually valuable (Sittenfeld and Villers 1994, 500–504). Based on a partnership of cooperative support and guidance with the Ministry of Natural Resources, Energy, and Mines (MIRENEM, 1992), and in collaboration with local and national institutions and individuals, INBio has agreed to carry out a variety of functions. They include the categorizing and inventorying of prospecting, information management, and dissemination of knowledge about Costa Rica's biological diversity in accordance with the existing legal framework. The institute believes that these programs—conducted at a local, national, and international level—will be decisive in preserving Costa Rica's biodiversity and promoting its sustainable use.

Bioprospecting contributes to INBio's institutional mission in two primary ways: facilitating society's sustainable economic use of biological diversity by establishing collaborative, market-driven research and development activities; and providing income to directly help cover the costs of activities supporting biodiversity conservation, through donating 10 percent of INBio's research budget, and 50 percent of any potential future royalties received by INBio, to the Ministry of Natural Resources (Sittenfeld and Gámez 1993). The program objectives move beyond providing the biological sample collection and distributor service that has tended to perpetuate inequitable agreements for natural resource transfer in the past. Activities and negotiations are geared toward incorporating a variety of innovative mechanisms that provide benefits for the country of origin; they also include an information infrastructure, technology access, scientific training and equipment, and new employment opportunities, while opening different avenues toward improved sustainable economic development.[1]

Recent experience in managing INBio's Bioprospecting Program demonstrates the multisectoral and multidisciplinary nature of this activity. Elevating the value of nature's biological resources by promoting their rational use within the context of sustainable economic development and continued biodiversity conservation is extremely complex. It requires the creation of bioprospecting frameworks based on cooperation among governments, intermediary institutions, private enterprise, and academia; the involvement and cooperation of local communities and entities; and the incorporation of scientists, lawyers, managers, and economists from developing and developed countries. At the same time, they require that macropolicy be intertwined with biodiversity inventories and information management, business development, and technology access, leading to an overarching bioprospecting framework (figure 1).

## Bioprospecting Frameworks

The fundamental point of departure for a bioprospecting framework is *macropolicy*: the set of governmental and international regulations, laws, and economic incentives that determine land-usage patterns, access to and control of biological resources, intellectual property rights regimes, technology promotion, and industrial development. Given a favorable policy environment, the framework should

FIGURE 1    Bioprospecting Framework

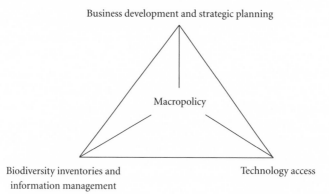

Business development and strategic planning

Macropolicy

Biodiversity inventories and
information management

Technology access

then focus on three essential components required for the rational use of biological resources, namely:

*Biodiversity Inventories and Information Management.* Development and management of information (including biological inventories and information management systems), regarding taxonomy, organism distribution, ecology, natural history, and the biological interactions between organisms and their ecosystems that add value to the raw biological resources.

*Business Development.* The definition of markets, market needs, and major actors; identification of in-country capabilities and needs as well as institutional goals and strategies; and the formation of collaborative arrangements to achieve institutional and national strategies.

*Technology Access.* Access to technology—through its development, transfer, or acquisition by other means—that converts raw biological materials into more highly valued products and promotes capacity building.

## Macropolicy

On the international level, macropolicies such as agreements and conventions establish the relationships and protocols for conduct between countries. For example, the Convention on Biological Diversity (CBD), calls on the 171 countries that have signed and ratified the treaty to create a framework for regulating access to and control of biological resources, intellectual property rights, environmental protection, and commercial laws that must be harmonized with the goals of development, conservation, and the fair and equitable sharing of benefits derived from the sustainable use of biological/genetic resources (appendix, this volume). Article 3 of the CBD recognizes an important principle of sovereign rights over biological diversity, dramatically changing previous assessments made by international organizations, such as the Food and Agriculture Organization, who in the past considered these resources to be the "heritage of humankind" and freely accessible (Blixt 1994; see also Reid, this volume).

On the national level, sovereign governments determine macropolicies. They will deal with issues such as landownership, the creation of conservation areas, access and control of biological resources, nationally recognized intellectual property rights, and whether or not biodiversity constitutes a resource in the public domain. They will also address the creation of incentives or deterrents for undertaking research and industrial activities. Such policies create a national context in which bioprospecting can either develop favorably or be effectively stifled.

In Costa Rica, the establishment of a National System of Conservation Areas—clearly defined, private and public, protected areas encompassing 27 percent of the national territory (Costa Rica 1992b)—was the macropolicy that supported the creation of INBio as an instrument for conducting a biological inventory and encouraging intellectual and economic use of biodiversity. Knowing that biological resources are "saved" and regulated within a given area makes it easier to justify the major investment required to undertake biological inventories. From there, a systematic and careful collection process helps to build the information infrastructure needed to carry out activities that increase biodiversity's economic and intellectual value (Raven and Wilson 1992).

The protected status of these resources, coupled with the clarity of laws that regulate access to and ownership of the conservation areas where they are found, substantially reduces the risks for potential industrial and academic partners desiring to conduct research and development. There are numerous examples of pharmaceutical and agrochemical companies that have identified lead chemical compounds from an initial plant sample only to discover that the source material can no longer be found because of the conversion of forest into pastureland. Similarly, a lack of explicit laws regarding landownership makes collecting from those areas inherently risky since uncertainties arise as to who has the authority to grant legitimate access to materials, and under exactly what conditions. In December 1992, a Wild Life Protection Law was approved by the Costa Rican Congress. Under the new law, any entity that wants to collect or manage biodiversity samples from a conservation area for research or other uses must obtain a permit or sign a concession agreement with the Ministry of Natural Resources (Costa Rica 1992b, arts. 2–4, 17). In this way, resources and their use are tightly regulated and safeguarded from overuse and exploitation. This legislation declares all wild animals (including invertebrates) to be "public domain" and "national patrimony," regardless of whose property they inhabit. Wild, nontimber plants as well as the conservation, research, and development of genetic resources are viewed as "public interest"; that is, they are considered to have public utility and, therefore, their use is specifically regulated. Bylaws regulate the conditions related to the production, management, extraction, commercialization, industrialization, and use of genetic resources (Costa Rica 1993).

Still other sets of macropolicies affect the dynamics of bioprospecting. For example, Costa Rica's heavy investment in education has created a base of highly competent and skilled technicians and scientists—in addition to well-developed laboratories within the universities. They have supported, and will continue to do so, the country's bioprospecting. Social macropolicies are extremely important in providing the human resources needed to implement bioprospecting frameworks. Nevertheless, much needs to be done in Costa Rica to strengthen its human, physical, scientific, and research infrastructure, and to provide industry with incentives that will enable it to compete on par with developed and some developing countries.

It is clear that without favorable macropolicies, bioprospecting programs are bound to fail. Yet countries that dedicate themselves to learning about stronger and more dynamic incentives are better positioned to realize their goals for balancing the economic use of biodiversity with conservation.

Biodiversity Inventories and Information Management
Given a favorable policy environment, particularly the organization and management of conservation areas, biological surveying and biodiversity information management systems become a feasible and crucial first step in creating bioprospecting frameworks. Biodiversity inventories—through the development and management of biological, ecological, taxonomic, and related systematic information on living species and systems—increase the value and promote the sustainable use of raw biological resources (Reid et al. 1993; WRI, IUCN, and UNEP 1992; Aylward et al. 1993a). This fundamental information provides the knowledge of what species are available, and where they may be collected in sufficient numbers without damaging ecosystems (Raven and Wilson 1992; Caporale and Dermody 1994).[2] As emphasized in the biodiversity conservation strategy, data are critical for conducting a wide variety of activities related to sustainable use and conservation, of which bioprospecting is only one.[3]

The biodiversity inventory of Costa Rica's wildlands builds on a long history of specialized inventories of fauna and flora conducted by national and international researchers (Janzen 1983). INBio's present program is intended to complete the inventory and induce broad national participation in the process of gathering such information (Schweitzer et al. 1991; Janzen 1992, 27–54). The basic fieldwork is being conducted by a group of laypeople trained in the vocation of "parataxonomists" (Gámez 1991). Based out of INBio's biodiversity offices, located in the country's conservation areas, a parataxonomist not only collects specimens for the national inventory, but also initially catalogs the biological material before transporting it to INBio's headquarters for identification and storage. They receive feedback, planning, and guidance from the institute's staff of curators, who work within a larger network of national and international curators and taxonomy experts.

Parataxonomists play an important role at the local level by forming a more immediate and stronger link between institutions and local entities, and by bringing INBio—and biodiversity as a whole—into the communities where they live, which are located in and around the conservation areas. In the process, they are becoming local educators, teaching nearby communities about their environments (Aylward et al. 1993b).

The inventory's first goal, for national and international purposes, is to accumulate the specimens necessary to "clean up" the taxonomy of Costa Rica's biodiversity (completing and taxonomically organizing already existing collections), including information about at least one site for each given organism (Sittenfeld and Gámez 1993). "Taxonomic cleanliness" will take the form of identified reference collections, field guides, biological species catalogs, and electronic identification services, such as expert systems. In the long term, the inventory will establish species' ranges in more detail and initiate the process of understanding their natural history and other characteristics.

To facilitate the management and manipulation of species and conservation information accumulating in the institute's databases, INBio has designed a computerized biodiversity management information system capable of handling traditional geographic information systems, text, and numbers. This initiative was made possible through a collaborative agreement with the Intergraph Corporation. The design and management of this new system forms a part of the recently created Division of Biodiversity Information Management at INBio. While large amounts of information will be generated and distributed by INBio, it will not necessarily become an information center. Highly detailed information regarding local institutions and projects will remain available at the local level.

As a first step toward sharing information with national and international users, data from the institute's botany database are now available worldwide on the Internet. The institute is also in the process of making its biodiversity information available to a larger cross section of society, both inside and outside of Costa Rica, through the newly established Program for Biodiversity Information Dissemination. Among other activities, the program focuses on biodiversity education for children of all ages and community outreach. In addition, the past two years witnessed the institute's first international workshops, aimed at sharing the INBio experience with different regions and countries, including Latin America, the Caribbean, Cameroon, Madagascar, and Ghana.

## Business Development

Business development, which covers an entire process—from identifying markets and strategic planning, to negotiating collaborations and contractual agreements—is often neglected in the emerging bioprospecting literature, but it forms a critical component of any successful bioprospecting program. This component builds on the knowledge of what biological resources exist, and begins to identify and develop collaborative arrangements with potential economic users of biological diversity.

Based on the results of market research, as well as a keen awareness of in-country research skills and capacity, a bioprospecting program can develop institutional and financial goals and objectives that include: facilitating the acquisition and/or development of information, technology, and products that increase the value of samples used by academia and local private enterprise for research purposes. If they are capable of being transferred to private enterprise to promote sustainable economic development and create jobs, this will work toward increasing the overall value of biodiversity in the eyes of society. The objectives should also contemplate generating income through direct research fees, donations, and potential future product royalties to support the country's conservation areas and conservation management activities (Roberts 1992; Sittenfeld and Gámez 1993; Touche 1991).[4]

To initiate collaborations, INBio conducted an informal audit of in-house and in-country strengths and weaknesses addressing the question of how Costa Rica could increase the value of its biological resources, and the level of in-country research capacity needed to manipulate those resources. From there, the institute set out to identify and understand the fundamental needs and characteristics of the resource's potential users. Through studying market surveys, INBio has been able to understand major industry trends, identify market needs, and locate potential company collaborators in the pharmaceutical, agrochemical, and biotech areas. Additional research into other industries that use biological resources—namely, the cosmetic, flavor, and fragrance industries—has provided INBio with knowledge concerning the basic characteristics of other potential industrial-user markets.

## Technology Access

Successful bioprospecting requires access to appropriate technologies, and the CBD underscored the significance of this factor. Access to technology, through acquisition or development, and its transfer to industry, permits developing countries to convert the raw materials of biological diversity into industrial inputs and products of greater value (Reid et al. 1993; Reid, this volume).[5] Conceptually, technology must first be accessed and mastered on an experimental, piloting scale before it can be successfully transferred and accessed on an industrial level (figure 2).

Capacity building for conservation depends on a preexisting knowledge base

FIGURE 2   Accessing Technology

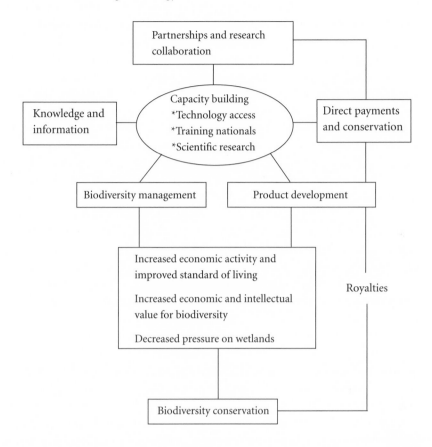

that attracts bioprospecting collaboration and partnership. These partnerships provide investment and funding (including equipment transfer, scientific exchange and training, and so on) that can be reinvested in increasing capacities, or contributed directly to national conservation efforts. In the long term, collaborative research investments not only advance scientific research, but also provide access to technology and technical training. In turn, this promotes a new avenue for economic development through innovative local product development, training and involvement of local communities, and higher employment rates (see INBio-British Technology Group agreement). Products developed from bioprospecting collaborations can also aid in biodiversity preservation; contracts can arrange for product royalties to be channeled toward conservation (see INBio-Merck agreement below). The expected benefit from these advances is a higher standard of living for society coupled with a decrease in the pressure to develop wildlands. Simultaneously, capacity building in the areas of conservation management and scientific training (both dependent on developing the initial knowledge base through inventories and other studies) will add value to biodiversity by improving resource management techniques, as well as providing information for eco-tourism and biodiversity protection, among other activities.

Negotiating successful technology transfers depends on complex and inter-dependent factors that begin with the existence and continual advancement of a solid foundation of knowledge and scientific expertise. National inventory programs are an example of a knowledge base that can serve as an excellent instrument for promoting institutional and national scientific capacity building.[6] For prospecting activities, it is clear that strong source-country scientific capacity can attract collaboration, effectively fostering incentives and reducing risks for these programs in the eyes of private enterprise. Such was the case for Merck and other companies and industries that were initially attracted and encouraged to enter into collaborative research agreements with INBio, which offered national, institutional, technological, and intellectual capabilities. Collaboration enables continued development of organizations or countries concentrating on biodiversity conservation and capacity building.

For biodiversity-rich countries, increasing the value of natural resources and promoting greater conservation incentives through an information infrastructure is only one aspect of technology transfer. Ensuring that funds are put toward conservation efforts and technology access (for equipment transfer, scientific training, and more) for the source country also constitutes an essential component of prospecting agreements. Investing in conservation or technology access will provide a number of tangible benefits, including a competitive technological edge for development (Reid et al. 1993). National capacities, beyond those of an individual organization or institution, need to be encouraged so that the country as a whole will benefit from any technology access resulting from prospecting agreements. During negotiations, then, it is to the mutual benefit of all parties—individual organizations, the source country, donor agencies, and private enterprise—to ensure that part of the investment is directed back into increasing those capacities that keep the wheels of this highly advantageous cycle in motion.

To date, INBio's efforts have been largely focused on this necessary first phase of technology access. With this, however, comes the recognition of a long-term goal to transfer environmentally sound and user-friendly technologies to promote sustainable economic development, and implicitly, the continuation of biodiversity conservation. In its collaborations with industry, government, and academia, INBio works hard to involve leading national laboratories and private enterprise so that this approach to accessing technology takes account of both static and dynamic national advantages, while keeping the future in mind.

FIGURE 3    Research Collaboration

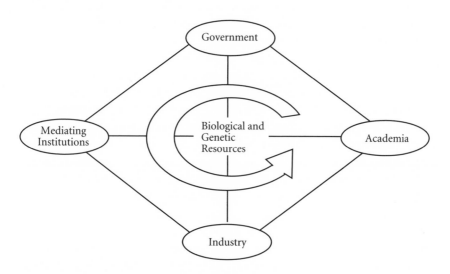

## Multisectoral Collaborations

In bioprospecting, public or private institutions act as mediators, playing a critical role as brokers between biological/genetic resources, industries, and other institutions interested in accessing and using those resources. INBio has negotiated several collaborative research agreements with the pharmaceutical and agrochemical industries, and has begun negotiations for additional partnerships in these and other industrial sectors. Each collaborative research agreement reflects a multisectoral approach based on the sustainable use of biological or genetic resources (figure 3).

Some of the primary actors that mediating organizations must approach and include are developing country governments, which frequently regulate access to, and control of, biological materials. Other key actors are private enterprises, such as particular companies or industrial partners interested in converting biological diversity into a biological/genetic resource. Often, access to the economic resources for financing the partnership is helped by private enterprises. They may frequently agree to access or transfer technology in exchange for in-country collaboration.

Current collaborative research agreements with the pharmaceutical industry reverse a centuries-old tradition of freely extracting raw plant and animal material from tropical countries for shipment to laboratories and companies in the North, while the country of origin receives no intellectual or economic returns (Reid, this volume; Joyce 1991). Raw biological samples have low market values (Reid et al. 1993; Simpson, Sedjo, and Reid 1994; Aylward et al. 1993a). Prospecting activities should seek to increase the value by moving beyond simple collection and distribution services. Research contracts can increase the value of biological materials by fostering knowledge about their ecology and taxonomy, and by carrying out some processing and research inside the source country. This enables mediating organizations and entities to offer research institutes and companies a consistent source of well-prepared and identified samples, while simultaneously augmenting the institute's bargaining leverage through the value this information adds to biodiversity (Lesser and Krattiger 1993b). An information package that provides ecological and taxonomic data along with the actual physical sample can significantly decrease resupply risks and increase the sample's value in the view of private enterprise. Additionally, a strong factor that can positively influence private enterprise's decision to collaborate with mediating institutions is the presence of "trained professionals, with whom . . . [companies] collaborate, provide identified samples, from recorded locations from which recollection is reliable" (Caporale and Dermody 1994).[7] Agreements can then include direct compensation for services rendered (such as sample collection, processing, identification, laboratory research, and resupply), contributions to support biodiversity inventories, direct funding for conservation activities to maintain national parks, and royalty payments to support conservation efforts if commercialization is successful.

There are those who express doubts about the value of payments contingent on successful product research and discovery because these payments are not only unlikely, but also long in coming. They suggest that it would be preferable for conservation activities to receive direct payments, which constitute a "greater [and more] certain total contribution to the preservation of biodiversity" (Lesser and Krattiger 1993b; Simpson, Sedjo, and Reid, this volume). While this argument is well placed, it should nevertheless be understood that the funds available for bioprospecting are not often available for direct payments to conservation. Both private industry and research institutes are not likely to donate funds directly to conservation efforts without receiving tangible short- or medium-term benefits in return. Indeed, straightforward donations are not nearly as attractive as funding programs that promise multiple solutions, even for bilateral and multilateral development agencies and foundations (Simpson, Sedjo, and Reid, this volume). Therefore, the investments made by Merck, Bristol Myers Squibb, the British Technology Group, the U.S. National Institutes of Health (NIH), the United States Agency for International Development (USAID), the National Science Foundation (NSF), and the MacArthur Foundation for prospecting activities could not have been put toward any other conservation or economic activities in Costa Rica. As INBio's research agreements indicate, however, portions of these otherwise unavailable investments *can* serve as "direct funding" for conservation if contracts are successfully negotiated to include this essential component.

Where research and processing (which add value to the resources) is involved,

INBio frequently seeks to make use of national and international university laboratories to assist in the research and development of technologies and new products derived from biological/genetic materials. Involving the academic institutions of developing countries in scientific training has the distinct advantage of ensuring that the technologies transferred or developed remain in the developing country, to be used by other university research programs and, eventually, by national industry. In addition, the advanced academic laboratories of developed countries are typically more disposed than corporations to sharing and transferring valuable technology with research counterparts.

## Contract Negotiations and Legal Issues

Contract negotiations are an essential aspect of bioprospecting frameworks, and give rise to three basic sets of issues: scientific, business, and legal. While the second two are related, it is often helpful to first outline scientific work plans carefully and then broadly define the business issues, later working to place them in a legal framework. To negotiate business issues, an organization needs to have a good sense of its own fundamental needs and those of its potential collaborator. The typical science-related needs that INBio focuses on in research negotiations are:

- generating income to support conservation areas and conservation management activities (including INBio's) through direct contributions as well as royalties;
- limited sample supply, to ensure that ecosystems and species populations remain uneffected and undamaged;
- the transfer of biodiversity processing technologies (equipment and know-how);
- creating opportunities for training Costa Rican scientists; and
- limited sample exclusivity, to allow for broad sample-screening exposure.

The private-enterprise collaborator needs and expectations include: access to new and diverse sources of biochemically rich materials; a high level of assurance that biochemically interesting materials can be resupplied; numbers of samples large enough to merit the effort of random screening; limited resource exclusivity; limited sharing of intellectual property rights; payment for resources commensurate with the estimated or perceived market prices; and assurance of legal, in-country practices for resource procurement, among others (Lesser and Krattiger 1993a). Once both sets of needs are defined, successful negotiation becomes largely a matter of how to maximize the benefits of collaboration for both parties. It has been INBio's experience that once needs and expectations have been discussed and understood, corporate lawyers with solid experience in licensing and joint venturing can assist in suggesting legal frameworks for achieving objectives.

The deliberation of appropriate royalty rates and intellectual property rights for biological materials is one of the major business issues discussed during bio-prospecting contract negotiations. In practice, the settlement of these issues generally comes down to arguing for the application of other industry precedents to the new situation of bioprospecting, such as precedents from the biotechnology industry, as well as market perceptions regarding resource supply and demand (Reid et al. 1993).[8] Therefore, bargaining power in these negotiations is considerably enhanced by a firm knowledge of the user's industry (for example, the pharmaceutical industry), the resource market (such as biological resources), and legal precedents in pharmaceutical and other industries, as well as an understanding of the needs of conservation activities.

When approaching legal issues, pairing up a team of institutional representatives and environmental lawyers from the developing country with developed-country management consultants and *pro bono* corporate lawyers, a scheme that INBio has adopted, has proven to be highly effective. INBio's recent experience in three such negotiations—with Merck, the British Technology Group/Hacienda La Pacífica, and Bristol Myers Squibb Corporation—that established favorable terms for technology transfer, royalties, and direct payments, among others, for INBio and Costa Rica's conservation areas, provide examples of different bioprospecting frameworks.

### INBio's Collaborative Research Agreements

The INBio-Merck Agreement
For years, ecologists and environmentalists have contended that the conservation of biological diversity is justified by its pharmaceutical and other commercial applications. In a concrete proposal to the MacArthur Foundation, Thomas Eisner, biology professor at Cornell University, argued that the "systematic screening of biological resources for uses, if it is done by developed and developing countries working together, could pay off for both while aiding conservation efforts" (Eisner 1989). With the initial support of the MacArthur Foundation, INBio and Cornell University began developing a Costa Rican research initiative on biochemical prospecting that had the long-range benefit of improving conservation. The initiative also provided for the organization of a small conference at Cornell University to set guidelines for the establishment of research links and contractual arrangements between INBio and specific university and industrial laboratories. The conference, held in October 1990, had one major consequence: it peaked the interest of Merck and Company, culminating one year later in the signing of a two-year INBio-Merck research agreement. In July 1994, INBio and Merck renewed the research project for another two years, followed by a second renewal in August 1996.[9]

According to the terms of the original two-year agreement, Merck paid INBio $1 million for a small, limited number of well-identified and documented biological environmental samples for microorganism isolation and plant and insect extracts and fractions for use in its drug discovery process. INBio has agreed not to provide these samples to any other organization for use in health or agricultural screening during this two-year period of initial evaluation in order to give Merck an opportunity to study them. All samples come from national parks or other kinds of conservation areas in accordance with the INBio-MIRENEM partnership, which was established in 1989 and formally ratified in May 1992.

INBio puts the ongoing funds from Merck toward research and certain start-up costs. Merck donated an additional $180,000 in laboratory equipment and supplies, and installed a "state of the art" extraction laboratory, to the University of Costa Rica (Sittenfeld and Villers 1993b). The project also included the training of four Costa Rican chemists at Merck or other prestigious international research centers. In addition, the collaboration supported scientific exchange for INBio's curators and taxonomists (Sittenfeld and Villers 1993b). The conservation areas and the Ministry of Natural Resources received 10 percent of the $1 million supplied by Merck, which was invested in the Island of Cocos National Park located off the Pacific coast of Costa Rica. The funds went toward augmenting the park's infrastructure, and financing park rangers to establish eco-tourism guidelines and enforce fishing regulations. Any royalties from products derived from the collaboration will be shared 50/50 between INBio and MIRENEM (Sittenfeld and Gámez 1992).

The project's bioprospecting framework was clearly understood by both parties, the end result of which demonstrates that companies can return part of the benefits of pharmaceutical development to the biodiversity-rich country where the chemical compounds originated. Both parties were conscious of the mechanisms needed to ensure that some of these benefits (that is, part of the initial funding) would directly finance conservation, while the remainder would indirectly finance it through investment in biodiversity inventories, biodiversity information systems, and bioprospecting in association with the conservation areas and national parks.

The negotiation process centered on the complementary nature of the activities that INBio and Merck carry out, which, once joined, would result in mutual benefits that could not arise solely from individual institutional efforts. INBio's experience and capability in biodiversity conservation and management, which rely heavily on the national biodiversity inventory, could only enhance Merck's drug discovery and development process, while Merck could support INBio's institutional mission and Costa Rica's conservation efforts. The collaboration even resulted in the additional benefit of scientific publications, which will be highly valuable not only for scientists, but also those interested in the advances of biochemical prospecting activities.[10]

The INBio-British Technology Group/Hacienda La Pacífica Agreements
In what could possibly result in the first commercial product from a Costa Rican conservation area since INBio's formation, INBio is collaborating with the British Technology Group (BTG), a major technology licensing company in the United Kingdom, and the Kew Royal Botanical Gardens in England, to isolate and test a plant compound for nontoxic, nematocide properties. Following the lead of a 10-year-old ecological observation indicating that field mice would starve to death before eating the seed of a certain Costa Rican plant (Janzen et al. 1990), biochemical research financed by BTG was initiated. Research on the seed revealed that a naturally occurring sugar analogue demonstrated a range of activities against several plant-parasitic nematode species (Birch et al. 1993). This resulted in the filing of a patent (no. GB 2 250 439 A in Great Britain) for the compound's use as a biodegradable, phloem-mobile, nontoxic nematocide, which controls the growth of root nematodes through foliar, seed, and soil application.

Once the patent was filed, BTG contacted INBio for assistance in obtaining the plant samples required for field testing, with the understanding that successful results would lead to large-scale plant domestication and production to promote sustainable resource use. Both INBio and BTG believe the product has major potential to eliminate the nematodes that plague bananas and other crops. And as the world's second largest banana producer, Costa Rica could only stand to benefit from INBio's partnership with BTG.

In agreeing to provide BTG with the desired quantity of the isolated compound, INBio, on the advice of Kew Botanical Gardens, first had to develop an effective protocol for isolating the compound from leaves (rather than seeds), which provide a more abundant and sustainable supply source. In return, BGT is funding INBio's initial research, with 30 percent of the budget proceeding to the Costa Rican National Park Fund. As part of the collaborative research agreement, INBio receives rights to the product's exclusive research and commercial development in Costa Rica—an activity that INBio is carrying out in partnership with local industry.

Support for field trials and other on-site research will come from the Costa Rican Association of Banana Growers (CORBANA), which has demonstrated a keen interest in the project. In addition, a major Costa Rican nematocide manufacturer has examined the feasibility of production on the industrial level. Further assistance comes from one of the country's better-administered conservation areas, which is donating space at its experimental forestry station for research in plant domestication. Presently, in collaboration with Hacienda La Pacífica (Costa Rica) and Ecos Holdings (Switzerland), studies are being carried out to determine the optimal conditions for creating large-scale, plantation production of the tree within the country. Based on positive results, the nematocide, to be produced in Costa Rica, will be available on the market within three years, providing royalties to the national parks and INBio. If all goes well, INBio, alongside national scientists and industry, will play a pivotal role in the development of the first environmentally friendly agrochemical that has the distinction of coming from a Costa Rican conservation area, and that provides tangible benefits (jobs and improved exports, in addition to direct financial returns from product development) for the country of origin.

INBio-Cornell-Bristol Myers Squibb Partnership

A partnership composed of INBio, Cornell University, the Guanacaste Conservation Area, the University of Costa Rica, and the Bristol Myers Squibb Company represents one of the world's five International Cooperative Biodiversity Groups (ICBG), supported by NIH, NSF, and USAID. All five ICBGs are comprised of private and public institutions, including pharmaceutical companies and environmental organizations, collaborating on projects that address biodiversity conservation and promote economic activity through drug development from natural products. The Costa Rican ICBG is seeking to discover new chemicals from arthropods for drug development in an attempt to introduce insects and arthropods into the marketplace as a new, potentially valuable source of pharmaceutical compounds. This five-year program involves the extensive biochemical and ecological training of field experts, termed *biodiversity ecologists*, who in addition to their regular collection activities are also trained to look for and record ecological indications of potentially powerful chemical compounds. An important component of the research effort focuses on developing expertise in chemical ecology, so that Costa Rican researchers will be better equipped to select, collect, extract, identify, and characterize promising chemical compounds proceeding from insects. INBio and Cornell expect that this consortium effort, and the knowledge resulting from it, will provide the impetus to formally introduce arthropods into the pharmaceutical market while generating valuable new ecological information that could increase the efficiency of drug screening in the long run.

The grant also supports an INBio/University of Costa Rica tropical drug discovery effort to test and identify chemical compounds developed from insect extracts for treating malaria. As part of the overall consortium agreement, Bristol Myers Squibb will directly contribute 10 percent of the research budget to the conservation areas to support biodiversity conservation efforts, and provide limited research funding to INBio, in addition to laboratory equipment and training for two Costa Rican scientists at its labs. If valuable drug compounds are identified and developed by Bristol Myers Squibb as a result of this collaboration, an intellectual property rights agreement signed among the consortium members guarantees INBio, Cornell, and the University of Costa Rica a share of potential future profits. Half of INBio's share of potential future royalties will be directly awarded to Costa Rica's National Park Fund through a parallel, bilateral agreement signed between INBio and the government of Costa Rica through MIRENEM.

## Conclusion

No single organization can advance bioprospecting. Rather, there should be multilateral collaboration to achieve the key components identified in our framework for bioprospecting. They include macropolicy, biodiversity surveys, business development, and technology access, creating dynamic advantages for bioprospecting. Throughout the journey taken toward building this framework, there exist a variety of issues, none of them simple, which require thorough consideration and evaluation on a case-by-case basis.

In scope, prospecting agreements should not be limited to the pharmaceutical industry or to the sole objective of generating profits from "gene hunting," which all recognize to be quite uncertain and long in their arrival. Further, research collaboration in this field is notably complex. It should incorporate: benefits, including capacity building and technology transfer, for the country as a whole; direct financial benefits for conservation in addition to potential royalties; the involvement of a country's institutions and entities on the local as well as national level; the creation of industrial incentives; and the attraction of industrial activities in general.

The nematocide project with BTG, a collaborative effort that has expanded to include other participants to locally produce the compound, is just such an example of how intricate, but carefully negotiated, prospecting agreements can achieve multisectoral benefits. It is a project that will create jobs for Costa Ricans, build scientific capacity and increase technology, improve economic stability by strengthening the country's largest export crop and lowering imports, and at the same time, provide returns for conservation efforts. With this type of collaboration, biodiversity can be valued beyond monetary profits resulting from successful product development; its benefits also reach individuals, communities, industries, institutions, nations, and biodiversity conservation in intellectual, technological, spiritual, as well as economic terms in a less-distant future. At present, Mexico, Kenya, and Indonesia, among other nations, have seen Costa Rica and INBio as an example of what is possible, an example not fully developed, yet one that can serve as a pilot project to be adapted or adopted to the individual circumstances of other nations (Sarukhan 1992, 317–43; Reid et al. 1993).

As INBio moves forward in its efforts to explore and conserve Costa Rican wildland biodiversity in collaboration with industry, it faces a number of pointed challenges. Principal among these are the need to guide public expectations regarding research results, and the need to find funding for INBio's ongoing conservation activities and for the conservation areas themselves. As an organization, INBio incorporates a number of activities beyond its Bioprospecting Program. Consequently, financial support from private enterprise represents only a small portion of the institute's funding, allowing INBio to fund its activities through a number of different channels. While the search for funding will always continue to a certain extent, the institute is attempting to diversify its sources, increase its endowment, and generate its own funds through providing a variety of services to other national and international organizations. The need to acknowledge that wildland biodiversity must become a sustainable generator of income if it is to survive underlies all of INBio's objectives (Sittenfeld and Villers 1994).

In general, successful bioprospecting programs face the challenge of expanding into markets besides the pharmaceutical and agrochemical industries, while also developing macropolicies that attract industrial investments beyond prospecting activities themselves, thus guaranteeing a solid industrial base for national economic development. Above all, we must control the expectation that bioprospecting will become a primary generator of conservation funds by keeping in mind that these activities represent only one means among many. Nevertheless, by continuing to integrate human and material resources in collaboration with highly capable industries, new products will hopefully be made available simultaneously with the development of new mechanisms for biodiversity conservation.

## Acknowledgments

This chapter has been adapted from a presentation to the conference Biological Diversity: Exploring the Complexities, 25–27 March 1994, University of Arizona, Tucson.

## Notes

1   There exist various perspectives concerning the exact premise of bioprospecting activities. On the one hand, some have questioned the validity of prospecting agreements as mechanisms that, in and of themselves, will preserve biodiversity and encourage its conservation through the financial benefits they may generate. Many conclude that such agreements may not provide the expected returns for conservation, given the low probability that successful products will result (Lesser and Krattiger 1993b; Simpson, Sedjo, and Reid 1994). Bioprospecting, however, may yield other benefits and advantages that have been overlooked. The point of view presented in *Biodiversity Prospecting* (Reid et al. 1993) recognizes "the enormous uncertainties involved in genetic prospecting [and] the exaggeration of benefits," as do Simpson, Sedjo, and Reid (1994), but also emphasizes that the "substantial genetic and chemical resource flows . . . from developing countries to pharmaceutical companies . . . are currently not resulting in either an equitable sharing of benefits or a return to conservation." The bottom line is that this new breed of bioprospecting agreement may rectify this situation above and beyond uncertain profits. Presently, agreements such as those negotiated by INBio are changing this inequity while generating direct funding for conservation, capacity building, and other benefits for the country (Reid 1994; see figure 2 for a discussion of capacity building).

2   It should be noted that at INBio, inventory sample collection is an activity separate from bioprospecting sample collection. Nevertheless, taxonomic and ecological data on which prospecting activities rely are shared from the inventory, while prospecting activities generate funding to support the inventory process.

3  As royalty payments resulting from pharmaceutical agreements are not guaranteed financial returns (Reid et al. 1993; Simpson, Sedjo, and Reid 1994), it is essential that prospecting collaborations
   negotiate elements that provide short-term contributions, such as funding, employment opportunities, information, and so forth (Reid et al. 1993).

4  For example, carrying out part of the sample research and processing procedures before shipment
   to local universities or private enterprise. See Lesser and Krattiger (1993b) for a more detailed discussion of value-added sample processing.

5  In addition to expanding scientific knowledge, INBIO's inventory has created economic benefits by
   providing employment for local communities, increasing local economic returns, and advancing
   intellectual capacities through the parataxonomist program.

6  Over the past four years, local Costa Rican universities have been the beneficiaries of close to
   $600,000 worth of laboratory equipment, scientific training, payment for services, and other forms
   of capacity building and technology access as a result of INBIO's efforts to help build national scientific and technological capacities, while also including national organizations and institutions in its
   scientific research collaborations.

7  For a detailed discussion of INBIO's informational service, see Sittenfeld and Gámez (1993).

8  INBIO ensures that partnerships include a guarantee of future potential profit sharing (typically
   through royalties based on net income) if commercial products are forthcoming.

9  The renewed agreement contains almost the identical provisions as the first collaboration in terms
   of the research budget, equipment and supplies, scientific training, and donations to conservation
   areas.

10 Two examples of publications resulting from the INBIO-Merck collaboration are: Bills, G.F., A. Ross-
   man, and J.D. Polishook. 1994. Rediscovery of *Albosynnema elegans* and *Sleimia costapora. Sydowia-*
   Band 46 (1):1–11; and Bills, G.F., and J.D. Polishook. 1994. Abundance and diversity of microfungi in
   leaf litter of a lowland rainforest in Costa Rica. *Mycologia* 86 (2):187–98.

## References

Aylward, B.A., J. Echeverría, L. Fendt, and E.B. Barbier. 1993a. The Economic Value of Pharmaceutical
   Prospecting and Its Role in Biodiversity Conservation. Discussion paper 93–05. London Environmental Economics Centre, London.

———. 1993b. The economic value of species information and its role in biodiversity conservation:
   Costa Rica's National Biodiversity Institute. Discussion paper 93–06, London Environmental
   Economics Centre, London.

Birch, A.N.E., W.M. Robertson, I.E. Groghegan, W.J. McGavin, T.J.W. Alphey, M.S. Phillips, L.E. Fellows, A.A. Watson, M.S.J. Simmonds, and E.A. Porter. 1993. DMDP: A plant-derived sugar analogue with systematic activity against plant parasitic nematodes. *Nematologica* 39:517–30.

Blixt, S. 1994. The role of genebanks in plant genetic resource conservation under the convention on bi-
   ological diversity. In *Widening Perspectives on Biodiversity*, edited by A.F. Krattiger, J.A. McNeely,
   W.H. Lesser, K.R. Miller, Y.ST. Hill, and R. Senanayake. Gland, Switzerland:World Conservation
   Union.

Caporale, L.H., and M.F. Dermody. 1994. *Drug Discovery and Biodiversity: Collaborations and Risk in the
   Discovery of New Pharmaceuticals.* Presented to the PAHO/IICA Symposium on Biodiversity,
   Biotechnology, and Sustainable Development at IICA headquarters, 11–14 April, San José, Costa
   Rica.

Costa Rica. 1992a. Ley de la Vida Silvestre. In *La Gaceta, Diario Oficial 253,* 7 December.

———. 1992b. Ley de conservación de la vida silvestre. In *La Gaceta, Diario Ofical 7317.*

————. 1993. Decretos 22545–MIRENEM, reglamento a ley de la vida silvestre. In *La Gaceta, Diario Oficial 195*, 13 October.

Eisner, T. 1989. Prospecting for nature's chemical riches. *Issues in Scientific Technology* (Winter):31–34.

Janzen, D.H. 1983. *Costa Rican Natural History*. Chicago: University of Chicago Press.

————. 1992. A North-South perspective on science in the management, use, and development of biodiversity. In *Conservation of Biodiversity for Sustainable Development*, edited by O.T. Sandlund, K. Hindart, and A.H.D. Brown. Oslo: Scandinavian University Press.

Janzen, D.H., L.E. Fellows, and P.G. Waterman. 1990. What protects *Lonchocarpus* (leguminosae) seeds in a Costa Rican dry forest? *Biotropica* 22 (3):272–85.

Joyce, C. 1991. Prospectors for tropical medicines. *New Scientist* 132:36–40.

Lesser, W.H., and A.F. Krattiger. 1993a. Facilitating South-North and South-South technology flow processes for "genetic technology." Working paper R7W, International Academy of the Environment, Biodiversity/Biotechnology Programme, September, Geneva, Switzerland.

————. 1993b. Negotiating terms for germ plasm collection. Working paper R8W, International Academy of the Environment, Biodiversity/Biotechnology Programme, September, Geneva, Switzerland.

MIRENEM (Costa Rican Ministerio de Recursos Naturales, Energía y Minas). 1992. *Biodiversity Country Study*.

NRC (National Research Council). 1992. *Conserving Biodiversity: A Research Agenda for Developmental Agencies*. Report of a panel of the U.S. Board of Science and Technology for International Development. Washington, D.C.: National Academy Press.

Raven, P., and E.O. Wilson. 1992. A fifty-year plan for biodiversity surveys. *Science* 258:1099–100.

Reid, W.V. 1994. Biodiversity prospecting: Strategies for sharing benefits. In *Biodiplomacy: Genetic Resources and International Relations*, edited by V. Sánchez and C. Juma. Nairobi: ACTS Press.

Reid, W.V., S.A. Laird, C.A. Meyer, R. Gámez, A. Sittenfeld, D.H. Janzen, M.A. Gollin, and C. Juma. 1993. *Biodiversity Prospecting: Using Genetic Resources for Sustainable Development*. Washington, D.C.: World Resources Institute.

Roberts, L. 1992. Chemical prospecting: Hope for vanishing ecosystems? *Science* 256:1142–43.

Sarukhan, J. 1992. The coordination of action on biodiversity in Mexico: A proposal of national priority [in Spanish]. In *Mexico Confronts the Challenges of Biodiversity*, edited by J. Sarukhan and R. Dirzo. Mexico: Comisión Nacional para el Conocimiento y Uso de la Biodiversidad.

Schweitzer, J., F.G. Handley, J. Edwards, W.F. Harris, M.R. Grever, S.A. Schepartz, G. Cragg, K. Snader, and A. Baht. 1991. Summary of the workshop on drug development, biological diversity, and economic growth. *Journal of the National Cancer Institute* 83:1294–98.

Simpson, R.D., R.A. Sedjo, and J.W. Reid. 1994. Valuing biodiversity: An application to genetic prospecting. Discussion paper 94–20, Resources for the Future, Washington, D.C.

Sittenfeld, A., and R. Villers. 1992. Building partnerships to save tropical biodiversity: The INBio-Merck agreement. *Network* 18:4.

————. 1993a. Biodiversity prospecting by INBio. In *Biodiversity Prospecting: Using Genetic Resources for Sustainable Development*, edited by W.V. Reid, S.A. Laird, C.A. Meyer, R. Gámez, A. Sittenfeld, D.H. Janzen, M.A. Gollin, and C. Juma. Washington, D.C.: World Resources Institute.

————. 1993b. Exploring and preserving biodiversity in the tropics: The Costa Rican case. *Current Opinion in Biotechnology* 4:280–85.

————. 1994. Costa Rica's INBio: Collaborative biodiversity research agreements with the pharmaceutical industry. In case study 2, *Principles of Conservation Biology*, edited by G.K. Meffe et al. Sunderland, Mass.: Sinauer Assoc.

Touche, R. 1991. *Conservation of Biological Diversity: The Role of Technology Transfer*. A report for the

United Nations Conference on the Environment and Development and the United Nations Environment Programme Intergovernmental Negotiating Committee for a Convention on Biological Diversity. London: Touche Ross and Co.

WRI (World Resources Institute), IUCN (World Conservation Union), and UNEP (United Nations Environmental Programme). 1992. *Global Biodiversity Strategy: Guidelines for Action to Save, Study, and Use Earth's Biotic Wealth Sustainably and Equitably*. United States: World Resources Institute, World Conservation Union, United Nations Environment Programme.

# V Moral Responses

# Biological Resources and Endangered Species: History, Values, and Policy

by Bryan G. Norton

This chapter examines the role of the U.S. Endangered Species Act (ESA) of 1973 in the broader context of protecting biological resources by considering its historical, valuational, and policy aspects. Three historical "phases" of policy thinking on the environment are identified: single-species protection (1800–1980), biodiversity (1980–1988), and sustainability of ecosystem health (1988–present). The ESA is representative of the thought behind the first phase and, in this sense, is anachronistic; but the protection of biological resources is supported by many important human values. So the question remains whether the ESA is the best approach to protecting biological resources, given that current thinking goes beyond individual species. It is argued that, given our state of knowledge and managerial abilities, the ESA remains the most viable policy tool for protecting biological resources, but it should be applied so as to improve managerial knowledge and encourage more eco-systematic approaches.

\*

The protection of biological diversity has become a central goal of environmental protectionism. On one level, this goal is noncontroversial, a motherhood-and-apple-pie issue—there are no advocates for species destruction or for accelerating species loss. Nevertheless, there remain important differences regarding how much we should do, what we should do, and even what is of ultimate value. Because the protection of biodiversity—no matter how strongly supported publicly —will certainly proceed within a context of uncertainty, this chapter advocates an adaptive management approach to protecting biological resources.

The argument for a more adaptive, reactive approach is well summarized in an analogy offered by William Ruckelshaus, who contends that the search for a sustainable society in the future ought not to be likened to a crossroads—an all-or-nothing choice between two alternatives—but rather, to a canoeist shooting the rapids. Survival and sustainability will depend more on a willingness and ability to react to new information than on a single and forever-binding choice (Ruckelshaus 1989). This adaptive approach emphasizes the importance of choosing policies that fulfill two criteria: chosen policies should, given the best available science at the time of their implementation, protect both species and the ecological processes associated with them; and chosen policies should be designed to increase our information base to support further policy actions (see Walters 1986; Lee 1993). Philosophically, these two policy criteria are supported as minimal requirements for fairness to the future, although it will also be noted that there remain considerable disagreements regarding how to formulate such foundational values in environmental ethics.

This chapter attempts to put these disagreements into perspective, given the history, competing values, and immediately pressing policy issues faced by conservation biologists and environmental managers. Part 1 explains how the conceptualization of the problem of protecting biological resources in the United States has evolved in three stages since early efforts to protect particular game species from overhunting and overfishing. Against this historical backdrop, part 2 discusses whether the ESA, as currently written and interpreted, represents a reasonable response to threats to biological resources in the United States. Finally, part 3 will speculate on the extent to which lessons learned in the United States can provide guidance in addressing the more complex problems of protecting biodiversity worldwide.

## Three Phases in the Protection of Biological Resources

Phase 1: Single-Species Protection (1800–1980)
As early as the eighteenth century, local governments in New England began protecting particular food, fur, and game species against the depredations of overharvesting (Cronon 1983). This species-by-species approach led to legislation at the local, state, and federal levels to protect biological resources, and eventually, to protect habitats for protected species (Fox 1981). These early attempts to regulate hunting and fishing, and later attempts to protect habitats in the main flyways for migratory waterfowl, remained "atomistic" in the sense that particular species were singled out as valuable and protection was conceived in terms of separate management plans for each protected species. This approach to protection reached its culmination in 1973 with the ESA, which provides for a process of identification, listing, and recovery of species, both game and nongame, that are endangered or threatened. While the ESA retains the atomistic conception of species, it nevertheless represents a landmark in the development of biological resource policy. The act protects, at least to some degree, *all* species (rather than just a few, valued game species), and it also pays attention to habitats, subspecies, and sometimes even populations. It is, then, important for its comprehensiveness.

Two intellectual-scientific aspects of this approach require comment. First, from a philosophical viewpoint, single-species management emphasizes protecting the elements of nature, and is therefore in keeping with long philosophical and historical traditions in Western civilization. It is often noted that Western philosophical traditions have stressed 'being'—what exists now and how to understand it—over 'becoming'—the processes of change and development that bring about and support those entities that exist at any particular time—(see Prigogene and Stengers 1984). Historically, this represented the triumph of the concerns of Plato and Aristotle for 'substance' over the ideas of Heraclitus, who, in 500 B.C., declared that "All is in flux." This triumph was embodied in modern science, which emphasized explanation at all levels of nature that can be "reduced" to the motion of elementary particles. This concept of reduction to elementary particles as the essential form of explanation is ultimately inseparable from the mechanistic view of nature that has held sway throughout the modern period of Western civilization (Prigogene and Stengers 1984). For this reason, the "atomistic" idea underlying a focus on elements is so deeply ingrained in Western thinking that it was difficult even to imagine alternative conceptualizations of nature until relatively recently.

With the publication of Charles Darwin's evolutionary theory, however, the importance of systemic change and irreversible developments of complex, dynamic processes has reasserted itself (Dewey 1910). At first, this revolution was limited to the biological sciences and, even there, the full implications have still not been worked out. Now, the revolution has extended to physics, and physicists are leaders in an interdisciplinary effort to develop a more dynamic worldview. But again, the full implications of a dynamic world, one that creates diversity and complexity, apparently at the edge of chaos, are only just now being felt. It may be decades before these concepts are well understood, but creative work in nonequilibrium systems dynamics is already leading to new insights in ecology (Pimm 1991; Lewin 1992), and this direction holds promise for applications to environmental policy (Norton 1992). This much we know for sure: the full absorption of new thoughts on evolving systems into environmental management will have far-reaching impacts on the policies we advocate, and will almost certainly require more attention to interspecific relationships and system-level characteristics.

Second, the single-species approach to environmental protection is based in *autecology*: the study of individual species within their habitats. Autecology has been a powerful force in applied biology, since it was dominant in forestry and applied entomology. It represents a research program attempting to understand populations within their environments; this approach often sought to understand populations and their characteristics in terms of the physiology of its members. In contrast, *synecology* concentrates on the relationships between, and systems of, species. Because of the dominant place of autecology in management agencies and practical sciences, concern for the protection of biological resources naturally turned into a concern for species and, by extension, other taxonomic groups (McIntosh 1985). Synecology has been stronger in academic ecology, and today, has given rise to "systems" ecology. Its continued study has, in turn, affected environmental management by encouraging more holistic and ecosystematic managerial experiments, giving rise to the current tension with species-oriented approaches.

Despite the growing popularity of ecosystem management, the single-species approach has several advantages: species are relatively easy to identify, they have a basis in biological fact, and they can be counted. Emphasis on species, therefore, provides fairly clear-cut management goals and at least the hope of achieving measurable success. Critics of autecology and single-species management, however, have mounted strong scientific and managerial arguments against this approach. Scientifically, autecology tells only one side of the story—certainly, the survival of species partly depends on the characteristics of individuals, but it also depends on a complex and changing set of relationships called their *habitat*. Managerially, emphasis on particular species in the short run can set in motion changes at the system level, resulting in strongly counterproductive management in the middle and long run.

This point was first made by Aldo Leopold in his trenchant criticisms of forest service management in the southwestern territories of the United States. In his brilliant metaphor of "thinking like a mountain," Leopold regretted his successful efforts to exterminate wolves in order to increase deer populations in the wilderness areas of the Southwest, and argued for managing nature on multiple scales, at the ecological system scale of time and space (the mountain) as well as at the economic, short-term scale (Leopold 1949). In short, Leopold contended that attempts to manage biological resources from the perspective of autecology would fail, especially in fragile ecosystems, because autecology does not pay enough attention to relationships among species at the community level, and second, because it falsely assumes that nature maintains a static balance. Leopold, therefore, criticized traditional, single-species management as being too atomistic and not sufficiently dynamic in its understanding of nature (Norton 1991). Leopold's work has been corroborated and developed by C.S. Holling and his associates, who have shown that large-scale systems that are managed for regularized production of one or a few resources become more "brittle," susceptible to collapse or gradual degradation to a new, less-desirable stable state (Holling 1992; Norton 1991). Based on these broad intellectual and scientific trends, our strategies to protect biological diversity must pay more attention to system-level characteristics and the dynamic processes that they represent.

Phase 2: Biodiversity (1980–1988)
By the early 1980s, it had become clear to policymakers that species diversity was only one aspect of life's variety. Species, genetic variation within species, and populations of species that are adapting to varied habitats were all recognized as components of *biological diversity*. This broader approach culminated in the Smithsonian Symposium on Biodiversity, which was followed by an important book and

a traveling exhibition (Wilson 1988). The biodiversity phase represented a distinct advance in conceptualization because the introduction of multiple layers of diversity, and the emphasis on varied dynamics and habitats as well as species, significantly expanded our understanding of how complicated the process of protecting biological resources really is.

But this phase was conceptually unstable because the third element of diversity —the multiple, dynamic processes within which different populations adapt and evolve—begins to shift attention away from the elements of nature, toward the processes that create and sustain those elements through time. Practically, this development is appealing; it promises to turn managerial attention away from a few desperately small populations of endangered species toward safeguarding processes on the theory that, if ecological processes remain intact, populations and species can take care of themselves. It is surely easier to keep species off the endangered list than to save them once they are in trouble. But the practical promise of a more dynamic approach has not led to decisive action or even very specific management proposals.

The problem is that the brilliant theoretical insights of Leopold have proven frightfully difficult to operationalize (Norton 1991). To manage an ecological system, one must know what processes are essential, and which elements depend on each other. Unfortunately, dependency relations represent one of the most difficult and least understood areas of ecological study. Worse, the dynamic, nonatomistic approach is very information intensive. Unlike physics, which has achieved remarkable generality, biology and ecology are studies of particularity (Ehrenfeld 1993). The information demands necessary to model impacts on ecological systems are not only heavy, but the relevancy of information is highly dependent on local conditions. Most commentators hold out little hope that there will be a grand breakthrough in ecology, such as the discovery of some simple and general laws that will make understanding ecosystems and their complicated interactions causally transparent (McIntosh 1985; Sagoff 1988). Consequently, while ecosystem management will require a great deal of detailed and locally relative information about species and their interrelationships, ecosystem ecology offers neither quick nor easy models. Ecology applies to ecosystem management, but the applications must be at the local and particular level. The strong trend toward locally motivated attempts to articulate "ecosystem management plans" for areas and regions may represent also a trend toward local responsibility in understanding and managing ecological systems.

Arguably, the ESA, even as glossed with the multilayered emphasis of the biodiversity concept, remains element oriented. The act attempts to identify, list, and recover species, subspecies, and so on. Its friendly critics may justifiably maintain that the act is not sufficiently protective of processes, and that we need an "Endangered Ecosystems Act" or an "Endangered Processes Act." This line of reasoning has spawned the current phase of environmental policy debate regarding how to protect biological resources.

Phase 3: Sustainability of Ecosystem Health (1988– present)
The biodiversity phase—given that it steps on an apparently slippery slope leading toward more emphasis on processes and less on elements—may therefore prove to be only a transitional phase, as policy and management will endeavor to protect communities of species by formulating strategies for protecting essential processes. Species remain important in this phase, but with the recognition that species in varied habitats anticipate quite diverse ecological and evolutionary trajectories. The importance of these trajectories apparently implies that a system representing a significant habitat is an important management unit. This approach, which remains highly speculative, argues that policies to protect biological diversity must monitor and protect larger ecological units, such as ecological systems. This management program requires introducing descriptive/normative concepts, such as *ecosystem health* and *ecosystem integrity*, which apply at the ecosystem level and that emphasize processes rather than elements.

It is not yet clear whether this new emphasis will require efforts in addition to or instead of the ESA and the efforts it currently mandates. Indeed, it can be argued, as we shall see below, that doing a better job of achieving the goals and objectives of the current act may be the best we can do for the foreseeable future. There is continued support for protecting species; there is growing support for ecosystem management. The question remains: Can the key concepts of dynamic, ecosystem-level management—ecosystem health and integrity—be defined with sufficient clarity to guide biodiversity policy?

## Why We Need the (Anachronistic) Endangered Species Act

We are embarking on a new adventure in environmental management by integrating traditional element- and species-oriented management into a multitemporal, multiscalar, ecosystematic approach. While the ideas that spark the adventure remain somewhat speculative, the goal must be to create a new paradigm of environmental management. Given the lack of reliable data and scientific models, the only reasonable path to pursue is *experimental* or *adaptive* management (Lee 1993), focused at local levels (Norton and Hannon forthcoming). Adaptive management is designed to function in an uncertain world; it devises managerial and other plans to make them learning experiences. We must make management proposals, try them out in carefully controlled situations, and then design pilot projects to determine the results of various methods.

There are two issues that are important to address in the uncertain context described above. First, we should examine the human values that drive the search for a better strategy for protecting species. Then we should discuss the practical, policy situation in the uncertain context that currently exists. At each reauthorization of the ESA there are attempts to rewrite the act. How should unflattering assessments of the act as a vestige of earlier thinking on the protection of biological resources affect concrete questions such as whether, and in what form, the ESA should be continued? What is the value of endangered species and biodiversity, and what kind of protection is appropriate given these values? Any discussion of the value of biological diversity should start with the recognition of the breadth of consensus favoring the protection of biological resources among serious scholars from every relevant discipline, and embracing virtually the entire intellectual landscape. If the present generation fails to protect biological resources into the middle and distant future, we will have committed a serious wrong.

The solidity of this consensus is sometimes obscured because the various academic disciplines and policy players often disagree on how to characterize that wrong, as well as how to weigh it against other obligations in comprehensive environmental decisionmaking. There are at least four quite distinct analyses of the values involved:

1. *The Economic/Utilitarian Approach.* According to what may be the dominant view, the values involved in protecting biodiversity are fully represented in an accounting of the welfare of humans in the present and the future. Protection of species, genetic, and habitat diversity is unquestionably important as a source of future resources. Protection of biological diversity can be justified because of the many ways in which species and ecosystems provide services that we would otherwise have to supply. It is important to realize that human values derived from nature are broader than economic values; many people value nature for its positive impact on the human spirit, and studies show a significant willingness on the part of consumers to pay for the protection of species and ecosystems, simply because they want to know they exist (Norton 1987).

2. *The Intergenerational Equity/Stewardship/Sustainability Approach.* The economic approach is often supplemented by an emphasis on intergenerational equity and an insistence that each generation should act sustainably. What exactly this emphasis adds to a longsighted application of economic or utilitarian reasoning is a matter of dispute, however. At issue is whether the obligation to the future is specific—there are particular resources that are essential and these should be protected—or general—we only owe the future a *just savings rate*, a rate of investment sufficient to ensure that future generations will have the opportunity to be as well off as we are. Economists and preference-based utilitarians tend to favor an *unstructured bequest package*, which aggregates all forms of capital—natural and human-made—into a single accounting system. They sum individual welfare within a generation and

then compare it across generations. The future has no reason to blame us, so goes this line of reasoning, if we leave a world capable of maintaining a nondeclining stock of undifferentiated capital. Under this view, sustainability becomes a mere afterthought to economic growth theory (see Solow 1974, 1993).

Most advocates of sustainability argue that present actions are constrained not only by an obligation to maintain a fair savings rate, but also by more specific obligations to protect essential resources, such as tropical forests and the oceans' fisheries. These advocates of a more structured bequest package as a guide to intergenerational fairness—often called *ecological economists*—contend that we should keep separate accounts for natural and human-made capital, and that we have obligations to protect natural capital in an intact state. Intactness of ecosystems is generally referred to as ecosystem health or ecosystem integrity, so the ecological economics movement complements the call for a stronger and more specific criterion of intergenerational equity, usually associated with "strong sustainability" (Daly and Cobb 1989). This approach has as its defining intellectual task the problem of determining what resources are essential elements of natural capital.

3. *The Biocentric Approach.* According to another line of moral reasoning, all living organisms are of equal value intrinsically, and humans are obliged to share resources equally with all other species (Taylor 1986). This approach, which is individualistic, focuses on the organism level; it values ecosystems and species, but only because they are made up of individuals. Denying human moral superiority, this approach argues that all living individuals, at least individuals who are morally considerable, have equal rights to the world's resources. This approach cannot, then, be applied to policy unless there are clear rules for defining which members of which species may exploit others in the struggle to survive. Various attempts to specify such rules have failed to gain broad consent. Until these problems are resolved, this approach is of little help in policy matters.

4. *The Ecocentric Approach.* Ecocentrists believe that ecological communities have inherent value and should be protected for their own sake. This approach to environmental valuation avoids many of the problems of biocentrism. It has a general answer to the question of which species are to be given priority: a species is valuable insofar as it contributes to the well-being of the ecosystem of which it is a part. This approach has been criticized for its "fascist" tendencies, in that it apparently sacrifices the interests of individuals to those of larger systems; it also appears, given the damage modern, technologically powerful humans now inflict upon ecosystems, to imply misanthropism. But most precise formulations of the principle are careful to limit the dominance of systems over individuals and to provide some special protections for human individuals (see, for example, Callicott 1989).

These various formulations are referenced not so much to emphasize the intellectual differences among them, although they are considerable, but rather to acknowledge and then consider them in policy contexts. While there is much discussion of nonanthropocentric approaches to environmental values (including three and four, above), what is often not recognized is the extent to which policy prescriptions converge between ecocentric approaches and the less-controversial commitment to protect biological resources for future generations (Norton 1991). If one adopts formulation three or four, in addition to one and/or two, most of the goals of the nonanthropocentric approaches will be at least approximated by efforts to fairly discharge duties implied by requirements of intertemporal fairness. For example, whatever content is given biocentrism, it apparently requires protection of habitat for living organisms in the future. But that is also what is required by the obligation to protect ecological processes for the benefit of future generations. One could not, that is, fulfill the goals of biocentrists without, in the process, fulfilling obligations to maintain essential ecosystem processes (Norton 1991). While there may be some differences regarding how far policies must go to protect biological systems and species, advocates of all four approaches, with few exceptions (such as Kahn 1982; Simon 1981; Simon and Kahn 1984), believe that we should be doing much more than we are currently.

The important point at a public policy juncture, such as the periodic reauthorization of the ESA, is that we agree on what is valued and what our goals should be. In this context, it is not significant that various scholars and activists use different languages to characterize the values involved in protecting biological diversity—the important thing is that there exists a strong consensus of high value and urgent priority for the protection of biological resources, and that this policy consensus is based on the broad agreement of experts plus considerable public opinion. In order to focus more sharply on the policy questions, three related but crucially different questions arise:

1. Do we need an *Endangered* "Biological Resources" Act?
2. Do we need an Endangered *Species* Act?
3. What kind of protection should be mandated in an Endangered Biological Resources Act?

The answer to question one is implicit in the analysis just completed: We do need legislation to protect biological resources. This conclusion follows from the wide intellectual and scientific consensus that we have obligations to the future. Confusion regarding how to characterize values derived from ecological systems does not undermine the overwhelming consensus regarding the need/value of protecting biological resources at some level and scale.

Question two, however, is much more difficult to answer, especially if one seeks a *scientific* response, because the ultimate importance of species in the overall picture is a matter of scientific disagreement and uncertainty. For example, it could be argued that, because of the significance of cross-population genetic variation, we need an "Endangered Populations Act"; or it could be argued, on the grounds that processes are really more important than species, that we need an "Endangered Ecosystem Processes Act." These difficult problems are really problems of scale. Once we have agreed that we need to save biological resources, there is still the question about the scale at which "protection" should be directed. These scalar problems have both a scientific and policy aspect (Norton and Ulanowicz 1992). Scientifically, it is certainly important to know how species extinctions and extirpations on large scales affect ecological processes; but there may be no general solution to this question. It appears that some extinctions are extremely significant ecologically, while others have very little ecosystemic impact (because their populations are rare, for example).

We cannot, however, afford the luxury of waiting for a full resolution of ongoing scientific differences. When the ESA comes up for periodic reauthorization, for instance, we face a *policy* question in the sense that a decision is forced, and must dominate the desire for further information regarding this decision. From a policy perspective, question two is actually easier to answer, once one accepts that the decision will be made with less-than-ideal amounts of scientific information. For example, the suggestion that we have an Endangered Populations Act can be ruled out because, while we might sometimes protect a population, it is too expensive to protect every population. Worse, saving every population would require "deep-freezing" nature, halting its constant dynamism.

Choosing between the ESA and an Endangered Ecosystem Processes Act is more difficult, however. As noted above, environmental management is entering a period of great flux, as the concept of "whole ecosystem management" is touted more and more in environmental protection and planning. The shift of focus to ecosystems involves the introduction of system-level characteristics—characteristics emergent on the ecosystem level—as measures of how well we are doing at protecting biological resources. We have noted that popular candidates include protecting the "health" and/or the "integrity" of ecological systems. One might ask: should we perhaps give up the attempt to save all species, and put more emphasis on ecosystems and their protection? While this ecosystems-sensitive approach has broad appeal, it might only be because nobody knows for sure what, exactly and operationally, would be necessary to protect the health or integrity of an ecological system. Yet there are interesting attempts in this direction (Costanza, Norton, and Haskell 1992). Advocates of economic development, for example, have shown interest in ecosystem management as a way of providing more "flexibility" to our efforts to protect nature. Protectors of nature, on the other hand, are beginning to realize that it is often better to concentrate on protecting ecosystems—which can harbor many species, including those that are rare and endangered—rather than making last-ditch efforts only after a species reaches the brink of extinction. There is no doubt, however, that any concrete measures to legislatively mandate the protection of ecosystem health and integrity, or descriptively/normatively characterize the changing state of ecosystems, would either be vague or controversial. While both Manuel Luhan, former Secretary of the Interior, and the deep ecologists have expressed an interest in more ecosystem management, it does not follow that they would agree on specific guidelines and actions to protect ecosystem-level characteristics.

For the present, there is simply not a detailed scientific-policy consensus regarding how to "define" ecosystem health and integrity, and it would be folly to dismantle the comprehensive endangered species protection system until there is a scientifically acceptable and politically viable definition of these ecosystem-level characteristics. The policy should be to protect species even as we continue to search for a more precise measure of exactly what we need to emphasize in the effort to protect biological resources over the long term.

This conclusion—that saving species may eventually play a less-central role in biodiversity policy and that we adopt the species-protecting policy as a temporary expedient—is not nearly as damaging an admission as it might seem. While the health/integrity movement emphasizes ecological processes, its advocates never question that species are necessary agents of those processes. An ecosystem that is rapidly losing species is most likely unhealthy; if steps are undertaken to reduce species losses, they will also contribute toward the goal of restoring ecological health to the system.

Indeed, the congruence of ecosystem and species-level objectives has a scientific basis. For example, some species are considered to be "keystone," since their loss would threaten other species and begin a cascading effect. Second, the plummeting population of one species can serve as an "indicator" for a whole ecosystem, as has been argued regarding the spotted owl and marbled mirulet in the Northwest. In this case, again, good ecosystem management will be identical to good species-level management. Third, virtually every definition of ecosystem level, descriptive normative characteristics includes as an important condition that a "healthy" system, one maintaining its integrity through time, must be capable of sustaining biological diversity.

Most important, the ESA functions as an extremely useful "working hypothesis" within an adaptive management strategy to protect biological diversity. As currently amended, the ESA's goal is to protect all species; however, it contains an "exceptions" clause, whereby a high-level governmental committee—the "God Committee"—can judge that steps to save a given species are too costly if they conflict with overriding regional and national goals. Given this content, the act is well suited to encourage many experiments in species protection, while also allowing an escape hatch if the social costs of saving some species prove to be too great. The system has not worked nearly as well as it should (Tobin 1990), either in gaining information or at protecting species. Nevertheless, it has the potential to encourage a steep learning curve regarding how to protect elements, when it is important to protect elements, and also regarding how processes and elements interact in many specific situations. This is the heuristic value of having an act that focuses attention at a policy level that is manageable. Given current information, we usually have at least some idea of actions that can improve the survival chances of a species. While it is manageable, however, it does stretch current information-gathering techniques and capabilities, creating a scientific and managerial environment that encourages relevant learning.

Despite some scientific uncertainty and the disagreements about direction, there is no question that we should continue to protect species for the foreseeable future, even as we try to become more sensitive to ecosystem-level management. This adaptive approach to management would be furthered if the ESA were made more flexible—providing more latitude to introduce "experimental populations"—or if it encouraged protecting "suites" of species that tend to survive or fail together: for example, HB 2043, introduced by Congressperson Gerry Studds in 1994, provided a listing category of species that would make it less likely that other species "dependent on the same ecosystem" would require listing. This modest improvement would encourage more attention to and research about species interrelationships.

Ecological science—both established principles and accepted uncertainties—supports this sort of incremental experimentation with new ecosystem-level management. We know what does not work—it does not work to allow development, at great cost to biological systems, and it does not work to try to freeze ecological systems to protect every population in a community. Unfortunately, we do not know what does work, and so it is often best to protect species. As Leopold (1949) says, "The first rule of intelligent tinkering is to save all the pieces." Once we admit we're tinkering, the policy choice should clearly be to avoid irreversible losses, if at all possible. For this reason, it may be concluded—at least for now—that there is only one reasonable policy choice: to protect species while continuing to explore ways to be more sensitive to, and even regulate to protect, ecosystem-level processes and characteristics.

The reasoning behind this chapter strongly conflicts with the proposal to amend the ESA to include a "no-surprise" clause, a proposal put forth by critics of the current law. A no-surprise clause would allow negotiated settlements between landowners/developers, on one hand, and regulatory agencies on the other. The settlements would be binding for a long period of time, such as 50 to 100 years. Once in place, the owner/developer would be immune from further regulatory burdens or liabilities for the period of the agreement. Critics of the current law do validly argue that some private landowners in some situations may be required to accept an unjust burden to protect biological diversity. It would, however, be better to address this problem with a system of compensation for the real losses of harmed landowners than to lock regulatory agencies into inflexible protection and recovery plans. The increasing emphasis on dynamic processes in management, as well as the growing importance of flexibility and adaptive management in the face of uncertainty, imply that a no-surprise clause would be a big step in exactly the wrong direction—a step toward inflexible management that is unable to respond and adjust in the face of new information.

We can now turn to question three, formulating it more precisely, in keeping with the foregoing argument: What kind of protection should be mandated in the ESA? Some environmentalists, such as David Ehrenfeld (1978), and deep ecologists have advocated a very strong and uncompromising line—that every species has a right to exist, and that we are bound to do anything necessary to prevent every possible extinction. Others would argue that we cannot save every species, but that we should try to save as many as possible, recognizing that species protection is not the only worthy environmental and social goal. Still others would consider the costs and benefits of protecting each species—calculating each case independently and, in essence, sorting through species, then saving only those that can pull their own economic weight.

It is difficult to defend either of the extremes. The hard line, that we should never give up on a species, is difficult to defend as a universal rule because there will, in some cases, be other competing environmental goods that override species protection. For example, it was reported in *Science* that an impasse was reached between efforts to restore water regimes to the Everglades ecosystem and attempts to save the Everglades kite, an endangered species (Alper 1992). Whether or not this case is an authentic clash of policies to protect ecosystems versus species, it does suggest that there may arise serious conflicts between those who emphasize species and those who emphasize ecological systems as the units of management. It is also possible to imagine rare instances in which some competing social good would override our obligation to protect a species. These uncertainties, far from undermining a commitment to adaptive management, support that approach precisely because it will bring these issues to light, encouraging the targeted study of particular systems and the feasibility of various conservation goals.

So any middle-ground position regarding the ESA moves significantly toward some balancing of the advantages and disadvantages of protection versus development in particular cases. But even the advocates of economic/utilitarian analysis have strongly advocated a presumption in favor of protection, because they recognize how difficult it is to identify and place dollar values on the benefits of species protection. Therefore, they argue, if the cost-benefit analysis is even close, policy should favor protection because of the greater difficulty in specifying the benefits of protection over those of development (Fisher 1981). The middle ground would seem to be the following position: there is a presumption of species protection, provided that the cost of protection is bearable; species will normally be saved, but interest groups have a right to challenge this presumption if important interests are at stake. This approach, which represents roughly the status quo in the ESA as amended to include the God Committee, can be theoretically formulated as an application of the "Safe Minimum Standard of Conservation," (SMS) due to Ciriacy-Wantrup (1952) and elaborated by Bishop (1978; also see Norton 1987, 1991). In situations where a resource may be irreversibly lost, always save the resource if the cost is bearable.

The matter of bearable costs, of course, is highly negotiable; one advantage— some may consider it a disadvantage— of this middle-of-the-road approach is that it makes justifiable costs a matter of degree. This means that most cases must be considered on their individual merits and that the question of protection is usually a matter for political compromise. Perhaps the scientific recognition that some species are much more valuable "ecologically" than others may enter into calculations regarding which species should be saved. Some species may deserve high priority because of their ability to support complexes of species; others may have little system-level significance. It may, therefore, make ecological sense to give up on some species and concentrate on those that are truly important for human or ecological reasons. But it is important that we not confuse ecological uncertainty with economic interest. Prior to amendments added in 1984, the ESA made a declaration of critical habitat for a listed species (which was required to complete the listing process) dependent on an analysis of the economic impacts of the listing. This requirement was misused early in the Reagan administration to virtually halt the listing process, requiring an amendment to separate economic from scientific considerations in determining whether a species should be listed. There have been several attempts to reestablish this linkage, but experience shows that this would be a mistake.

## Applying the U.S. Experience to Global Biodiversity

While this chapter has mainly focused on policies to protect biological resources in the United States, it may nevertheless be useful to briefly discuss the implications of the evolution of U.S. policy for attempts to protect species worldwide. Indeed, the lesson of adaptive management is generalizable to most situations. In many cases, efforts to protect biological resources are initially undertaken in a context of relative scientific ignorance, where little is also known about the social values associated with either the protection or continued degradation of biological systems. Policies should be instituted not only to deal with immediate, consensually accepted problems (since it is possible to muster social resources to address them), but also to increase the flow of information between scientists, policymakers, and the public. This lesson surely applies worldwide.

It may be futile, however, to attempt to export the species-by-species approach embodied in the U.S. ESA to many developing countries, especially tropical countries with rapidly increasing human populations, high degrees of biological diversity and endemism, and rapid deforestation rates. Under these conditions, so few species have been studied, and so many are threatened by rapid development, that policies directed at species would direct scientific study and management efforts at an unrealistically small scale. It makes more sense to focus immediate attention on "biodiversity hotspots" (Forey, Humphries, and Vane-Wright 1994) in order to protect areas with demonstrably high diversity and high rates of endemism. While this strategy implies a shift to a larger scale in the system, it is still possible to design policy responses to deal with immediate and recognizable problems *and* to design those policies to increase our knowledge base. This recognition may lead, eventually, to complementary scientific approaches—we maximize our understanding of systems in general by studying many of them, and by studying them at multiple levels of organization, encouraging the supplementation of autecology with a more system-level approach.

It is important to be realistic about what developing countries can afford to undertake on their own. Many industrial nations have created their wealth by destroying crucial elements of their own resource base. This wealth can then be used—provided there is public support and a political will to do so—to protect some areas as pristine representatives of historical systems that have evolved with few impacts from human activities. In countries where the development process has begun more recently, and especially where development has been retarded by international exploitation and colonialism, there may be little public support for the goal of "total" protection of areas in their pristine state. If environmentalists from the developed world hope to export this goal, they should recognize that they will be accepting responsibility to provide financial and scientific support as well. Assuming there will be limited funds to support such international efforts, it follows that most local and national strategies to protect biological resources will emphasize wise and sustainable use of resources, rather than protection of resources from any human use. Sustainable development may not lead in these cases to large, pristine reserves in developing nations; but it is becoming more and more widely recognized that some uses, such as sustainable eco-tourism, can protect natural areas from the utter destruction of clear-cutting and other highly destructive practices.

Other authors in this volume discuss various approaches to sustainable development, and a detailed discussion would be beyond the scope of this chapter, but it is important to ackowledge the importance of local context in developing management plans to protect diversity. The local context includes features of ecological communities—such as diversity and endemism—but it also must take into account local social and cultural conditions (see Norton 1994).

There is another apparent implication of the trends discussed above for international efforts: there will have to be experimentation with many new international institutions and perhaps the development of new concepts of property in biological resources, since public resources are often lacking in developing nations (see Vogel 1994; Stone, this volume). Again, the applicable lesson is local sensitivity to ecological and social conditions. One of the most important problems in developing an effective international policy is to create incentives to protect biological diversity on the part of private corporations. But here the international problem clearly outstrips the lessons that can be learned by examining the fledgling attempts of the United States to protect endangered species and other aspects of biological resources. These pressing issues must, then, be left to other authors in this collection, as well as those who are actively engaged in global biodiversity protection.

## Conclusion

The protection of biological resources/diversity must proceed amidst considerable uncertainty. It has been argued that—despite uncertainties regarding the comparative role of elements as opposed to processes, confusion over which evaluative conceptualization to use in characterizing these resources, and significant recent changes in these conceptualizations—there remains considerable consensus in favor of protecting biological resources. It is, therefore, possible to chart a reasonable policy course by preserving and improving current practices as mandated by the ESA, even while striving to develop more comprehensive methods for protecting ecological processes as well as elements. First, it is important to maintain the ESA in something like its present form; second, it is wise to experiment with more ecosystem-related management initiatives, testing the new ideas of ecosystem management. These actions rest on the overwhelming support for protecting biological resources from many philosophical and disciplinary perspectives. While it may be necessary to modify and extend current efforts to protect species, this should be accomplished through incremental change, experimentation, and pilot projects, while maintaining the protection currently mandated by the ESA. Given current knowledge, it is almost always better to protect species because this goal contributes strongly, for a variety of reasons, to both the traditional and more speculative goals of ecological management. Whether or not the experiences in the United States—at state and federal government levels—can provide a useful template for developing a response to the biodiversity crisis, one thing is certain: as long as there is uncertainty in our knowledge base for managing biodiversity, a high premium should be placed on an experimental approach that is both adaptive to local conditions and designed to increase knowledge even as it deals with pressing problems of biodiversity loss.

# References

Alper, J. 1992. Everglades rebound from Andrew. *Science* 257:1852–54.

Bishop, R.D. 1978. Endangered species and uncertainty: The economics of the safe minimum standard. *American Journal of Agricultural Economics* 60 (1):10–18.

Callicott, J.B. 1989. *In Defense of the Land Ethic.* Albany: State University of New York.

Ciriacy-Wantrup, S.V. 1952. *Resource Conservation.* Berkeley: University of California at Berkeley.

Costanza, R., B. Norton, and B. Haskell. 1992. *Ecosystem Health: New Goals for Environmental Management.* Covelo, Calif.: Island Press.

Cronon, W. 1983. *Changes in the Land: Indians, Colonists, and the Ecology of New England.* New York: Hill and Wang.

Daly, H., and J. Cobb. 1989. *For the Common Good.* Boston: Beacon Press.

Dewey, J. 1910. The influence of Darwinism on philosophy. In *The Influence of Darwin on Philosophy and Other Essays.* New York: Henry Holt.

Ehrenfeld, D. 1978. *The Arrogance of Humanism.* New York: Oxford University Press.

———. 1993. *Beginning Again: People and Nature in the New Millennium.* New York: Oxford University Press.

Fisher, A.C. 1981. *Economic analysis and the extinction of species.* report no. ERG–WP–81–4. Berkeley: Energy and Resources Group, University of California at Berkeley.

Forey, P.L., C.J. Humphries, and R.I. Vane-Wright. 1994. *Systematics and Conservation Evaluation.* Oxford, U.K.: Clarendon Press.

Fox, S. 1981. *John Muir and His Legacy: The American Conservation Movement.* Boston: Little, Brown.

Holling, C.S. 1992. Cross-scale morphology, geometry, and dynamics of ecosystems. *Ecological Monographs* 62 (4): 447–502.

Kahn, H. 1982. *The Coming Boom: Economic, Political, and Social.* New York: Simon and Schuster.

Lee, K.N. 1993. *Compass and Gyroscope: Integrating Science and Politics for the Environment.* Covelo, Calif.: Island Press.

Leopold, A. 1949. *A Sand County Almanac and Sketches Here and There.* London: Oxford University Press.

Lewin, R. 1992. *Complexity: Life at the Edge of Chaos.* New York: Macmillan.

McIntosh, R.P. 1985. *The Background of Ecology: Concept and Theory.* Cambridge: Cambridge University Press.

Norton, B.G. 1987. *Why Preserve Natural Variety?* Princeton, N.J.: Princeton University Press.

———. 1991. *Toward Unity among Environmentalists.* New York: Oxford University Press.

———. 1992. A new paradigm for environmental management. In *Ecosystem Health: New Goals for Environmental Management,* edited by R. Costanza, B. Norton, and B. Haskell. Covelo, Calif.: Island Press.

———. 1994. On what we should save: The role of culture in determining conservation targets. In *Systematics and Conservation Evaluation,* edited by P.L. Forey, C.J. Humphries, and R.I. Vane-Wright. Oxford: Clarendon.

Norton, B.G., and B. Hannon. 1997. Environmental values: A place-based theory. *Environmental Ethics* 19:227–45.

Norton, B.G., and R.E. Ulanowicz. 1992. Scale and biodiversity policy: A hierarchical approach. *Ambio* 21:244–49.

Pimm, S.L. 1991. *The Balance of Nature? Ecological Issues in the Conservation of Species and Communities.* Chicago: University of Chicago Press.

Prigogene, I., and I. Stengers. 1984. *Order Out of Chaos: Man's New Dialogue with Nature.* Toronto: Bantam Books.

Ruckelshaus, W.D. 1989. Toward a sustainable world. *Scientific American* (September).

Sagoff, M. 1988. Ethics, ecology, and the environment: Integrating science and law. *Tennessee Law Review* 56:77–229.

Simon, J. 1981. *The Ultimate Resource*. Princeton, N.J.: Princeton University Press.

Simon, J.L., and H. Kahn, eds. 1984. *The Resourceful Earth: A Response to Global 2000*. Oxford: Basil Blackwell.

Solow, R.M. 1974. The economics of resources or the resources of economics. *American Economic Review Proceedings* 64:1–14.

———. 1993. Sustainability: An economist's perspective. In *Economics of the Environment: Selected Readings*, edited by R. Dorfman and N. Dorfman. New York: W.W. Norton.

Taylor, P.W. 1986. *Respect for Nature*. Princeton, N.J.: Princeton University Press.

Tobin, R.J. 1990. *The Expendable Future: U.S. Politics and the Protection of Biological Diversity*. Durham, N.C.: Duke University Press.

Vogel, J. 1994. *Genes for Sale*. New York: Oxford University Press.

Walters, C.J. 1986. *Adaptive Management of Renewable Resources*. New York: Macmillan.

Wilson, E.O. 1988. *Biodiversity*. Washington, D.C.: National Academy Press.

# On the Uses of Biodiversity

by Mark Sagoff

This chapter maintains that ethical, aesthetic, cultural, and religious arguments justify efforts to prevent species extinction. At the same time, it urges that instrumental or economic arguments for preserving biodiversity, such as those presented in Agenda 21, be used with caution because, if overstated, they will bring efforts to protect nature into disrepute.

*

"The history of wild salmon," according to a recent article in *Sierra Magazine*, "is a sorrowful one" (Lord 1994, 63). South of Alaska, wild salmon stocks in the Pacific Northwest have declined by 80 percent, leading authorities to ban ocean salmon fishing off the coasts of Oregon and Washington States. The coho salmon in California nears extinction. Only a few of Idaho's Snake River sockeyes survive conditions—such as dams, pollution, habitat destruction, and overfishing—that presage their demise. As wild salmon disappear, so does the ethos to which they belong—one that "includes nature's design of a particular fish for a particular place, within an overall culture of water, gravel, and shade; copepod, bear, and people" (62).

## The Distinction between Morality and Prudence

The rapid decline in the population of wild salmon may concern us for two kinds of reasons—moral and prudential. First, we may regard salmon as wonderful creatures whose association with the ethos of a particular place compels us to mourn their extinction, even if the dams, highways, and industries that usurp their habitat are economically more valuable than the fishery. On these ethical or religious grounds, society may be justified in making great economic sacrifices—removing hydroelectric dams, for example—to protect remnant populations of the Snake River sockeye, even if, as critics complain, hundreds of dollars are spent for every fish that is saved.

Second, one could offer economic, prudential, or instrumental arguments in favor of protecting the habitat of the Idaho sockeye. Admittedly, to restore the salmon habitat, the eight dams along the Columbia River that support the entire industry of the area may have to be dismantled (*Washington Post* 1994b). Further, the use of agricultural chemicals may have to be curtailed, making many farming operations uneconomical. Commercial forestry may be curtailed to keep stream habitat clear. Moreover, far too many people may inhabit the area to maintain a *modus vivendi*—much less an ethos—with salmon, copepods, and bears. If prudence requires us to protect ecological systems, however, we may have to dismantle our industrial civilization, which has altered these systems everywhere. Actions that threaten the habitat of the salmon or the bear may arguably undermine ecosystems that sustain human beings as well. At any point in the "course of the empire" in the West, from the time of the first settlers, we might have gone too far; perhaps every change that humans have made to nature is risky.

Nevertheless, whatever instrumental arguments may be given for protecting salmon habitat, the economic value of the wild fishery is not one of them. The *Sierra Magazine* article points out that while the world production of salmon today stands at a historical high, and is expected to continue rising, prices continue to fall no matter what happens to the wild Pacific stocks. As aquaculture becomes more efficient, the capture fishery may not be economically competitive no matter what we do to keep it viable or sustainable in ecological terms. *Sierra Magazine* explains:

> Thirty years ago about 873 million pounds of salmon—Atlantic and Pacific combined—were going to consumers. Fifteen years ago the amount was 850 million pounds. In 1993, it was 2.5 billion pounds. There is so much salmon currently available that supply exceeds demand, and prices to fishermen have fallen dramatically. (Lord 1994, 63)

One important reasons for falling prices, of course, is fish farming and ranching, which would have rendered the wild fishery in Oregon and Washington nearly obsolete on economic grounds, even if it were not threatened ecologically. (This is true even though prohibitive tariffs against imports, especially from Norway, keep salmon prices artificially high in the United States.) Just as people moved thousands of years ago from hunting and gathering to farming and ranching for most agricultural products, the world relies more and more today on aquaculture and silviculture to satisfy demands for fish and lumber. The Consultative Group on International Agricultural Research has predicted that in 15 years, fish farming and ranching could provide 40 percent of all fish in the human diet (*Deutche Press–Agertur* 1995). The time may soon come when domesticated fish—engineered biologically and raised in artificial ponds and raceways—replace wild varieties in commerce just as domesticated cattle, chickens, and pigs have replaced their wild ancestors.

The prospect that fish, like chicken, could become an inexpensive agro-industrial product is not a fantasy. Here are some facts. Fish require less feed per pound to produce than chickens (Brown, Kane, and Roodman 1994, 34). If inks were not toxic, recycled paper, an organic product, when laced with synthetically produced proteins, might be useful as a basis for fodder. Fish food need be no more expensive than chicken feed. Better still, human excrement—presumably an abundant commodity—can feed algae that, in turn, feed fish. "In Calcutta, a sewage-fed aquaculture system now provides 20,000 kilograms of fresh fish each day for sale in the city" (Brown, Flavin, and Postel 1991, 72).

China leads the world in aquacultural production, with India and Japan as runners-up. Even Norway, a "boutique" operation, can now produce 400,000 metric tons of salmon annually, in spite of tight governmental controls. "The capacity of checked and approved marine areas for fish farming in Norway is 700,000 tons a year" (*Fish Farming International* 1994, 22). This figure—which equals almost 1.5 billion pounds—exceeds the total annual wild catch from the Atlantic and Pacific combined during the 1960s and 1970s (Lord 1994, 63). Prices of salmon are falling as more and more nations—Canada, the United States, Chile, and Scotland are also major producers—increase aquacultural production. "We must realize that what is happening to the salmon industry in Europe now is similar to what happened in the chicken industry decades ago," one trade journal editorialized. "Salmon is becoming a low-cost food, and we shall just have to find ways to live with this" (*Fish Farming International* 1994, 23).[1]

To be sure, aquaculture requires fairly clean conditions in those specific areas—from fishponds to fjords—where it is practiced, but its overall effect is to domesticate nature. Aquaculture, like agriculture, succeeds because economically it is more efficient than hunting and gathering. With respect to coastal seas, fish farming leads to the same kind of world that ordinary farming has produced on land—a world, according to John Stuart Mill (1987, 750),

> with nothing left to the spontaneous activity of nature; with every rood of land brought into cultivation, which is capable of growing food for human beings; every flowery waste or natural pasture ploughed up; all quadrupeds or birds which are not domesticated for man's use exterminated as his rivals for food, every hedgerow or superfluous tree rooted out, and scarcely a place left where a wild shrub or flower could grow without being eradicated as a weed in the name of improved agriculture.

### The Intrinsic Value of Nature

Insofar as we value objects such as salmon for reasons other than their use, we ascribe to them an intrinsic or noninstrumental value. We say that they have a dignity rather than a price. Attitudes toward children illustrate this familiar Kantian distinction. At the beginning of this century, when a million American children tended bobbins in sweatshops and "hurried" coal in mines, 12-year-olds were valued for economic reasons, for the useful work they did. Mine shafts were built to accommodate young children. Opposing child labor laws, mine operators argued that the economic costs of hiring adults would be prohibitive, since mines would have to be completely retunnelled to make room for adults. Child labor legislation restricted the functioning of free markets and, therefore, might not pass an honest cost-benefit test.

When Congress ended the exploitation of children, people did not stop having them. On the contrary, it is possible to think of children in noneconomic terms, that is, as objects of love and respect—even to think that they have a right to grow up, to be educated, and so on. (My wife will not even discuss sending our 12-year-old son out to work so that I can retire.) Americans have somehow gotten beyond thinking of children in relation to their instrumental value. This suggests we could move beyond arguments of utility and prudence with respect to salmon, whales, bears, and other natural creatures as well.

Nature is valuable—we take an interest in it—because of its aesthetic, historical, and expressive qualities rather than simply on the basis of its utility, that is, the benefits it offers us or the contribution it makes to our well-being. We value nature because of what it means to us, not just for what it does for us. The earth's evolutionary and ecological heritage is the object of religious, aesthetic, and cultural contemplation and appreciation; it is also the subject of the science of natural history. As part of that heritage, salmon have enormous historical, aesthetic, and scientific value, which is why the extinction of these magnificent creatures in the wild fills us with dismay. The extinction of species is shameful for human beings; it is the moral crime, as naturalist E.O. Wilson (1980) once said, for which our progeny are least likely to forgive us.

An example from another field may illustrate the crucial distinction between intrinsic and instrumental value. President Ronald Reagan once said in a speech that morality requires us to be celibate before marriage and monogamous after it (Fullinwider 1994, 14). Many of us would disagree with this injunction—especially the part about celibacy, given how late in life people marry these days—but we understand the moral and religious traditions on which the president might base his view. Rather than mention these traditions, thus emphasizing the intrinsic reasons for marital fidelity, President Reagan went on to argue that freer sexual activity might lead to unwanted pregnancy, AIDS, or some other disaster. In short, what has love to do with it? President Reagan attempted to base a moral proposition on an instrumental argument.

Sometimes, religious people have worried that the availability of technology that makes sex safer—the contraceptive pill, for example—encourages people to be more promiscuous. Indeed, it should, if the only reason for sexual abstinence is prudential rather than moral. Similarly, advances in biotechnology have made extinction "safer" insofar as human well-being is concerned. Rather than depend on nature to come up with useful organic compounds, industry can design its own. If instrumental values alone concerned us, every advance in genetic engineering might be an argument for a retreat in environmental protection, and every new organism dreamed up in the laboratory might be an excuse for allowing a plant or animal to go extinct in the wild.

## Prudential Reasons for Preserving Wild Salmon

Why should we be concerned about the extinction of wild salmon or, for that matter, the extirpation of any wild creature that might stand in the way of economic development and improved agriculture? As we have seen, we might answer this question in either of two ways—ethical or instrumental. An ethical argument offers a principled reason for a position; for example, it may assert that creatures who have endured the labor of millions of years of evolution have a moral standing and right to remain on this earth, even if that course is economically disadvantageous to us. Ethical arguments usually draw on principles that teach us to respect goals beyond our own personal welfare or well-being.

Prudential, instrumental, or economic arguments, in contrast, connect efforts to preserve wild salmon and other creatures with human well-being. One might argue, for example, that the extinction of millions and millions of species in many different environments represents incalculable damage to the ecological life-support systems on which human survival is based. This argument, however, warns us only against large-scale extinctions over a variety of environments. It does not demonstrate that every species should be protected or, indeed, that it is dangerous to destroy occasional species-rich habitats, say, for agricultural purposes. To be sure, we cannot survive without some species, but this does not show that we could not do without many others. No one would argue that the demise of the passenger pigeon—once the most conspicuous creature on the American continent—imperiled human life on earth. Certainly, many events, both human-made and natural—such as storms, earthquakes, and the invasion of exotic organisms—have altered ecosystems greatly without threatening human survival.

Human beings have greatly altered the function of many ecosystems; little remains of the ecology of pre-Columbian times, for example. One might suppose that ecosystems were in better shape in 1492 or 1620 than they are today, since human beings arguably had disturbed them less. Would human beings survive better under those wilderness conditions than, say, in built-up places, such as Brookline, Berkeley, and Bethesda, where nothing remains of the original ecosystems? When the Pilgrims came to these shores, nearly half of them died the first winter from cold, hunger, and disease. Does prudence argue, then, that we restore the frontier conditions in which these settlers perished? Is untouched nature always better for us? Should we resurrect natural and original ecosystems because they support all life? What other conclusion would follow from the premise that any change we make in the way ecosystems operate, we make at our peril?

In 1600, the carrying capacity of what is now the U.S. continent for people like us—that is, essentially urban people—was close to zero. If 250 million of us were suddenly blessed with pristine and healthy ecological systems, such as existed in the colonies many centuries ago, very few of us would survive more than a few weeks. The economic development of this continent hardly respected or protected the integrity and authenticity of natural ecosystems. Rather, humanity fundamentally had to alter these ecosystems to eke out a living—for example, to plant amber waves of grain and then decide whether passenger pigeons or human beings would harvest them. The question arises, then, whether every change we make to nature that threatens species habitat—every dam we construct, road we pave, field we plow, condominium we build—also threatens the ecological systems on which human life is based. Plainly, some changes in the environment do threaten human welfare—for instance, global warming, for which hydroelectric dams offer a partial solution. But changes to nature may promote well-being, not simply undermine it, and so an argument is needed to show why a particular alteration may be either advantageous or dangerous, all things considered.

From an economic point of view, changes to the natural world must be evaluated "at the margin" on an instance-by-instance basis. Plainly, biodiversity in general, like water in general, being essential to human survival, is of infinite value. The absolute value of biodiversity as a whole tells us nothing, however, about the marginal economic value of any particular species. (The argument that every part is valuable if the whole is valuable—for example, that the seepage in your basement is valuable because water is essential to all life—is the fallacy of division.) Markets sort out the exchange, marginal, or economic value—these are equivalent terms—for natural objects. It is not clear whether markets fail to correctly determine the economic value "at the margin" for the next gopher, beetle, skunk, or darter that is threatened with extinction. If individuals or businesses are willing to spend millions to build housing in the habitat of a rare butterfly, but nothing to protect the insect, what can we say about its economic worth?

To argue that since one or another species is economically valuable all are, is to commit the fallacy of existential generalization. One of the key tools of biotechnology, the polymerase chain reaction, depends on an enzyme isolated from a microbe, *Thermus aquaticus*, discovered in a hot spring in Yellowstone National Park. This fact provides no information at all about the value of any other creature. By analogy, one may note that Bill Gates is worth billions of dollars. Does this suggest that everyone else has the same net worth? Shall we produce every child possible because some children grow up to make great contributions to industry or the arts? Shall we build every possible shopping center because some make millions? As an ethical matter, we may be obliged to protect every species. Whether we are to spend millions to protect a threatened beetle, one among 600,000 species of beetles, as an economic matter, is best left to markets where buyers and sellers can consider the benefits and costs.

Those who emphasize economic grounds for preserving species of wild salmon may point out that ranched or farmed fish could succumb to a disease for which defenses might be found in the genome of wild stock. (It is customary and usual to refer to the Irish potato blight as an example of what disease can do to a monoculture.) Yet artificial selection, genetic engineering, and other techniques have greatly added to the genetic variety and diversity of plants and animals. A biotech firm in Waltham, Massachusetts, has developed a commercial salmon genetically engineered to grow up to six times faster than wild stocks. There is no scarcity at all of genetic material to swap around among fish.

Consider an analogy with cattle. All cattle are descended from the *Bos primigenius*, a single species that disappeared from Asia and Africa thousands of years ago. The last wild cow is thought to have died in Poland in 1627 (*Washington Post* 1994a). If the absence of wild stock imperils agricultural production, then the beef and dairy industry are in a precarious situation indeed. For the past 400 years, cattle breeders have been unable to rely on the genetic material of wild populations, and this accounts for the touch-and-go character of dairy and cattle operations. Oddly, the Cattlemen's Association has never bitten its fingernails over the idea that the wild relatives of cows long ago became extinct. Similarly, kennels, without relying on any wild stock, seem capable of breeding as many kinds of dogs as anyone will buy. Geneticists at Perdue and Holly Farms show no interest whatever in the fate of the Atwater prairie chicken, an endangered species. And genetic engineers at Ciba-Geigy, Monsanto, DuPont, and other corporations are hard at work on programming basic crops for pesticide resistance, frost resistance, and more, without giving a thought to the opportunities foreclosed to them because of the loss of the passenger pigeon.

The engineering of the chicken genome would advance along essentially the same lines whatever the fate of wild avian species. A research effort to map the chicken genome is identifying the alleles that influence production traits. Genetic engineers are developing ways to improve these traits by manipulating genes and by introducing genetic material from other species. (A tomato containing a flounder gene, to use a different example, is about to be marketed.) "In addition, much of the information and technologies will be readily transferable within poultry since the genomes of other avian species such as turkeys and ducks are similar to the chicken genome" (Cheng 1994, 24).

Advances in biotechnology allow us to store genetic material *in vitro* often more cheaply and effectively than to preserve species *in situ*. Even if chicken producers have no interest in the fate of the Atwater prairie chicken, for instance, someone may prevail on them to freeze a few specimens or even raise them in captivity against the day that their genetic material could conceivably come in handy. To be sure, genetic engineers are more interested in introducing genes from other animals into chickens—such as flounder or pigs—than from other chickens. And they may be more interested in mapping and manipulating genes than in breeding new domestic varieties from wild stock. But who knows? It is cheap to store all kinds of genetic material—which can easily be replicated in cultures—in the freezer.

One might argue that wild species of salmon (as opposed to domesticated breeds) may contain some chemical that could cure AIDS or cancer. This is often presented as a reason for protecting any and every of what may be scores of millions of species. One problem with this argument, however, is that to maintain the wild habitat of some species of salmon, such as the endangered Idaho Snake River sockeye, society must make great sacrifices, which may involve dismantling the hydroelectric dams that impede the path the anadromous fish takes from the ocean to its breeding grounds. The aluminum industry—a major employer in the Pacific Northwest—depends on those dams. The question then arises whether the added chance of discovering a life-saving drug in the Snake River sockeye preserved *in situ* rather than *in vitro* justifies the elimination of much of the industry of this large area.

What is more, there may exist as many as 100 million species of organisms, each with a unique biology, containing thousands of chemical mysteries, which await effort and ingenuity to unlock. The most likely organisms to contain antibiotic and antiviral activity are not the charismatic megafauna, such as salmon, but lowly microbes. For example, Merck and Company has made useful discoveries in nature, not among plants and animals in Costa Rica, where a few years ago it set up a highly publicized $1 million screening project (Sittenfeld and Lovejoy, this volume), but in soil samples taken from places like Rahway, New Jersey. It is plausible to look in toxic waste dumps for microorganisms that might, for example, degrade pollutants. Pristine natural areas are *prima facie* no more likely than cesspools to merit protection for the useful chemicals its creatures may contain.

Once again, biotechnology undercuts the prudential arguments (though not the moral ones) that might be offered for protecting biodiversity. Pharmaceutical companies are not found in the forefront of contributors to conservation activities, in part, because they find it more profitable to design new drugs on computers than to prospect for them in the wild. They are looking at viruses, bacteriophages, and other microscopic critters rather than at elephants or tigers. As technology makes it possible to design new chemicals to meet specific criteria based on analyses of deleterious organisms, scientists are likely to proceed by synthesizing drugs they design for specific purposes rather than by prospecting at random in nature. This is not to deny, of course, the enormous public relations windfall a company like Merck may realize as a consequence of investing a tiny fraction of its research budget in highly visible prospecting efforts. The major research thrust, however, will involve rational design, not random discovery.

One hesitates to question standard economic arguments for protecting biodiversity, however, because to do so suggests that one's heart is in the wrong place, and subjects one to the censure of those who cannot understand you when you say that all the most valuable things are quite useless. They may believe that instrumental arguments are justified at least on instrumental grounds, to protect biodiversity. And so they may question the motives of those who subject prudential and instrumental arguments to honest criticism. We should advance, where we can, credible instrumental arguments for protecting biodiversity. Yet we may bring the cause of preserving biodiversity into disrepute if we shield these arguments from criticism or rely too heavily on them.

The idea that a herb may have genes that can counter potato blight does not impress those who believe, with Walt Whitman, that a leaf of grass is no less than the journeywork of the stars. A mouse is miracle enough to stagger sextillions of infidels, whether or not it contains useful compounds. We may fail to see the difference between instrumental arguments for protecting species and the kinds of reasons that Jonathan Swift advanced for valuing poor Irish children. They may be good to eat. The more we rely on instrumental arguments for protecting species, however, the more we vindicate the economic approach to valuation. This may be a dangerous strategy, if our true reasons for valuing biodiversity are ethical or spiritual rather than economic.

## Why Preserve Biodiversity?

Many voices within the environmental community, not to mention the agrochemical industry, assert that biotechnology has opened the door to greater use of biodiversity (Horsch and Fraley, this volume). This economic rationale linking biodiversity with biotechnology is clearly reflected, for example, in Agenda 21, which emerged from the Earth Summit (the United Nations Conference on Environment and Development). Such an important expression of international consensus deserves examination, especially if biotechnology may hinder rather than advance the protection of habitat.

The biodiversity of Agenda 21 observes that "recent advances in biotechnology have pointed up the likely potential for agriculture, health, and welfare of the genetic material contained in plants, animals, and micro-organisms" (UNCED 1992, 439). It calls for the "development and sustainable use of biotechnology" as a way of exploiting biodiversity, on the grounds that biodiversity is valuable as a storehouse of raw materials for genetic engineering.

What may be most striking about Chapter 15, "Conservation of Biological Diversity," in Agenda 21 is the strong emphasis it places on the development of biotechnology. This is especially remarkable because Chapter 16, "Environmentally Sound Management of Biotechnology," is itself devoted to issues in biotechnology. There is nothing wrong, of course, with wanting to promote biotechnology. But what has this got to do with—and why is it so central to the chapter on—ending the mass extinction of species and the destruction of their habitats?

Agenda 21 makes the answer to these questions all too clear. Its authors apparently assume that the principal reason to protect biodiversity is to maintain an enormous inventory of raw materials for eventual economic applications, for example, in biotechnology. The chapter on biodiversity (chapter 15) acknowledges that "essential goods and services depend on the variety and variability of genes, species, populations and ecosystems." Its concern with "the conservation of biological diversity" is a logical consequence of its larger concern with "the sustainable use of biological resources." The chapter does not point out that the storehouse of biological resources is so vast—including thousands of compounds in each of millions and millions of unknown species—that marginal losses might hardly be noticed. It would take decades to even identify the world's species, much less perform biological assays of them all.

Now there is nothing wrong with finding uses for biodiversity; indeed, in some instances, we might need to do so in order to preserve it. The question arises, of course, as to how much is reasonable to spend per species—there are millions and millions of them—to find some that are useful. What should be done with those plants and animals that have been screened and found to be useless, like those screened in Costa Rica that have no pharmaceutical application? If the point of the exercise is economic, then it would seem that we should not sink a great deal more money into the enterprise than we expect to earn as a result. What is more, as species are screened, the reasons for protecting them might diminish.

If the real point of bioprospecting is to find some sort of use for these animals that might then serve as an economic pretext for protecting them, however, then no investment may be too great. Even a trillion-dollar investment for a dime's worth of utility could be justified—if it is really a moral rather than an economic goal we are pursuing. This is because the trillion-dollar investment could be written off as a sunk cost, and the organism itself would still be worth the dime. To be sure, on a strict cost-benefit calculation, bioassays of microbes in soil samples from Rahway may merit more investment than organisms such as the Snake River salmon or a prairie chicken. Yet with a large enough investment, some use for these more charismatic species could conceivably be found, and if the point is a moral rather than an economic one, perhaps that would justify such an investment.

It may be cheaper if we simply recognize that the principal and sometimes only reason for protecting a species such as the Snake River salmon is essentially moral or cultural rather than economic. Then we could spend our resources immediately on protecting it, rather than exhaust them many times over on futile bioassays that seek to find in its genome a cure for some dread disease. The primary reason to preserve biodiversity, in other words, does not lie in the uses that we find for it. The reason to find uses for biodiversity, rather, is to preserve it, a course we wish to take and ought to take on ethical grounds. If it becomes too expensive to find economic uses for endangered species, we should not bother with the task. We should just protect them for their own sakes.

Naturalist such as S.J. Gould (1989, 284) remind us that the minute particulars of plants and animals at specific times and places—their natural histories—"are endlessly fascinating in themselves, in many ways more intriguing to the human psyche than the inexorable consequences of nature's law." Wilson (1992, 345) elegantly takes up this theme in arguing that every organism, large and small—the flower in the crannied wall—"*is* a miracle," but one that makes sense—is explicable—in the context of a rich historical narrative. "Every kind of organism has reached this moment in time by threading one needle after another, throwing up brilliant artifices to survive and reproduce against nearly impossible odds."

To study these "artifices"—to appreciate the toil each species endures to prevail in the vast labor of evolution—is to be moved to more than economic arguments for protecting plants and animals. And to live close to nature, to have an intimate regard for the native flora and fauna that characterize a place one calls home, is to identify with, to respect, and love—not simply want to use—biodiversity.

Where, then, beats a heart so cold and dead as to never think that the creatures with whom we share the earth are valuable and wonderful in their own right? Who would have thought that the only or even the primary reason to value them lies not in their magnificent qualities, but in benefits they may confer on us? The answer, perhaps, is the authors of Agenda 21. They appear to regard biodiversity mainly as a warehouse of information for biotechnology and genetic engineering—a kind of "natural capital."

The principal interest we have in preserving the last vestiges of the natural world lies in what Agenda 21 (chapter 15) lists casually as "spiritual nourishment"—that is, in ethical or religious motivations. Since these values compel us to protect species, we may be willing in some instances to make the necessary investment to find economic uses for them. It could be cheaper in the long run, however, to do the right thing for the right reasons, rather than forcing ourselves to sink all kinds of money into attempts to identify uses for these wonderful plants and animals. Some of us may worry that moral values carry little weight, or that public officials and authorities respond only to economic arguments. This would be a self-fulfilling prophecy and a counsel of despair.

## Biodiversity and Biotechnology

"Biological resources," so Agenda 21 reasons, "constitute a capital asset with great potential for yielding sustainable benefits" (chapter 15). Biotechnology, according to this view, is supposed to release the economic potential "of the genetic material contained in plants, animals, and micro-organisms." The problem, though, is that biotechnology and biodiversity are not always complementary yet substitutable forms of "human-made" and "natural" capital. Indeed, advances in biotechnology may make it less and less important economically to protect the "library" of creatures that nature provides. We can start writing our own books.

Three reasons suggest that advances in biotechnology—absent policy direction—will not support but hinder efforts to protect biodiversity. First, genetic engineers create a lot of their own materials, for example, by the computer-assisted design of molecules. Second, biotechnology may encourage the domestication of nature, for instance, by replacing wild habitats with bioindustrial systems of aquaculture, silviculture, and agriculture. Third, it seems impossible for biotechnology firms to prospect—nearly at random—among the trillions and trillions of bits of unknown genetic information in millions and millions of unidentified kinds of organisms. Biotechnology can "complement" and thus provide a reason for preserving only a very small part of the biodiversity we ought to and want to preserve. If we took instrumental approaches seriously, many creatures would have to succumb to the same economic rationale that leads us to protect others.

Research budgets in the pharmaceutical industry are stupendous: Two pharmaceutical houses based in New Jersey, Merck and Johnson & Johnson, spent $1.1 billion in 1992 on research alone (Unger 1993). SmithKline, another large house, has an estimated annual research budget of $800 million (*Canada NewsWire* 1994). No matter how urgently environmentalists stress the value of prospecting for chemicals in tropical forests, these pharmaceutical companies count on "biorational," computer-assisted drug design. Environmentalists quite rightly emphasize that useful drugs, such as aspirin, have been derived from nature. Pharmaceutical companies, however, apparently think that computers now offer a more likely path to discovery.

Recently, space station advocates, rather than relying on cultural and moral persuasion, have argued that research carried out by scientists in the weightless environment of space might lead to a cure for cancer. Environmentalists counter that protecting rain forests will lead to wonder drugs and wonder crops. Who could turn down a plea of this sort?

Critics nevertheless scoff at these self-serving arguments. They carry no weight with researchers at the National Cancer Institute, for example, who compete with space buffs and environmentalists for scarce funds. Only if one believes that good things never conflict—that money put into one worthwhile project is not taken from another—may one attempt to justify the space program, endangered species policies, and other valuable projects wholly or even primarily for the contributions they might make to agriculture, health, and welfare. If one sees that there are trade-offs between investing in different research agendas—for example, to cure cancer—then we might as well leave it to the experts at Merck and SmithKline to decide where the best prospects lie. To argue for species protection on economic grounds is to call for the market to resolve the matter. A market resolution, however, may not go well for biodiversity.

## The Effects of Biotechnology on Natural Ecosystems

The new biotechnologies give us the power to domesticate or even "industrialize" nature—that is, reconstitute, manipulate, and control for economic purposes "wild" ecosystems, such as forests and estuaries, that have so far escaped humanity's conquest of nature. The new technologies empower us to force nature's favors, even in the most recalcitrant environments. They invite us, then, to simplify or otherwise alter wild and natural ecosystems to convert them to quasi-industrial, aquacultural, or agricultural management. It can also work the other way. The various forms of biotechnology, by adding enormously to the efficiency of production, may increase agricultural surpluses, while creating a glut of those commodities, like fish, that do not already exceed effective demand. This glut may cause lands that are now intensively farmed and seas that are intensively fished to revert to "wild" or "natural" conditions. (One hears discussion of the "buffalo commons" to replace the farms on the prairies and plains that agricultural surpluses have made uneconomical.)

Consider the fate of tropical rain forests. The primary argument against cutting, burning, and clearing these wonderful ecosystems for agriculture has been economic or broadly utilitarian: few if any crops will grow in their place. Conservation biologist Daniel Janzen (1987) noted in a letter to *Science* that recombinant DNA techniques are likely to provide varieties of economically valuable plants and animals that will do well on land that rain forests now occupy: "When genetic engineering gives us crops, plants, and animals that thrive in the various tropical rain forest habitats, it is 'good-bye, rain forest.'" If this is true, then to protect rain forests, we should have to make agricultural production elsewhere so efficient that there is no incentive to bring these environments into cultivation. We could as well rely on biotechnology to save the rain forest as subject it to the axe and the plow (Horsch and Fraley, this volume).

Even if human beings, with the aid of biotechnology, to a large extent create our own niche, rather than find one already implicit in nature, we still value the remnants of our evolutionary and ecological heritage. People want to live in harmony with nature, not just triumph over it. One might even say that Americans identify themselves as a nation that cares about the environment—that considers nature an object of moral attention rather than economic subjugation, or as a condition of our consciousness as human beings, not merely as a vehicle to satisfy consumer wants or demands. Thus, we may reject the idea that biodiversity and biotechnology serve the same end, which is to perfect nature for human wants and needs.

Agenda 21, in contrast, views nature as a source of profit, which it sees biotechnology extracting from biodiversity. It seeks to distribute the profits equitably. Little in its analysis, however, suggests any reason for protecting the ecological and evolutionary heritage of humankind for any reason but as a source of genetic and other resources, or as a sink for wastes. Yet if we regard nature wholly in instrumental terms, we might value engineered or domesticated species more than wild ones. We might then conclude that the basic problem is not the risk bioengineered species pose to natural ones, but vice versa. The goal in agriculture, silviculture, and aquaculture is to replace nature with high-tech and industrial production systems. Why would biodiversity have a place, then, outside the zoos, specimen collections, and freezers where seeds and tissue samples are kept?

## Nature and the Environment

The distinction between *nature* and the *environment* does not receive enough attention. Nature is Creation, fresh from the hand of God or, if you prefer a secular or scientific account, it is the amazing surviving record of the billion-year-old toil of evolution. Our evolutionary and ecological heritage beggars our ability to understand or even imagine its possibility; as F. Scott Fitzgerald (1953, 182) wrote, nature provides for the last time in history an object commensurate with our capacity to wonder.

Nature is whatever humanity did not make; something is "natural" to the extent that its qualities owe nothing to human beings. The values nature inspires in us— many of them fundamental to our culture—tend to be religious, moral, and aesthetic. What distinguishes humanity, perhaps, from other creatures are those moral ideals that impel us to appreciate and protect nature for its intrinsic qualities, rather than only for the properties that may benefit us.

The natural environment, in contrast to nature, is valued only in terms of the benefits we may extract from it. It is the infrastructure we find, as distinct from that we build. The environment comprises those aspects of nature that are useful to us and, therefore, that we value for welfare-related reasons as long as technology does not render them obsolete (consider salmon, for example). The environment is what nature becomes when we cease to believe in it as an object of cultural, religious, and aesthetic affection, and instead, come to view it as a prop for our welfare—a source of materials and a sink for wastes.

Human beings have enjoyed higher and higher standards of living by making economic production ever less dependent on the vagaries of a fickle and stepmotherly nature. Progress has followed more from the conquest than the conservation of natural environments. From this point of view, the AIDS virus may represent the sort of thing we can expect from the rain forest, for example, no less than the miracle medicines that environmentalists hope to find there (Webber 1984, 115–16). The approach to nature that regards it as a frontier to conquer is familiar in the medical sciences. Plague, smallpox, malaria, heart disease, and polio are all part and parcel of nature; one by one, medicine has pushed back these limits to life—and is at work on genetics. Biotechnologists may regard nature *per se* as the enemy of humanity; indeed, our mortal enemy, since it kills us all.

The image of nature as raw material to be formed by us—a kind of genetic frontier for human manipulation—can now be found in the ecological sciences. For example, Frank Forcella (1984), writing in the *Bulletin of the Ecological Society of America*, called on his colleagues to embrace the "biotechnologist" credo "to engineer and produce plants, animals, and microbes that better suit the presumed needs and aspirations of the human population." He continued:

> Ecologists are the people most fit to develop the conceptual directions of biotechnology. We are the ones who should have the best idea as to what successful plants and animals should look like and how they should behave. . . . Armed with such expertise, are we going to continue investing nearly all our talent in Natural History? . . . Or should we take the forefront in biotechnology, and provide the rationale for choosing species, traits, and processes to be engineered? I suspect this latter approach will be more profitable for the world at large as well as for ourselves.

Similarly, the underlying message of Agenda 21 may argue that by transforming nature, technology expands the resource base (for example, by domesticating species) and enhances waste removal (for example, by building landfills and sewer lines). By controlling, subduing, and eliminating nature, under this view, we push back its barriers, and thus we improve the environment. The prudent course, then, is the proven course—that is, to continue to try to conquer nature or it will conquer us.

Those who believe that economic growth is "unsustainable," on the other hand, tend to think of nature as a benevolent mother rather than a cruel adversary. Thus, while they are pessimistic about technology, ecologically minded economists are optimistic about the utility of the natural world. They would preserve endangered plants and animals for uses that may yet be found. What is more, they urge us to preserve the habitats of these creatures, not just specimens or tissue samples for future replication and study. The prudent or "precautionary" approach would save "all the parts," that is, whatever vestiges of nature are left.

Not much difference in the end separates these positions—since each is concerned not with the love of nature, but only about the use of the environment. It comes down to what you are optimistic about. Ecologically minded economists are optimistic about the utility of nature; their opponents tout the prospects of technology. Since each approach, in its own way, seeks to make a better—safer, more comfortable—environment for human beings, neither gives us any reason to protect nature or to value it for its own sake.

## Biotechnology and the Reversion of Land to Nature

"My biggest fear is not that by accident we will set loose some genetically defective Andromeda strain," then Senator Albert Gore (1987) has written, but that the nation will drown in a sea of surplus agricultural commodities. "The Green Revolution made America the world's breadbasket, but it has also brought on an age of intractable overproduction," Gore noted. "Unless we plan more carefully, the Gene Revolution could do the same—on an even grander scale." Gore observed that scientists "are working on Supercows, Superpigs, even supersized salmon." Other experiments concern improved reproduction, disease resistance, and growth. "Unless we can somehow find a way to create very hungry Superhumans," Gore correctly warned, "each of these advances may produce nothing but glut."

Agricultural biotechnology affects the environment indirectly by making production less and less dependent on the functioning of natural ecosystems. Phosphate-solubilizing and nitrogen-fixing plants now in the experimental stage, for example, can compensate for poor soil conditions; why worry, then, about erosion? Likewise, drought-resistant crops, also being developed, substitute for water. Advances in biotechnology—the gene revolution is poised to take over where the green revolution left off—may rob environmentalists of important utilitarian or prudential arguments for protecting natural ecosystems. As we find cheap technological substitutes for nature's free gifts, instrumental arguments for protecting nature become harder and harder to defend.

The main reason that farms may go out of production is not erosion or lack of genetic diversity; it is surplus or glut. The possibility arises, indeed, that the factory will replace the farm as the location where food and fiber are produced. The principal problem, in other words, may not be surpluses of farm commodities, but the industrial substitutes that biotechnology will create for those commodities. In the past, most agricultural products have been grown on farms and then processed in factories. In the future, factories may "grow" as well as process food and fiber—or the two functions will be one. As one scientist has said: "We have to stop thinking of these things as plant cells, and start thinking of them as new microorganisms, with all the potential that implies" (Curtin 1983).[2]

The substitution of industrial for agricultural products is a familiar story; for example, the replacement of cane sugar by artificial sweeteners, such as Nutra-Sweet, and high fructose corn syrup. Plant-produced milk proteins, known collectively as casein, may soon do with cheese what oleomargarine has done with butter. Likewise, synthetic fabrics like nylon and orlon have captured much of the textile market. Spices, fragrances, and flavoring agents are now being produced *in vitro*: vanilla is one example. Corporations competing to culture major crops *in vitro*—such as coffee, tea, rubber, cocoa, cotton, and so on—eventually may replace Third World growers as the principal suppliers of these commodities.

Agricultural products that the United States still imports, such as coffee and cocoa, are following the path of sugar and vanilla. Multinational firms, like Nestle and Plantek, have engineered varieties of rust-resistant coffee plants with higher yields that will grown even in temperate climates. World markets are soon to be flooded with coffee engineered to grow in various climates with such characteristics as frost resistance, herbicide tolerance, flowering uniformity, low-caffeine content, and so on. Eventually, easily grown crops such as peanuts and lentils, into which coffee- or cocoa-flavored genes are inserted, may replace the original plants as the primary source of coffee and chocolate.[3]

The major effects of biotechnology on the environment are likely to be twofold. First, many "wild" or "natural" ecosystems may be converted to species and processes suitable to large-scale, highly controlled aquacultural or silvicultural production. Second, as agricultural surpluses begin to be seen as infinite, and as the factory replaces the field as the location where food and fiber are fabricated, the agrarian economy must shrink, and many farms will go out of production. Highways, commercial strips, and suburban "sprawl" might absorb some of this land, but a great deal may revert to "nature." This is already happening; farms in the eastern United States have been reverting to woods as the agricultural sector there shrinks. In developed parts of the world, where political stability and institutional capacity permit knowledge and technology to be applied to agricultural production, yields may be expected to increase, decreasing the need for land (Waggoner 1994).

In other parts of the world, alas, chaos and civil war may continue to produce mass starvation, no matter how much food is available worldwide. We must distinguish, in other words, between there being enough food versus people having enough food. The first problem, to produce enough food at today's historically low prices to feed everyone, will take further technological advances, but few agronomists doubt that these are in the cards. To make sure that people have enough food, in contrast, requires us to solve not technical but social problems, such as our inhumanity toward one another. No solution to this kind of problem—moral or economic—is in sight.

## The Inherent Contradiction in Agenda 21

The biodiversity chapter of Agenda 21, as we have seen, celebrates the application by biotechnology of "the genetic materials contained in plants, animals, and micro-organisms" (chapter 15). The chapter also, at least ostensibly, seeks to protect biodiversity—minimally, as a source of these materials.

To collect and screen these materials, however, does not require us to protect rain forests or other habitats and natural environments. It may suffice to send prospectors to remove specimens and tissue samples from nature for storage and examination *in vitro*. If useful materials are found, the natural environment can then be domesticated for the production of whatever it is that may be sold profitably. This is not necessarily a recipe for protecting nature; it may only pave the way for the further industrialization of the natural world.

Agenda 21 reflects an influential view that public policy should be concerned primarily, or indeed exclusively, with improving the overall efficiency of the economy. According to this philosophical approach, regulatory actions are justified insofar as they bring market "externalities" or "spillovers" like pollution into the pricing mechanism, so that the private and social costs of production will not diverge. In this way, regulation will help markets to allocate materials and resources to their most valued employments, which is to say, to those producers and consumers who are willing to pay the most for their use.

The major environmental impacts of biotechnology, such as those described here, may have little to do with spillovers, externalities, or inadvertent side effects. The major impacts generally will be intentional and profitable, as public officials use the weapons of biotechnology purposefully to defeat and, one might say, humiliate nature.

Why worry about these impacts? Why worry about altering ecosystems—for example, converting forests to tree farms—if these changes are socially profitable over the long run? Our present emphasis on prudential or instrumental values suggests we do not intend to protect the environment from changes that are economically beneficial. We may happily put nature into the hopper of biotechnology as long as it is economically efficient—in other words, as long as there is a market, a price paid, for everything that comes out.

We environmentalists have to confront the distinct possibility that the blind forces of evolution did not perfectly fashion nature for our economic purposes. We must then consider the likelihood that recombinant DNA organisms—fish that bite better, trees that grow faster, birds that sing prettier, microbes that fix nitrogen more efficiently, and so forth—will be superior economically to the natural species they may be intended to replace. While these artificial species may outcompete their "wild" counterparts economically, however, they may not outcompete them ecologically. We will have to eliminate wild species to make room for domesticated ones—as we have always done. Should we then say that "unimproved" native species pose a risk to the "improved" engineered organisms that managers may wish to substitute for them? Which kinds of species—natural or artificial—is it in our interests to protect?

Since our welfare—indeed, our entire agricultural economy—is likely to depend on genetically engineered organisms, it may be these domesticated organisms, rather than their "wild" counterparts, that really constitute the basis of the "carrying capacity" of the earth or the systems that support human life. (Remember that we would die in droves, like the Pilgrims, in a really natural ecosystem.) If environmental regulators take economic profitability and productivity seriously, then they will not necessarily attempt to protect nature from the risks posed by biotechnology. It may be more rational, in many instances, to protect domesticated plants and animals from the risks posed by naturally occurring organisms. Why protect nature from humankind when it is more rational to protect humankind from nature? Why not welcome totally engineered, artificial surroundings, if that is the way to maximize consumer surplus, or efficiency, or profitability—even if we say "good-bye" to the natural environment?

This is exactly the prospect, of course, that J.S. Mill contemplated in England more than a century ago. He saw that every animal his fellow citizens did not domesticate for their own use, they exterminated as their rivals. Mill understood that the economy of England could not have been sustainable—it could not have fed many millions of people—if land were left to the spontaneity of nature. Yet Mill, a great utilitarian philosopher, also understood the aesthetic, religious, and cultural importance of that kind of spontaneity, and he regretted its loss. Those of us who share his utilitarian philosophy will not find reasons for protecting species like wild salmon unless we share his reverence for nature as well.

### Acknowledgments

The author gratefully acknowledges support for this research from the National Science Foundation, grant no. SBR9422322. The views expressed are those of the author and not necessarily of any funding agency.

### Notes

1   Salmon prices in the United States are kept artificially high by steep tariffs against imports (*Fish Farming International* 1994).
2   See also Chaleff (1983, 679): "With recognition of the similarities between cultured plant cells and microorganisms came the expectation that all the extraordinary feats of genetic experimentation accomplished with microbes would soon be realized with plants." Chaleff enumerates the difficulties that must be resolved before this expectation may be fulfilled.
3   For more information about product substitution, see the World Bank (1989), especially page 31. Cocoa, already grown commercially in Hawaii, may be engineered to grow in even more temperate climates.

# References

Brown, L., C. Flavin, and S. Postel. 1991. *Saving the Planet*. New York: W.W. Norton.

Brown, L., H. Kane, and D. Roodman. 1994. *Vital Signs 1994: The Trends That Are Shaping Our Future*. New York: W.W. Norton.

*Canada NewsWire*. 1994. Arthritis Society receives funding for clinical fellowship program. 11 April.

Chaleff, R.S. 1983. Isolation of agronomically useful mutants from plant cell cultures. *Science* 219:676–82.

Cheng, H.H. 1994. The chicken genetic map: A tool for the future. *Poultry Digest* (June): 24–28.

Curtin, M.E. 1983. Harvesting profitable products from plant tissue culture. *BioTechnology* 1:657.

*Deutche Presse–Agentur*. 1995. Rapid increases in fish farming needed to boost food supply. 12 May.

*Fish Farming International*. 1994. Norway's salmon capacity is now nearly 300,000 tons. 22–23 July.

Fitzgerald, F. 1953. *The Great Gatsby*. New York: Scribner.

Forcella, F. 1984. Commentary: Ecological biotechnology. *Bulletin of the Ecological Society of America* 65:434.

Fullinwider, R. 1994. Chastity, morality, and the schools. *Report from the Institute for Philosophy and Public Policy* 14:14–18.

Gore Jr., A. 1987. Gene revolution's progress must be aimed toward helping small individual farmers. *Genetic Engineering News* (May): 4.

Gould, S.J. 1989. *Wonderful Life: The Burgess Shale and the Nature of History*. New York: W.W. Norton.

Janzen, D.H. 1987. Conservation and agricultural economics. *Science* 236:1159.

Lord, N. 1994. Born to be wild. *Sierra Magazine* 79 (6):60–65.

Mill, J.S. 1987. Bk. 4, chap. 6, sec. 2 in *Principles of Political Economy with Some of Their Applications to Social Philosophy*. 1848. Reprint, Fairfield, N.J.: Augustus M. Kelley Publishers.

UNCED (United Nations Conference on Environment and Development). 1992. In vol. 1, *Agenda 21 and the UNCED Proceedings*, edited by N.A. Robinson. New York: Oceana Publications.

Unger, M. 1993. Drug money. *Newsday*, Nassau and Suffolk edition, 2 May, 88.

Waggoner, P.E. 1994. *How Much Land Can Ten Billion People Spare for Nature?* Ames, Iowa: Council for Agricultural Science and Technology.

*Washington Post*. 1994a. In search of wild cows, 9 September, D1.

———. 1994b. Plan to save salmon roils the Northwest; Change seen causing ripples in economy, 15 December, A3.

Webber, H. 1984. Aquabusiness. In *Biotechnology in the Marine Sciences*, edited by R. Colwell, A. Sinskey, and E. Pariser. New York: John Wiley.

Wilson, E.O. 1980. Resolutions for the '80s. *Harvard Magazine* (January–February): 22–26.

———. 1992. *The Diversity of Life*. Cambridge, Mass.: Belknap Press of Harvard University Press.

World Bank. 1989. *Sub-Saharan Africa: From Crisis to Sustainable Growth*. Washington, D.C.

# VI Legal Implementation

# What to Do about Biodiversity:
# The Earth's Biological Riches in Law
# and Economics

by Christopher D. Stone

Among the earth's natural resources, some of the most important are biological—which, unlike coal, are capable of regeneration, even increase, if we temper our rate of harvest. Yet not all harvest (perhaps even extinction) is unwise. Humankind has regularly converted some biological assets into other assets, without regret. The question is, what can we do to ensure that the rate of consumption/conversion is efficient? The first part of this chapter looks at an array of policy options, including privatization, trade restrictions, and subsidies, in the context of ordinary, consumable biological assets. Next, the chapter examines an emerging category of interest, genetic resources, against the background of the new U.N. Convention on Biological Diversity (CBD). The exploitation of genetic information puts a positive value on conserving as genetic "libraries" many of the same natural ecosystems that are under pressure for conversion to other uses, such as agriculture and lumber. The chapter discusses the mechanisms for tempering the rate of conversion, including "prospecting contracts" and intellectual property protection alternatively for countries of resource origin and for processing firms. The conclusion reached is that, even were the value of the biological asset to be enhanced by these and all other legal devices, the full social benefits of the resource lie beyond appropriation by markets, and the case for international subsidization of selected conservation areas appears powerful.

★

## Biology, Biodiversity, and the Law

The CBD (appendix, this volume) is only the latest expression of the law's concern with the planet's biological capital.[1] Hammurabi's Code dedicated 11 sections to agriculture (Saggs 1989, 158). The Old Testament enjoined belligerents to spare their enemies' fruit trees even in laying siege to their cities (Deuteronomy 20:19–20). International agreements providing for the conservation of fish, migratory birds, and marine mammals go back at least a century (Birnie and Boyle 1992, 425; Lyster 1985).

In recent years, legal attention to biological issues has been expanding. Intellectual property law is groping with the patentability of engineered life-forms. Over 100 nations have ratified the Convention on International Trade in Endangered Species of Fauna and Flora (CITES) (United Nations 1973). There are international agreements on wetlands preservation, salmon fishing, the use of drift nets, the life of the Antarctic, and living resources that migrate across international boundaries.

The social goals underlying this broad swath of legal activity are not exactly congruent. Many laws and treaties, such as fisheries regulations, aim to shield regenerative resources from unsustainable levels of consumption. The CBD was motivated largely by the concern for the portfolio value of genetic material; the idea is a sort of long-term insurance for human health and welfare (Arrow and Fisher 1974, 312; Fisher and Hanemann 1986, 169). Laws such as the Endangered Species Act (1995) and international conventions such as CITES aim to protect certain living things that find favor for moral, aesthetic, or symbolic reasons, quite apart from their utility as "insurance" or products and services.

On the other hand, while the exact motivations vary, all these efforts have a common theme and intricacy. Among the resources we inherit, some of the most important to us are biological—hence, unlike coal, capable of regeneration, even increase, if we choose to temper our rate of harvest. Yet not all harvest (perhaps even extinction) is unwise. Humankind has regularly converted some biological assets into other assets, without regret (Swanson 1994). Forests have not only been cut for timber (consumption), but because the underlying soil is more productive—promising a higher income stream from other uses, such as farmland. Alternatively, the cash from the timber, and the physical site, united into investment in a factory, may yield a higher social rate of return producing machine tools. The question for a general theory is, how do we ensure that the rate of consumption and conversion is not too much? In the first part of this chapter, some general features of the ideal framework for the conservation (and consumption) of biological assets are examined; next, a special category (insofar as it is a special category) of the newest major challenge, genetic resources, is addressed.

### Protecting Biological Assets: The General Framework

For a general theory, a good starting point is to mark the traditional distinctions among private goods, public goods, and externalities. All three categories can be illustrated in the context of a tropical forest.

*1. Private Goods.* Wild berries are (potentially) private goods.[2] The consumption of the berries is *rival*: the more I pick and eat, the less are left for you. Berries, latex, and lumber can be parceled, and the parcels priced in markets.

*2. Public Goods.* The forest's trees and other biomass draw carbon out of the atmosphere and pump back oxygen. In contrast with the berries, the benefit I derive from my "use" of the forest—to restore a congenial carbon balance—does not make it any less valuable in the same way to you. To put it otherwise, we benefit in a way that makes our consumption *nonrivalrous.* Indeed, whatever contribution to atmospheric quality that the forest provides, the benefits are also *nonexclusionary*: the quality is supplied to each and every person on the globe, whether or not it is the precise quality they would prefer, and whether or not they contributed to the costs of its provision. This means that the contribution of the forest to the enhancement of the atmosphere, unlike its contribution in berries, cannot be parceled and priced in markets.

*3. Externalities.* Even when goods (or activities) have been privatized, or brought within the sovereignty of a particular nation, external effects may be displayed. Sometimes the externality is positive. Nation $N$'s forests may decrease the risks of flooding in neighboring nation $Z$ or bolster $Z$'s wildlife. Other externalities are negative. $N$'s forests may breed mosquitoes that spread malaria across the border to the detriment of $Z$.

Privatizable Biological Assets
Against this background, let us begin with those biological assets, whether in the wilds or in agro-ecosystems, that are potentially private goods—that is, whose use is rival, the control over which can be made exclusionary, and that have marketable commercial value.

For these goods, the ideal is to ensure that the resources are priced at true marginal social costs. Governments can advance toward this ideal by establishing and enforcing property boundaries around resource areas,[3] and by deploying various measures to internalize externalities.[4]

The point cannot be overstated. Across the world, environmental (and not merely living) inputs are commonly being priced imperfectly. Some, such as forests (and almost universally, water), tend to be underpriced and, therefore, overconsumed. Other assets, such as the waste storage capacity of air and water basins, are not merely underpriced, but all too commonly elude pricing systems altogether. In these situations, polluters treat aqua systems and the atmosphere as cost-free waste dumps. Even worse than that, some countries have adopted resource policies that actually subsidize the destruction of biological assets.[5] Ocean fishing is so heavily subsidized that the costs of sweeping the seas reportedly exceeds the value of the catch ($92 billion in costs, $72 billion in value). It is no surprise that fishery yields are declining (United Nations 1994, 1–3).

Hence, the advance of privatization (or, at least, of cost-based pricing where the

FIGURE 1   Provision for Supply of Public Goods

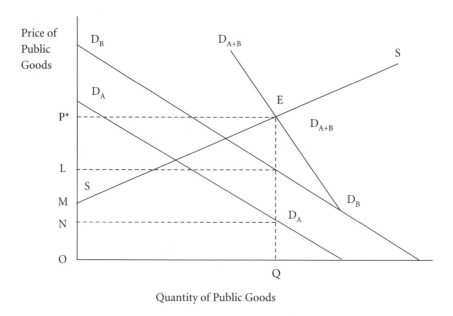

Quantity of Public Goods

asset is government managed) would, in a single stroke, advance human welfare and restrain the plunder of biological (and other) assets worldwide.

Biological Assets That Are Public Goods
In some situations, however, a privatization strategy is unavailable or falls short of achieving social ideals. Most obvious are circumstances in which the biological asset provides a public good. Consider the great algae colonies of the high seas. They are of value to the world, first, by "supporting" elements higher in the food chain, and second, by drawing carbon out of the atmosphere, thereby reducing risks of climate change. Yet, unlike fish, that value cannot be captured and expressed in private markets. Even if the colonies were placed in the hands of an owner, the owner could not exclude others from, and collect for, the benefits. It is easy to see that any natural condition threatening the "collapse" of the algae colonies—a virus, for example—would merit international response. It has even been proposed that our responses to climate change should include expanding algae colonies by "seeding" them with iron ferrules. As with U.N. peacekeeping missions, however, financing such efforts—the medication or augmentation of the colonies—would require raising public contributions from the world community. But how much of a contribution? And how would it be exacted among the benefiters? By definition, markets will not produce the answer for public goods. What will? The ideal is represented in figure 1.

Figure 1 expresses the demand for the conservation of members of a commercially valueless species—say, penguins—in a world of two persons, $A$ and $B$. Although the penguins (like the algae colony and ozone shield) have no appreciable commercial value, they do have "existence" value (Freeman 1979). People across the world (represented by $A$ and $B$) are willing to pay to support the penguins—just to know that they exist (or perhaps to hold an option to see them some day, or to reserve them as a bequest to their progeny). $D_A$ is $A$'s demand for penguins; $D_B$ is $B$'s. (Observe that $B$ gets more benefit—is willing to pay more—for any given number of penguins than is $A$.) S–S is the supply schedule reflecting the marginal costs of conserving and perhaps expanding the number of penguins. The intersection of the aggregated demand curve, $D_{A+B}$ [6] with S–S yields the efficient level of species members, $Q$; $OMEQ$, the cost of providing $Q$ penguins, is the appropriate subsidy for maintaining the habitat.

Biological Assets That Afford Positive Externalities

Externalities effects may or may not be regarded as a *tertium quid*, depending on how broadly the external effects radiate, and perhaps on the costs of internalizing them. For purposes of analyzing legal options, it is useful to treat the three categories separately, and call attention to the many situations in which the same biological asset provides all three types of good in what might be called separate layers. The asset provides some benefits appropriable by the owner in private markets. A second layer of benefit spills across so broad, even global, a community as to present a public good. And we may find a third layer of effects that are externalized across a discrete set (some but less than all neighbors), and that therefore are potentially internalizable through legal instruments.

The situation is illustrated in figure 2, below. Nation $N$ possesses a rich forest area. Whether it retains the land under forest cover (withholding it from alternative uses, such as for cropland, logging, and human habitation) or disinvests it depends on a number of factors, including: the expense of deforesting; the asset value of the goods and services that the land produces if conserved as forestland, such as nuts, latex, and sustainable levels of timber harvest; and the costs of managing the area, such as shielding the ecology from natural catastrophes.

Assume (i), (ii), (iii), and such to be exogenously given and constant. $DD$ is the demand for the land in forest-stable condition. To appreciate the influence of various legal regimes, imagine first that the government is subsidizing a competing use of the forest area, for example, agriculture or logging. Under those circumstances, disinvestment accelerates and deforestation (and loss of habitat) will reach $OQ_S$. It is evident that such a degree of consumption (conversion) is too high, and that the perverse subsidies that induce them need to be eliminated.

Suppose, now, that the subsidies accelerating alternative investments have been abolished, but ownership rights for the resource area are poorly defined. Poor de-

FIGURE 2   Consumption (Conversion) of Forest under Five Legal Regimes

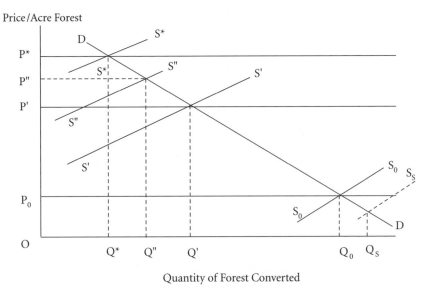

Quantity of Forest Converted

finition may take the form of the forests being held in common; or if the title is al-located, owners may have little assurance that the land will not be nationalized without compensation; or the owner's right to exclude may be protected inade-quately—by, for example, undependable enforcement against trespassers and poachers. Under these circumstances, the quantity of covered forestland converted to alternate uses is $S_o-S_o$; $P_o$ represents the price an investor would pay for the asset under those fragile ownership conditions, and $OQ_O$ represents the quantity con-verted at equilibrium.

If, however, property rights are defined and protected, the supply curve shifts left, with $S'-S'$ replacing $S_o-S_o$. This shift reflects the intuition that as the protec-tion of ownership interests in the property (such as the rights to exclude and to convey good land title) become more dependable, the landowner's opportunity costs of quick cutting the asset increase. The value of the land in forest-covered condition rises to $P'$ to reflect the additional security of the expected income stream from the future sale of forest products, and conversion is accordingly di-minished to the more efficient level $OQ'$, reflecting the lower rate of reinvest-ment—via deforestation—to other uses.

Now, imagine that the forest provides an external benefit: it is so situated as to mitigate the risks of flooding in neighboring nation $Z$. That benefit can be inter-nalized. $Z$ may respond to the owner's threat of deforestation by offering a "bribe" (a transfer payment to the landowner directly or through $N$) not to deforest to $OQ'$.[7] The amount of the payment is $P''-P'$. Under those conditions, the supply of the acreage in covered condition shifts left to $S''-S''$, with the corresponding diminution in deforestation to $OQ''$.

Finally, the forest, by sequestering carbon and thereby tempering the risks of climate change damage, provides a benefit that radiates across the globe. That layer of benefit is a pure public good. So, too, is the benefit it provides in harboring flora and fauna—rare forms, in particular. The value of the public good layer would be determined as illustrated in figure 1—by the willingness of all benefiters to pay. One can imagine circumstances where, if the public good benefit were appropriately consolidated in the asset price—that is, if the owner of the forest could collect not only the value of the conventional forest products (the income stream they represent), but in addition, from each benefiter, the value of the benefit received from existence values and the reduced risks of climate change-induced damage— then the conversion of the uncut acreage would shift further left again ($S^*-S^*$). The price of the acreage rises and deforestation decreases to $P^*$ and $OQ^*$, respectively—the global optima.

Consider how the analysis applies to the preservation of noncommercial species, even those creatures whose preservation is motivated by moral considerations— such as, that we owe it to future generations or nature itself to conserve them. $N$ harbors a habitat for an endangered species, for example, black rhino. While there is a commercial market for these animals and their parts, their market value often falls far beneath their existence value—essentially, what people across the world would be willing to pay to preserve rhinos in the wilds, if their payments could be collected. Privatization of the habitat as a paying tourist attraction, allied with creative incentives for neighboring indigenous peoples (such as guide jobs for those who might otherwise be poachers), can moderate the consumption from $OQ_O$ to $OQ'$. But considering that only a small fraction of those who want the habitat to exist will actually make the trip and pay entrance fees, the owners cannot receive as reward the full value to the world of the habitat, which is reflected, in figure 2, in the level of consumption $OQ^*$.

## Strategic Responses

In the circumstances depicted in figure 2, where the asset manager is rewarded for providing the private good only, the problems of conserving the nonmarket good are aggravated. The priced good is being overproduced. The asset manager is not merely indifferent about the nonpriced public good being sacrificed; distorted market signals are allied behind its elimination. What countermeasures are appropriate?

## Traditional Legal Remedies

If moral suasion fails, there are—at least theoretically—legal options. In circumstances where the impairment of the public goods supply is traceable to human activity, the response may take the form of charging those responsible in proportion to the harm they cause (as opposed to raising funds from benefiters in proportion to their benefit, discussed below). For example, if $N$'s penguin habitat were being obliterated not by natural causes (a virus or solar cycle), but by effluents traceable to coastal manufacturing plants, we could seek efficiency through a cost-internalizing instrument, such as a tax gauged to internalize on the offending nation $N$ the true long-term social costs. Or, we could consider internalization through damage actions: that is, try to arrange international institutions so that $N$, the destroyer of the penguins' habitat, could be sued in a world body.

But while such cost-internalization techniques are theoretically available, the concept of host-state sovereignty over its internal resources is so entrenched in international law that efforts to tax $N$, or to make it answerable to the world community for the destruction of its internal environment, are unlikely to succeed.

Perhaps one may be slightly more optimistic regarding the future of cost-internalization techniques when the aim is to protect global commons resources. A nation is apt to be less resistant to user charges when they are based on the use of commons areas, than when they are based on the nation's activities within, and affecting, its own sovereign territory. Elsewhere I have expanded on the idea of imposing charges on the "use" of commons areas—such as taxes on ocean pollution, high seas fishing, and atmospheric carbon emissions. The funds raised would be placed into a Global Commons Trust Fund, which in turn, would underwrite the costs of commons-area environmental repair (Stone 1993, 201–34). But the immediate prospects of institutionalizing even such modestly scoped charges (limited to impairment and repair of the commons) are not bright.

## Trade Sanctions and Other Special Measures

In general, then, as a practical matter, traditional legal measures such as taxes and torts have a limited role to play in protecting biological assets in the international arena. The more realistic, but nonetheless controversial alternative involves the deployment of trade leverage. Trade measures invoked pursuant to multilateral agreements such as CITES, which preserve threatened species by dampening the demand for them, will probably survive their chafing with the General Agreement on Tariffs and Trade (GATT) and other trade agreements. A group of nations can refuse to buy rhino horns from $N$. But trade pressure that takes the form of a refusal to buy unrelated products (say, the produce of land $N$ converted from a rhino habitat to farmland), particularly if imposed by $Z$ unilaterally, is more vulnerable to attack. Essentially, for a nation to condition its trade in order to influence another nation's environmental policies is not a violation of traditional international law, because no unlawful force is involved, and each nation is free to deal or not to deal with others as it chooses (Damrosch 1989). But such a use of trade measures (particularly unilaterally) is increasingly disapproved of in the world community, appears to conflict with the United Nations Conference on Environment and Development's (UNCED) Principle 12, and would certainly be challenged by any target nation that was a member of the World Trade Organization (WTO). [8]

My own view is that those who decry the deployment of trade measures to influence the behavior of other nations toward the environment are expressing legitimate concerns, such as a hesitancy to appear eco-imperialistic. But particularly because of the dearth of alternatives, it is an option not to be lightly discarded, although it should only be exercised judiciously in selective circumstances.

For example, there is a strong argument that we may legitimately invoke our purchasing power to protect international common-pool resources, such as whales and dolphins, of which we are (in a sense) "co-owners." And we should not hesitate to invoke it (as we have in the past)[9] to discourage—even merely to protest—truly egregious, even macabre environmental practices, for example the decimation of tigers to produce tiger penis soup at $320 a plate (*New York Times* 1993). Trade pressure—and in some cases, trade pressure alone—is capable of dissuading or at least bringing to the bargaining table environmental miscreants and those who support them. (Compare the threats denying China Most Favored Nations status if it fails to improve on human rights.)

Compensation to Reduce Consumption of Biological Assets

No one can predict how the bounds between legitimate and illegitimate trade measures will evolve. But it is safe to say that the practices to which environmentalists typically object will continue to lie beyond the reach of trade "sanctions," just as they will continue to lie beyond the reach of any truly toothy international law mechanisms, such as "tort"-style relief.[10] Converting a species-rich habitat to more remunerative uses is not the same as permitting tiger slaughters. Nor is it the same as impairing the ozone shield, which is common property, the loss of which threatens intraterritorial damage.

To illuminate the options in the more typical case, let us return, then, to $N$ and its penguin habitat. If $A$ and $B$'s moral suasion fails to convince $N$ to save the habitat, and taxes, torts, and even trade sanctions are practically unavailable, $A$ and $B$ may have no choice but simply to pay $N$ for the penguins' maintenance. That is, many international disputes over biological assets may have to be approached as problems of public goods provisions. In terms of figure 1, if $A$ and $B$ want $Q$ penguins at the price $P^*$, they may have to come up with $OMEQ$, the cost to nation $N$ of providing the $Q$ penguin habitat (or the cost to an international body, if the penguins lie—on the Antarctic, say—outside the jurisdiction of any nation).

As between the payers, $A$ and $B$, inherent in the notion of the penguins being a public good is that each party "takes" the same supply. But from that supply, each party receives, as we saw, a different level of benefit. One presumes that each nation's contribution should equate with its marginal benefit, $A$ paying $ON$; $B$, $OL$ (where $ON + OL = OP^*$).

There are, however, many impediments, well reviewed in the literature, that hinder cooperative solutions to public goods provisions—these are solutions where, as in the international arena, an authoritative central authority with taxing power is lacking.

Even if all parties are represented and know their preferences, cooperation can be frustrated by strategic behavior. For example, $A$ might well underrepresent the true value to it of $Q$ penguins ($ON$ in figure 1) in the hope of a "free ride," namely, that enough other nations will cooperate even if $A$ defects so that the preserve will be underwritten at no or minimal ($<ON$) cost to $A$.

The nation on the other side of the deal—the country whose cooperation in maintaining the penguins is being purchased—may disturb negotiations with strategic behavior of its own. If the benefiters' ($A$'s and $B$'s) aggregate willingness to pay exceeds $N$'s true opportunity costs, $N$ may be tempted to overstate what it needs to receive to maintain the habitat (its costs), hoping to induce a higher price. And then, even if $A$, $B$, and $N$ can agree on a price, the parties still have to devise some institutional way of ensuring that $N$, especially as governments change, will continue to apply the funds for the agreed-on purposes.[11]

As a consequence, arranging subsidization to achieve the ideal level (corresponding in figure 1 to $OQ$ penguins, and in figure 2, to retaining $OQ^*$ forests) is no easy task. Yet, even with all its impediments, subsidization would appear to have at least as significant a role to play in the management of biological assets as its alternatives.

Let us summarize up to this point. No single approach is perfect—nor are we forced to rely on any one strategy to the exclusion of others. The best beginning is to recognize—and publicize—the institutional and economic structure that underlies and shapes the problem. Many nations host biological assets that are either pure public goods, or, as joint products with marketable goods, retain an appreciable layer of public good benefit, even after the domestic legal regime has gone as far as it can to appropriate benefits and internalize costs through available legal instruments. This would destine many biological assets to undersupply in the best of circumstances. But we are not in the best of circumstances. Many important biological assets lie within the jurisdiction of less-developed countries; the pressures that many are under to pay down loans and simply survive from day-to-day is destined to accelerate the consumption of wildlife areas.

There are some "stick" devices (such as taxes, tort suits, or trade sanctions). Yet, their employment is almost certainly going to remain marginal, reserved for a narrow band of quite egregious situations. As a consequence, we face trends that cannot be righted without subsidies in some form. It is not that we get no public goods unless we pay for them. We can get some police protection without a formal police force. The alphabet, algebra, the concept of zero, and various recipes for beer—public goods all—made their appearance without subsidies (or even patent protections). The point is, the supply of biologically rich areas is doomed to be suboptimal while consumption remains superoptimal absent a concerted effort by the world community to make conservation an attractive option to those with jurisdiction over the resource.

## Genetic Resources

Against this general backdrop, let us turn now to the question of genetic resources. The CBD, together with much of the literature, singles out these resources as a special case requiring specially tailored approaches.

In some ways, genetic resources do display intriguing properties. But the most important issues they raise can be brought under the same headings we have already discussed in connection with other biological assets. To put it another way, the overall theoretical framework is the same, but some distinctive properties of genetic resources influence the selection of specific policy options.

What distinguishes genetic resources from most other resources can best be introduced with a contrast. In exploiting a typical asset—say, the bark of the cinchona tree to extract quinine—what the exploiter is after is a substance, a raw physical input for production. The bark is a classic private (or privatizable) good, much like lumber. Its use is rival: if I appropriate the tree's bark, it is not available for you. More medicine depends on more bark. The more trees, the better.[12]

But when our concern shifts to the genetic resource value—the possibility that there is a synthesizable secret waiting to be discovered in the bark of the $z$ tree—what we are interested in is not substance so much as information. If $F$ is a pharmaceutical firm prospecting for genetic material, $F$ intends nothing that will threaten the forest. What $F$ is after is fundamentally little more than a few seeds or leaves—or, more exactly, the genetic information they encode. The information could be supplied from a single sample, if need be, and copied.[13] Once copied (and eventually, hopefully, synthesized), the value is shifted from the original specimen to its synthesized end product. In that process, the use of the information is nonrivalrous. That is, $F$'s carting off of sample leaves and seeds for information does not diminish the availability of the same information to the country of origin (hereinafter COO). Nor does $F$'s use of the information to produce a better seed or medicine conflict with anyone else's similar use of it. In the same way, one person's enjoyment of a Shakespeare play does not conflict with another person's enjoyment of the same work.

In its nonrivalrous aspect, genetic material, conceived as information, displays one distinct marker of a public good. As Fisher and Hanemann (1986, 170–71) point out, there is an appreciable option value in retaining biodiversity in the following sense. A decision to extinguish a biologically rich area (to convert it to another use) has the characteristics that the decision is (practically) irreversible, and that we may at a later time acquire better information about the costs and benefits of the decision to conserve or convert. (That is, with the passage of time, we better appreciate the value of any element in a biological "hot spot.") In this context, the world enjoys a sort of "flexibility premium" by postponing development.

But the information cached in a forest, including its option value, is not a pure public good, like atmospheric quality or climate. In the instance of a pure public good, recall, there is no feasible mechanism to prevent each member of the community from receiving the same output. The good in those (pure) cases is, in addition to being nonrivalrous, nonexclusionary. Genetic material (the information) resembles, in this respect, the wildlife preserve: its use is nonrivalrous, but, just as we can fence and charge admittance to the wildlife preserve, we *can* arrange various legal rules, such as intellectual property laws, so as to charge those who most immediately benefit from the information. The question is, should we?

To approach that question, we have to change focus slightly from the genetic information itself to the source of the information: the geographic area in which the genetic resources are naturally stored. Although the use of the information is nonrivalrous and can be made available to all to share at no cost, the same is not true of the resource area. To continue with the example of a forest (although certain marine areas, also valuable storehouses, would illustrate equally well), the genetic storage capacity is a joint product with some other forest products. A strategy of holding a genetic inventory at least roughly tracks a strategy of carbon sequestration and nut production: the more trees, the more nuts, the more genetic material. But with other uses of the land, gene storage is in clear competition: the more acreage we conserve for gene storage, the less acreage is available for wheat and cattle.

As the value of the competing uses rises, raising the opportunity costs of gene inventorying, it is possible to remove specimens from nature and store them *ex situ* in special gene banks and even museums. But *ex situ* storage, for all its advantages—like ready access and insurance against habitat loss—is not quite the equivalent (Benford, this volume). Removal of a specimen to storage "freezes" its evolution at a point in time. *In situ*, the organism continues to adapt to new stresses of viruses and climate (Grossman 1988, 261). In all events, keeping in mind that as the costs of *in situ* storage rise, some substitution by *ex situ* storage is appropriate and inevitable. It is clear that overall, the analytic framework for optimizing genetic resources is essentially the same as guides us to the optimization of any other biological asset with a public goods layer.

Figure 3, therefore, is simply a modification of figure 2. It considers the genetic

FIGURE 3    Consumption (Conversion) of Forest under Five Legal Regimes
(Modified to Account for Genetic Resources)

Price/Acre Forest

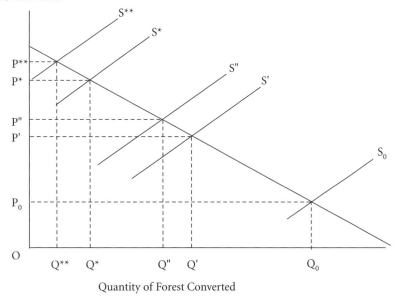

Quantity of Forest Converted

information to be one more forest resource, the exploitation and conservation of which varies with legal regimes. Illustration of perverse subsidy effects are deleted for simplicity. $S'$ is the supply curve with ordinary landowner protection, as before. $S''$ represents a further possible raising of the supply curve (retarding of deforestation) that might be achieved by special measures of added legal protection for genetic resources; the legal policy options range from securing the power of the asset manager, including the coo, to excluding "trespassers" from access to biological resources, to awarding intellectual property rights (IPRs), such as patents, on valuable species with the manager's borders (see below).

The main point—again—is that each additional measure of protection is aimed at raising the supply curve, dampening conversion of forest cover to the correspondingly reduced level $OQ''$. $S^*$ expresses the demand for the various public goods discussed above: carbon sequestration, wildlife refuge, and so on. (We can imagine transfer payments from neighboring nations for such benefits as flood protection to be folded into $S^*$.) $S^{**}$ represents additional demand that the forest be conserved and reflects the widely radiating public benefits of genetic resource portfolios. The law—even with its most favorable skewing of landownership, patent laws, and other devices—cannot deliver these benefits to the asset manager, otherwise they would be represented by $S''$.

## The Legal Options

Within a general framework, mentioned above, what are the legal options for achieving the desired shifts? The first goal of the law is to optimize the supply of genetic resource areas, to foster the shifts in supply of *in situ* storage areas so that the ideal (non)consumption of the genetic resources will be achieved $OQ''$—or even a more stringent $OQ^{**}$, as may be warranted by the opportunity costs of the land, robustly calculated.[14]

The second goal is to distribute valuable information, once it has been identified. There are, as we have observed, no marginal costs to the use (reproduction) of the information, and the *prima facie* inefficiency of charging people for a free good is patent: some consumers will be denied a product that costs nothing.

The principal proposals to accommodate genetic resources within the legal system, basically, involve three modes of rewarding the host for conserving genetic resources by reinforcing the coo's power to exclude from the resource area; empowering the coo to embargo exportation of resource samples; and giving the coo an intellectual property right in the resource (ordinarily a species) *and/or products derived from it.*

## Exclusion from Access to Resources

Start with the power to exclude. Nation *N* has or believes it has valuable genetic resources. Foreigners want to come in and exploit them. It is unlikely that *N*, in forbidding them from doing so, would violate any rules of international law. That would remain true even if *N* discriminatorily permitted its own citizens to engage in activities from which foreigners were restricted.

The CBD does not appear to change this picture in any way. It is true that Section 15 enjoins "each Contracting Party [to] endeavor to create conditions to facilitate access to genetic resources for environmentally sound uses by other Contracting Parties and not to impose restrictions that run counter to the objectives of this Convention" (United Nations 1992b, art. 15, sec. 2). This, by itself, would appear to undercut the right to exclude. But the same section provides that such access shall be "on mutually agreed terms" and "subject to prior informed consent" (United Nations 1992b, art. 15, sec. 2, paras. 4–5). In other words, *N* is confirmed in its power to exclude—but is urged to waive it for a price. The practical result is that *N* can negotiate genetic prospecting contracts with firms that want access to resource areas (Reid, this volume; Sittenfeld and Lovejoy, this volume; also, see below).

## Embargoing Exportation

The export restriction is a slight variant. A firm *F* (foreign or domestic) has identified a valuable resource. *N* wants to prevent *F* from taking it out of the country. Several countries have reportedly tried this tack. Ethiopia once embargoed export of its coffee plant germ plasm (Macfadyen 1985); India, the export of genetic material related to black pepper and turmeric; and Taiwan, the distribution of sugar-cane germ plasm (Sun 1986). And Brazil tried to close its borders to the export of rubber germ plasm (Reid et al. 1993, 21).

As a matter of law, the status of the export restriction is slightly less clear than the right to exclude. A total prohibition on exports would almost certainly raise conflicts among parties to GATT, because it is a quantitative restriction. Some em-bargoes, such as the United States's embargo of oil exports, can be validated under GATT's Article 20 exception for measures aimed at conserving "exhaustible natural resources." But no such resource-conserving defense would appear to be available to a nation that banned the export of a sample of its germ plasm. On the contrary, allowing some of its plant germ plasm to leave the country is a way in which a country can provide for restocking if the domestic stock is eliminated in a disaster. Yet it is not clear on what basis an export tax (even a very high export tax that ef-fectively throttled exports)—as distinct from a prohibition—would violate any-thing in GATT. The point is, just as $N$ could use its power to exclude to extract (share in) some of the value (above), so it has the theoretical legal, if not the prac-tical, power to levy an export tax to extract some of the value of genetic resources.

## Are These Powers Defensible?

Existing law gives the COO the power both to exclude and embargo, and so to ex-tract value. Is there any reason to eliminate or curtail these powers? The point to keep in mind is that the case to exclude access to genetic material, the use of which is nonrivalrous, is not as strong as the case for excluding access to ordinary private goods, such as timber. Open access, so serious a menace to efficiency in the context of rivalrous resources that we seek to eliminate it through privatization, would be, in the context of these nonrivalrous goods, a defensible improvement. Indeed, ge-netic resources have historically been free-access goods, with the result (as with all free trade) that the world has benefitted mightily.

The counterargument, of course, is this: while free access may have been desir-able at certain points in civilization, competing demands for the inventory-rich areas have never been higher. This dictates creating new and stronger incentives for the landowner (or sovereign) to withhold potential resource areas from in-compatible activities. The genetic information will not be produced at an efficient level unless we make it worthwhile to the asset manager to warehouse the genetic material. The right to exclude (and therefore license entry) and embargo (and therefore tax export) moves $S''$ above $S'$. This will retard consumption from $OQ'$ to $OQ''$.

However, does this represent an efficient incentive? Or are we providing mo-nopoly rents? The issue is a familiar one. The law can underwrite added invest-ment by manipulating a host of (broadly speaking) "property" laws. Whether it is wise to do so depends on a complicated conjecture as to the benefit of the incen-

tives so provided (inducing investment in the good) versus the risks of unjustifiable monopoly power (inducing restriction in supply).

As a practical matter, the risk of these powers generating rent is insubstantial—
but it is a matter on which better-informed judgment should be sought. For one
thing, the chances of any COO harboring the only cereal germ or even the *only* cure
for influenza are not large. And even in the infrequent cases where the COO has a
relatively unique resource, and there is therefore the potential for rent over some
time frame, legal measures to fortify a property interest through exclusion or export embargo are simply so hard to police and enforce that they would be unlikely
to make much difference.

To find a successful monopolization, one may have to go back in history over
1,000 years: the ancient Chinese long denied the West the secret of silkworms, but
all it took to finally break the monopoly (legend has it) were two Nestorian monks,
who bootlegged some eggs out of China in their pockets in 552 A.D. (*Encyclopedia
Americana* 1989, 816).[15] The reason for the enforcement difficulties (both of the
embargo and the exclusion, although perhaps more of the former) is worth noting. The owner of a timber stand gets a tangible advantage from property rights
because the ownership interests are relatively easy to monitor and enforce. That is
because any appreciable invasion of those interests—cutting and carting away a
timber stand—requires repeated trespasses, which will ordinarily provide the
owner ample opportunity to invoke enforcement mechanisms. By contrast, probably most genetic samples can, like the silkworm eggs, be poached surreptitiously
and with relative impunity—particularly from under the eyes of a national authority, where a local landowner or official is willing to deal with the appropriating
exporter. A single "leak" may be hard to detect even after it has occurred. And
should the smuggle eventuate in a commercial product, one cannot count on trac-
ing the innovation back to the true COO, particularly in a world with increasing
synthetization and blending with innovative steps.[16]

In fact, the practical problems are severe enough that there is less likelihood that
empowering $N$ to exclude or embargo will generate rent than that it will fall short
of moving the "new" property regime supply, $S''$, out to the desired $S^{**}$ margin.

Intellectual Property Rights
The possible shortcomings in the exclude/embargo tactics strengthen proposals
to award the COO an intellectual property right (IPR)—in the form of a patent,
copyright, or trade secret protection[17] on potentially valuable species (or molecules).[18] IPR protection would not only offer further reward; it would be a partial
response to the problems of policing that undermine the exclude/embargo options. Illicit smuggling would be policed not only at customs inspection points; in
tracking down the exploiter of a smuggled resource, the COO might enjoy the additional powers provided by local and international IPR rules.[19]

The question is under what circumstances would existing law allow such a thing—allow, say, Peru to acquire some form of IPR protection for its indigenous tomatoes—or, would no real reforms be required? The rules for each type of IPR vary, among themselves and jurisdictions. Patent laws, for example, characteristically require some sort of inventive step. The details of the innovativeness requirement vary from jurisdiction to jurisdiction, but universally would appear to rule out protection for the discoverer (or the COO) of the novel tomato. Similarly, to acquire a copyright, the applicant has to show that the work owes its origin to the author. This may leave room for the "author" of a novel gene sequence to argue for copyright, but certainly the COO cannot claim original authorship of the information naturally stored in a leaf or tree. The law as it now stands is therefore not congenial to COO acquisition of an IPR in any naturally occurring species.

Whatever the requirements of present law, they are, of course, open to amendment. Why should patents be available to encourage research and development that accrues to the advantage of the developed world, while the same laws do not encourage conservation, which would inure to the advantage of many species-rich, less-developed countries? The question, however, is not rhetorical. In demanding uniqueness and some equivalent of an "innovative step," traditional patent laws are not just being fussy. They are responding to important considerations.

Patent rights convey monopoly power that should not be doled out lightly. Consider a worst-case scenario. Reforms in the law favoring $N$, such as COO patents, have no effect on $N$'s domestic policies: rather than respond with special set-asides for conservation, $N$ continues to deforest as before. Timber from a ravaged forest is shipped to the United States where a research firm, $F$, working with a sample of tree $T$, produces after many years of work a valuable molecule. $F$ files for patent protection on the improvement, $T_1$. Informed of the development, $N$ files for a patent on $T$. $T$ is also found in Peru and Argentina—but they file after $N$.

If $N$ has a patent on $T$, vis-à-vis $F$, it potentially shares a bilateral monopoly position with $F$, although $N$ has expended no effort. "Rewarding" $N$, at a cost to $F$, would constitute a disincentive to efficient investment in research and development. $F$ would be required to divide its returns with $N$, which has made no deliberate investments at all. Moreover, vis-à-vis Peru and Argentina, is $N$ to be given a patent-term monopoly on $T$ and uses from it? Can $N$ restrain the rest of the continent from using $T$ because $N$ applied to some patent office first?[20]

These are all challenges to iron out, but they need not be fatal. Perhaps specially tailored IPRs could be formulated to cope with such problems, along the lines of the special provisions the law has made for sexually reproducing plants[21] through innovative plant breeder's rights (PBRS). For example, the worst-case hypothetical suggests that we might condition the award of a species patent on the demonstration of special efforts—for instance, conservation set-asides—as an equivalent of innovative effort. As with traditional homestead or mining claims, the patentability of a species might be made to turn on the applicant demonstrating a certain level of appropriate investment in the resource area. In regard to the problem of nonunique source—for example, a specimen that ranges across $N$, Argentina, and Peru—the system might provide for a division of royalties according to some formula, or at the least, a right of "free use" by countries other than $N$ that can show that the molecule is natural and indigenous to them. Indeed, perhaps there should be a division that provided some resources to an international organization that funded the conservation of key areas (see below).

The fact is, patents (or other IPRs) could be fashioned in such a way as to provide a further incentive to COOs that might be welcome in some circumstances. But it also adds to the risk of deadweight social loss: monopoly profits for genetic resources, the supply of which was inelastic to the law.

Hence, here too, whether to support IPRs in this area depends in some measure on a complicated conjecture as to whether the marginal benefit of the incentives so provided outweighs the risks of unjustifiable monopoly power. The calculations are complicated because it is all too easy to exaggerate the monopoly perils of patents. Many countries do not formally recognize trade secrets and patents, or do not enforce them in good faith. Even where they are enforced, the costs of proving infringement undermine their economic value. This is true even in the context of patents for normal mechanical contrivances and processes. Efforts to patent natural molecules and species invite novel, unmapped disputes, raising enforcement costs and eroding the value even further. For example, in regard to plant genetic resources, seed lines are so thoroughly blended that for most crops it would be difficult to determine which genes had originated in which country (Wilkes 1987, 215). The same problems of blended sources would vex IPRs for industrial applications.

In conclusion, IPR protection for the COO is worth further discussion—to examine, among other things, the significance of the range of cases where it might do more good than harm. The case for them is not as untenable as it first may have seemed. Indeed, it appears that IPRs are just as likely to fall short of our ideal—of inducing the efficient supply—as curtailing that supply through economic rent.

Genetic Resource Prospecting Contracts

In contrast with embargoing or patenting the resource, the COO can negotiate an offer to assist in its exploitation. As indicated above, the power of the COO to negotiate such a deal would be advanced by firming up the COO's power to exclude foreign prospectors and to embargo exports. But control over access and exports, while a sufficient condition of a management contract, is not necessary. Even in the absence of these powers, the COO can offer valuable services. The COO's strength lies in the fact that the exploiting firm's ($F$'s) position is improved if it can count on some continuing services that the COO is in a good position to supply or ensure—even though $F$ may be able to scuttle off with a few samples in any case.

Such services include smoothing relations with indigenous peoples and facilitating leads, often from traditional healers, as to which samples are most promising. Moreover, to enable $F$ to backtrack and recover fresher or closely related samples, specimens should ideally be dated, the exact place of origin and collection method recorded, and the location preserved. Obviously, all these continuing efforts can be facilitated by governmental patronage (Sittenfeld and Gámez 1993).

This is the route Costa Rica adopted to capitalize on its genetic resources (Sittenfeld and Lovejoy, this volume). The government established the National Institute of Biodiversity (INBio) as an autonomous, nonprofit organization. Then Merck and Company, in return for nonexclusive facilitation assistance, paid INBio $1.1 million up front and committed to an undisclosed royalty thought to be between 1 and 3 percent of any marketable products that may be developed.

This is an appealing and imaginative alternative well worth backing. It carries less risk of rent than the patents and other IPRs—even less than attaches to the fullest powers of embargo. Indeed, the management contract appears to be unexceptionable.

Its limits should not be overlooked, however (Simpson, Sedjo, and Reid, this volume). Given the nature of the business—the ease of a noncontracting competitor "poaching" or dealing directly with a local private landowner, and so forth—it is virtually impossible for a country to assure a $F$ access rights that are truly exclusive. In fact, the Costa Rica-Merck agreement makes no such pretense. Each $F$ knows that, in reality, other firms will still have the capability— reflected in the modesty of Merck's offer—to scout up resources on their own terms. Hence, the prospecting contract is a positive tool in promoting resource-area conservation. But there is no assurance that there is enough financial merit in bioprospecting that such contracts will raise the value of tropical forestlands appreciably.[22]

Modes of Ownership by Processing Firms
The debates over genetic resources have included the discussion of property rights in favor of the exploiting firm, no less than in favor of the COO. Much controversy has swirled over efforts by $F$ to secure a return on its investment through IPR instruments. These range from a full patent to a PBR, discussed above (Sagoff, this volume).

How far the law is prepared to extend IPRs to genetically engineered forms of life (plant and animal) remains an open question in industrialized nations, with the European Community and the United States moving along slightly different tracks (Cripps, this volume). There is much to be said about where the lines should come out. But in our context, there are only a few points that need to be noted.

First, some of the tension reflected in the literature—the suggestion that awarding an IPR to the exploiting firm works an injury to the COO—is simply unfounded. Awarding protection to $F$ benefits the COO as well (although third parties might object). This is so since the IPR, by enhancing the value of the specimen to $F$, thereby raises the offer that $F$ brings to the bargaining table. Moreover, it is hard for a debt-ridden COO to securitize the expected income stream from the biological portfolio tempting it to chop the trees for debt service. The COO's First World corporate partner, whose securities trade on global exchanges, is better positioned to monetize the future benefits of the biodiversity portfolio—and thereby to decelerate the pace of consumption. In fact, in any jurisdiction where the patent applicant has to truly identify the source of requisite materials under penalty of law, the system may reinforce the COO's ability to police poaching beyond its own boundaries. Thus, awarding a patent to $F$ for advances in the art, advantages both $F$ and the COO mutually.

Second, it follows from the value-enhancing effect of IPRs that strengthening IPR protection will lead toward a more welcome level of conservation. But, the magnitude of the shift (as well as the overall social benefit) is open to question. A broad, empirical survey of U.S. research and development officers suggests considerable differences in the importance of patents, industry to industry. Patents, however, were generally rated the least effective of various mechanisms of appropriating the value of innovation, below lead-time and learning-curve advantages (Levin et al. 1987, 794). This should be all the more true in the global context, where the patent claimant may face multinational litigation, even in jurisdictions that do not recognize IPRs.

The conclusion is that awarding patents for the improver has the advantage of moving consumption of resource-rich areas in the right direction: tempering conversion from $OQ_O$ to $OQ''$. But there is no assurance that IPRs for the improver (any more than IPRs for the COO) will succeed in raising the price/consumption to p**/$OQ$** in those cases where it would improve biological reserves. In summary, the practice can be endorsed, but it is destined to have less impact on conservation than one might suppose—and less than one would desire.

Subsidization Revisited
Each of the various instruments discussed above would, at their best, move $S''$ in the direction of $S$**. As figure 3 suggests, one might imagine, too, that in the course of arranging for subsidization of joint public goods *other* than genetic portfolio, such as carbon sequestration and wildlife refuges, we will move as far as $S^*/OQ^*$. But when privatization instruments and "first-tier" public goods subsidies have exercised their influence, we still may be overconsuming by $Q^{**} - Q^*$.

Hence, even when all the instruments we have examined thus far have exercised their sway, there will almost certainly remain a shortfall to be addressed. And the only means of addressing it (by definition, if all other major policy instruments have been deployed) is subsidy.

The CBD, of course, anticipates the need for subsidies from developed to developing country parties (appendix, this volume, art. 20 [2]). It has fallen to the restructured Global Environment Facility (GEF 1994) to collect and distribute funds to cover "the agreed incremental costs" of biodiversity-advancing projects.

Unfortunately, establishing a mechanism to play this role is fraught with difficulties. We have already alluded to the general problems that beset any attempt at multinational public goods provision; namely, overcoming the familiar impediments to collective action, like free riding and concealment of preferences. Worse, in the biodiversity context, there are several specific considerations that ally to exacerbate the ordinary difficulties.

### The Ambiguity of "Maximizing 'Biodiversity'"

Strategic posturing aside, negotiations over subsidization can be destabilized by honest differences over which assets are to be prioritized. Selecting which areas to set aside as genetic storage sites is not as straightforward a business as selecting areas to preserve gorillas or tigers. We know with a fair certainty which areas we have to "rent" to protect such endangered species. It is far more difficult to anticipate what needs humankind is going to have in the future, and which resource areas are most likely to be able uniquely to satisfy them.

One wants to offer the maximization of *biodiversity* as a good proxy for human needs (Norton, this volume). But even among biologists, the term is too open-ended to signal clear guidance for policymakers.[23] Some identify it with maximizing the sheer number of species, in which case, preserving the prolific rain forests would be the highest priority. But others champion grassland areas, which are more important custodians of "higher taxonomic categories"—notably mammals, which generally store, species for species, more genetic information than the lower life with which the wetter jungles teem. But this is not consistently true. While the number of genes per organism range from 1,000 in bacteria, to 10,000 in fungi, to 100,000 in the typical mammal, some flowering plants carry 400,000 or more (Swanson 1994, 4). If variety is referred to as the breadth of taxonomic categories—classes, divisions, phyla—fantastically exotic marine habitats may merit the highest priority (United States 1991). Others maintain that biodiversity is best measured not by the number of species we conserve, but by the evolutionary distance between them, and what we should aim to optimize is a function of that distance (Weitzman 1992; Solow, Polasky, and Broadus 1993).

To make collective action even less harmonious, some ecological disturbances can be viewed not only as inevitable, but as necessary to the health of an ecosystem (Lugo, this volume; Botkin 1990). "Disturbance ecology" reminds us that, at the extreme, biodiversity in the added-species sense may hinge on exposing life to stresses: it is, after all, the destabilization of ecosystems that furnishes new species (Pickett and White 1985; Mayr 1963). Consequently, it is not altogether clear when, as defenders of biological diversity, the negotiants would rationally buffer an ecosystem and when they would expose it. The Independent Evaluation of the GEF Pilot Phase called attention to the fact that "the GEF still lacks a good operational definition for biodiversity. . . . Thus, success is hard to define and even harder to measure" (UNEP, UNDP, and WB 1993, 55).

The Uncertain Value of Biological Diversity
Suppose that even without complete agreement on criteria, the negotiants are able to agree on a consensus list of high priority areas. Even then, there remains the problem of valuation. Everyone agrees that the total value of the world's genetic library is immense. Few are untroubled by the massive, swift conversion of forests and wetlands. The question for subsidy negotiants is subtler and more controversial: What is the marginal value of a promising species or acre? The range of current estimates is so broad as to constitute an independent negotiating obstacle: from $24 million per untested species *in situ* to $44 (Simpson, Sedjo, and Reid 1993, 8).

There are several reasons why the economic value of any "book" in the genetic library may be both hard to settle and closer to the low end of estimates than the high. For one thing, Simpson, Sedjo, and Reid (1993, 8) point out that current high-side estimates overlook a considerable redundancy factor—the high probability that any particular need (a cure for $x$) is likely to be satisfied by more than one untested wild organism. After all, the number of presently untested species is exceedingly high; the number of medical conditions and other applications is limited; and no one tree, for example, is likely to be the unique repository of a valuable "code" (that is, if we eliminate the Brazilian $x$ tree before we have unlocked its hidden cure for gout, the Argentinean $y$ bush—that evolved in response to the same environmental stresses—is apt to be reserving for us the same properties). Hence, Simpson, Sedjo, and Reid conclude that "if the set of organisms that may be sampled is large, the value of the marginal species [or marginal acreage] is small." If so, it is quite possible (generalizing from their argument) that the incentives the world community would be warranted to provide species-rich countries to conserve those areas on the basis of realizing the potential medicinal, industrial, and so forth, value of genetic specimens would be unlikely to bring forth a dramatic increase in acreage conserved. $S^{**}$ may not lie appreciably above $S''$ at the present.

The Institutional Quagmire Controversy
Finally, all efforts to underwrite repair and protection of the global environment are dogged by problems of institutional design. Problems have included the allocation of voting power between the prospective donor countries—the developed nations—and the prospective recipients, the resource-rich, less-developed countries. And there are unresolved tensions between efficiency—funding projects to

maximize global cost effectiveness—and equity—funding projects in a manner calculated to achieve regional "balance" (UNEP, UNDP, and WB 1993, 56).[24]

Moreover, as mentioned above, the GEF, the CBD's lead agency, is charged with underwriting the incremental costs incurred by a host nation in achieving global environmental benefits of agreed-on measures. The test, if left intact, could prove a three-stage nightmare of uncertainties. First, there has to be agreement that some proposed measure advances GEF priorities. Then, at the second stage, in determining the amount of the fund, there are the stock complications of "additionality"—untangling which expenses are readily attributable to the compliance effort. For example, nation $N$ proposes to expend $1 on a forest management project. How much (what baseline) is attributable to the achievement of nondiversity benefits that $N$ would have undertaken anyway? Only the investment above the baseline is "additonal." But the third stage, if it is insisted on, will be even trickier. Once the benefits of the biodiversity-enhancing component have been disentangled, the GEF is being urged to calculate $N$'s own domestic benefits and subtract them from the gross additional cost, on the view that only the portion of the project costs attributable to spillover global benefits should be eligible for funding (GEF 1993b).

Resolving all these complexities is destined to hinder subsidization efforts. Yet the conviction that at least some degree of subsidization is merited should give us determination to press ahead. Even if no single criterion—such as a straightforward maximization of species—can generate a consensus ranking of all resource areas, a well-informed advisory board could probably identify to general satisfaction a set of high priority areas within budget constraints. Uncertainties over special conservation incentives based on the genetic resources by themselves does not derail the case for subsidizing biologically rich acreage. Areas well endowed with genetic resources typically harbor other biological assets as joint public goods. In the case of many resource-rich areas, even if the argument for subsidizing each such product, considered on its own terms, is indecisive, the argument for subsidization will be strengthened for areas that display a bundle of complementary products.

Moreover, we are not reliant on the GEF alone. A host of other agencies and nongovernmental organizations are in the business of gathering and distributing conservation funds—from the Man and Biosphere Program (Batisse 1993), to the Food and Agriculture Organization, to the World Conservation Foundation.

Further, thought needs to be given not only to agencies and agency design, but to modes of wealth transfer. An agreement with a COO to set aside a resource-rich area in exchange for a front-end lump sum payment will not ensure continuing performance of the COO's obligation as effectively as periodic "rentals" or disbursement from a fully funded endowment. There have been suggestions that, in some areas, conversion of resources-rich areas could best be retarded by emphasizing investment in agricultural productivity of lands outside the biological hot spots, thereby relieving some of the pressures to expand inside the reserves (GEF 1993a). There are also major policy choices to be resolved regarding the division of financing between *in situ* and *ex situ* conservation, and how to divide emphasis between multiple-use and dominant-use strategies (Vincent and Blinkley 1993).

## Conclusion

The charge that the earth's biological riches are being plundered is not the product of a few environmentalists' overheated imaginations. Reports of accelerating species loss are all too genuine. And the imprudence of it for humankind—sentiments aside—is confirmed by the coolest verdict of economic theory. Across the world, biological (living) assets—and indeed, a whole range of environmental inputs—are being underpriced, altogether unpriced, or even subsidized out of existence. In those circumstances, overconsumption, even if we speak only in efficiency terms without entering the conundrums of environmental ethics, is inevitable.

We can relieve the stresses significantly through the privatization of resources and embrace of optimal-pricing strategies. Asset-protecting privatization can take many forms, from the government simply establishing and enforcing property boundaries to its abjuring uncompensated "takings." Optimal pricing involves renouncing subsidies that perversely foster overconsumption of the biological base, and ensuring that prices reflect true long-term social costs, externalities included.

By making conservation more profitable to the manager of the asset, each of the various internalizing mechanisms can temper the pace of destruction. There is a lot of good work to be done refining and refitting these traditional mechanisms in the context of new technology. It may even be defensible to adapt patent or other intellectual property laws to provide countries of origin an additional incentive to conserve especially promising acreage. A case has not been made for any single instrument; a range of measures may work comfortably in combination.

But even were the best of those measures in place, the pace of destruction would still be too fast. That is because there inheres in the earth's biological assets—tigers, lions, portfolios of germ plasm, and marine resources of the global commons areas—an uneliminable public good dimension. The full social benefits of these things are beyond appropriation by markets, even when market value is fully enhanced by all the devices of the law.

Theoretically, there are two approaches to the overconsumption of public goods. We could sanction the offending nation with sticks: taxes, torts, trade sanctions, and so on. But as both a practical matter and a matter of international law, the room for such measures is distinctly limited.

That leaves us with carrots: essentially, transfer payments. The case for an overlay of selective subsidization appears all the more imperative when one considers how much of the resource base lies in countries that are poor, debt-pressured, and operating, presumably, under steep discount rates. Any given forest may be a valuable trove of biological assets—some day. But clearing its trees for timber and crops will produce cash, and underwrite debt service—now.

Thus, there is every reason to fear that the earth's biological assets will continue to be "consumed," even from the most unsentimental point of view, at the wrong rate. Reasonable persons can differ over the right rate. Contingent valuations are involved, and people differ as to how these are to be measured. Of course, some people attach a higher shadow price on the existence value (or moral claims) of whales and snails than do others. Those debates, however, as absorbing as they are for moral speculation, implicate differences largely of detail and degree. Far more striking than the quarrels is the common ground of agreement: unless we modify our institutional frameworks, foresight, and most important, willingness to subsidize, the squandering of our biological base will continue.

## Acknowledgments

The author would like to acknowledge the comments of Scot Altman, J.P. Benoit, Howard Chang, Linda Cohen, Michael Knoll, and Edwin Smith, and the research assistance of Robert Tesler, Tony Christopolous, and Steve Harris.

## Notes

1    As of April 1997, 165 nations, including all the major industrialized ones other than the United States, had ratified the convention.

2    "Potentially," because privatization of the berries depends on the state's commitment to exclude others from the picker's (or landowner's) exclusive rights over them once he has lawfully appropriated them (or the land).

3    This is true of nonfugacious resources: rational management of fugacious resources, such as migratory animals and flowing water, requires more complex arrangements.

4    The litany of measures, some of which are discussed in the text following, range from a tax on effluents to damage suits against polluters.

5    "Subsidy" should be understood broadly. For example, local land laws can have the same perverse effect when they require would-be claimants of public lands to exhibit dramatic evidence of their possession, such as chopping or burning the forests.

6    Representing in a vertical summation the marginal benefits to $A$ and $B$ for each level of supply.

7    $Z$ could try to internalize the risks on $N$ through judicial action or diplomatic maneuvering, but under existing international law, success is not probable.

8    GATT, Part 2, Article 11 disallows all prohibitions and restrictions beyond the limited duties and taxes delineated elsewhere in the agreement.

9    The United States has, for example, denied import permits for marine mammals harvested extra-jurisdictionally in ways that would violate U.S. laws—actions that have been upheld by U.S. courts. See *Animal Welfare Institute v. Kreps*, 561 F.2d 1002 (D.C. Cir. 1977), a case in which the importation of South African seal skins was banned.

10 The reasons for this skepticism are presented in Stone (1993, 37–70).

11 One answer is for the donors to schedule periodic payments ("rent") in lieu of a lump sum cash payment up front; the rent is contingent on the leasing sovereign's continued compliance with the conservation conditions.

12 The same is true when we regard the trees as sequesterers of carbon (a public good). The marginal benefit of each additional tree is fairly constant over a colossal range of supply: climate-wise, the more trees, wherever in the world, the better.

13 While one small stand of $z$ trees will do, several scattered stands would be nice as "insurance" against fire and pests. In a pinch, we do not even have to store the samples *in situ*; we can, and do, store specimens *ex situ* in gene banks. As discussed below, however, there are some advantages to retaining fresh, multiple samples *in situ*.

14 As I have explained elsewhere, I am prepared to include in the shadow price for some resources a "morally corrected shadow price" over and above our ordinary willingness to pay—an extra measure reflecting our moral obligation not to destroy certain natural objects, such as planetary heritages (Stone 1987, 78–79).

15 Inasmuch as silk was discovered in the third millennium B.C. and the silk culture flourished no later than the Shang dynasty (sixteenth to eleventh centuries B.C.), this monopoly—uncomplicated by GATT and probably backed by capital punishment—enjoyed a 1,000-year life.

16 In the case of plant genetic resources, at least, the whole idea has been labeled "unworkable" because the final product, the crop, is typically a blending of genes from so many different sources. Which country or countries would a farmer be obliged to pay (Wilkes 1987)? Wilkes's analysis—a response to Kloppenburg and Kleinman (1987)—is excellent. The same problems of blended sources would vex IPRS for industrial applications.

17 As distinct from patents, trade secrets protection potentially lasts forever. Many countries provide no legal protection for trade secrets, and where the protection exists, enforcement is uncertain and costly, with the result that the role of trade secrets is almost certainly marginal in the biodiversity area (Gollin 1993, 164; Gollin 1991, 201).

18 I interpret Roger Sedjo (1989; 1988, 310) as favoring a patent-like right for natural species.

19 Under U.S. law, the applicant has to deposit a specimen; it is less certain that the applicant would have to specify the place of origin of the specimen. If the origin has to be identified, a firm that covertly poached and failed to disclose origin would run the risk of fraudulent procurement.

20 Even if we assume that the plant $P$ is unique to Peru, patent protection based merely on place of origin invites highly problematic disputes. For example, a British firm has genetically engineered a line of potatoes that produce polymers rather than starch. Imagine that Peru, as the COO of the original potato brought back to England, had the potato under patent and sued for infringement (ignore laches). Alternatively, suppose the COO has a patent on the species $P$, which it licenses to $F$; $F$ derives a marketable product from $P$, for which it pays the COO royalties. But $F$ quickly proceeds to develop a new product, $P'$, which the COO alleges was in some way made possible by the license on $P$. Should the law—can the law—successfully engage such controversies, particularly when the parties are, as they ordinarily will be, foreigners to one another, sometimes even nation states able to claim refuge in acts of state and sovereign immunity doctrines?

21 Innovators face the threat that when their development germinates in the hands of the first buyer, the first buyer can easily set aside some of the newly produced seed to sell in competition with the innovator.

22 The contentiousness over "equitable sharing" is almost certainly disproportionate to the likely booty. Simpson, Sedjo, and Reid (1996), using what they label highly optimistic assumptions to estimate the pharmaceutical value of the world's top 18 "hot spots," conclude that some areas of western Ecuador might be worth $20 a hectare as bioprospecting sites, but their median is only

about $2. Others, using a different methodology, conclude that the world's 3 billion hectares of tropical forest are worth, on average, 90¢ to $1.32 per hectare as pharmaceutical "mines" (Mendelsohn and Balick 1995, 223). No real rush of pharmaceutical houses has materialized to mimic the Merck-Costa Rica arrangement, which is not known to have generated anything for Merck to date except publicity. Many will recall how the Law of the Sea negotiations were complicated and prolonged by heated efforts to divide up the riches of seabed manganese nodules—riches that even today have yet to materialize.

23  In the CBD, *biological diversity* is said to encompass "the variability among living organisms from all sources including, *inter alia*, the terrestrial, marine and other aquatic ecosystems and the ecological complexities of which they are a part; this includes diversity within species, between species, and of ecosystems" (United Nations 1992a, art. 2).

24  The problem is endemic in all GEF focal areas. In the case of climate change expenditures, if the GEF's aim were simply to maximize carbon reduction per budgeted dollar, the lion's share would simply be expended, quite efficiently, in China and India (ibid., 66).

## References

Arrow, K.J., and A.C. Fisher. 1974. Environmental preservation, uncertainty, and irreversibility. *Quarterly Journal of Economics* 88:312–19.

Batisse, M. 1993. The silver jubilee of MAB and its revival. *Environmental Conservation* 20:107–12.

Birnie, P.W., and A.E. Boyle. 1992. *International Law and the Environment*. Oxford: Clarendon.

Botkin, D.B. 1990. *Discordant Harmonies: A New Ecology for the Twenty-first Century*. New York: Oxford University Press.

Damrosch, L.F. 1989. Politics across borders: Nonintervention and nonforcible influence over domestic affairs. *American Journal of International Law* 83:1–50.

*Encyclopedia Americana*. 1989. Vol. 2, international edition.

*Endangered Species Act of 1995*. U.S. Code, vol. 16, secs. 1531–44.

Fisher, A.C., and W.M. Hanemann. 1986. Option value and the extinction of species. *Advances in Applied-Economics* 4:169–90.

Freeman III, A.M. 1979. *The Benefits of Environmental Improvement: Theory and Practice*. Baltimore, Md.: Johns Hopkins University Press for Resources for the Future.

GEF (Global Environment Facility). 1993a. Economics and the conservation of biological diversity. Working paper no. 2, prepared by K. Brown, D. Pearce, C. Perrings, and T. Swanson, Washington, D.C.

———. 1993b. Issues to be addressed by the program for measuring incremental costs for the environment. Working paper no. 8, prepared by K. King, Washington, D.C.

———. 1994. Participants meeting report. 14–16 March. *Instrument for the Establishment of the Restructured Global Environment Facility*.

Gollin, M.A. 1991. Using intellectual property to improve environmental protection. *Harvard Journal of Law and Technology* 4:193–235.

———. 1993. An intellectual property rights framework for biodiversity prospecting. In *Biodiversity Prospecting: Using Genetic Resources for Sustainable Development*, edited by W.V. Reid, S.A. Laird, C.A. Meyer, R. Gámez, A. Sittenfeld, D.H. Janzen, M.A. Gollin, and C. Juma. Washington, D.C.: World Resources Institute.

Grossman, R. 1988. Equalizing the flow: Institutional restructuring of germplasm exchange. In *Seeds and Sovereignty: The Use and Control of Plant Genetic Resources*, edited by J.R. Kloppenburg Jr. Durham, N.C.: Duke University Press.

Kloppenburg Jr., J.R., and D.L. Kleinman. 1987. The plant germplasm controversy: Analyzing empirically the distribution of the world's plant genetic resources. *BioScience* 37:190–98.

Levin, R.C., A.K. Klevorick, R.R. Nelson, and S.G. Winter. 1987. Appropriating the returns from industrial research and development. In vol. 3, *Brookings Papers on Economic Activity: Special Issue on Microeconomics*, edited by M.N. Bailey and C. Winston. Washington, D.C.: Brookings Institution.

Lyster, S. 1985. *International Wildlife Law: An Analysis of International Treaties Concerned with the Conservation of Wildlife*. Cambridge, England: Grotius Publications.

Macfadyen, J.T. 1985. A battle over seeds: The third world asks for a share of gene stocks bred in northern laboratories from southern seed. *Atlantic*, November, 36–44.

Mayr, E. 1963. *Animal Species and Evolution*. Cambridge, Mass.: Belknap Press.

Mendelsohn, R., and M.J. Balick. 1995. The value of undiscovered pharmaceuticals in tropical forests. *Economic Botany* 10:49.

*New York Times*. 1993. A bowl of tiger penis soup sells for $320 in Taiwan. Only while supplies last. Advertisement, 17 November.

Pickett, S.T.A., and P.S. White. 1985. *The Ecology of Natural Disturbance and Patch Dynamics*. Orlando, Fla.: Academic Press.

Reid, W.V., S.A. Laird, R. Gámez, A. Sittenfeld, D.H. Janzen, M.A. Gollin, and C. Juma. 1993. A new lease on life. In *Biodiversity Prospecting: Using Genetic Resources for Sustainable Development*. Washington, D.C.: World Resources Institute.

Saggs, H.W.F. 1989. *Civilization before Greece and Rome*. New Haven, Conn.: Yale University Press.

Sedjo, R.A. 1988. Property rights and the protection of plant genetic resources. In *Seeds and Sovereignty: The Use and Control of Plant Genetic Resources*, edited by J.R. Kloppenburg Jr. Durham, N.C.: Duke University Press.

———. 1989. Property rights for plants. *Resources* 97:1–4.

Simpson, R.D., R.A. Sedjo, and J.W. Reid. 1993. The commercialization of indigenous genetic resources: Values, institutions, and instruments. Paper presented at the Conference on Market Approaches to Environmental Protection, 3–4 December, Stanford University, California (paper available from Resources for the Future).

———. 1996. Valuing biodiversity for use in pharmaceutical research. *Journal of Political Economy* 104:1.

Sittenfeld, A., and R. Gámez. 1993. Biodiversity prospecting by INBio. In *Biodiversity Prospecting: Using Genetic Resources for Sustainable Development*, edited by W.V. Reid, S.A. Laird, C.A. Meyer, R. Gámez, A. Sittenfeld, D.H. Janzen, M.A. Gollin, and C. Juma. Washington, D.C.: World Resources Institute.

Solow, A., S. Polasky, and J. Broadus. 1993. On the measurement of biological diversity. *Journal of Environmental Economics and Management* 24:60–68.

Stone, C.D. 1987. *Earth and Other Ethics: The Case for Moral Pluralism*. New York: Harper and Row.

———. 1993. *The Gnat Is Older Than Man: Global Environment and Human Agenda*. Princeton, N.J.: Princeton University Press.

Sun, M. 1986. The global fight over plant genes. *Science* 231:445–47.

Swanson, T.M. 1994. *The International Regulation of Extinction*. New York: New York University Press.

United Nations. 1992a. Conference on environment and development. Rio declaration on environment and development. UN Doc. A/CONF. 151/5/Rev. 1.

———. 1992b. Framework convention on biological diversity. *International Legal Materials* 31:818–41.

———. 1994. Food and Agriculture Organization. *Review of the State of the World Marine Fishery Resources*. Rome: United Nations Food and Agriculture Organization.

———. Treaty Series. 1973. Convention on international trade in endangered species of wild fauna and flora. *Treaties and International Agreements Registered or Filed or Reported with the Secretariat of the United Nations*, vol. 993, no. 243 (1975).

UNEP (United Nations Environment Programme), UNDP (United Nations Development Programme), and wb (World Bank). 1993. *Report of the Independent Evaluation of the Global Environment Facility Pilot Phase.* 23 November.

United States National Resource Council. 1992. *Conserving Biodiversity.* Report of a panel of the Board of Sciences and Technology for International Development. Washington, D.C. National Academy Press (Citing to Thorne-Miller, B., and J. Catena. 1991. *The Living Ocean: Understanding and Protecting Marine Biodiversity.* Washington D.C.: Island Press.)

Vincent, J.R., and C.S. Blinkley. 1993. Efficient multiple-use forestry may require land-use specialization. *Land Economics* 69:370–76.

Weitzman, M.L. 1992. On diversity. *Quarterly Journal of Economics* 107:363–405.

Wilkes, H.G. 1987. Plant genetic resources: Why privatize a public good? *BioScience* 37:215–17.

# Aspects of Intellectual Property in Biotechnology: Some European Legal Perspectives

by Yvonne Cripps

The relationship between biodiversity and intellectual property with particular reference to patents is outlined in this chapter. Granting patents on novel genetically engineered organisms may have both advantageous and disadvantageous effects on biodiversity. A comparison of American and European attitudes toward the patentability of genetically engineered organisms reveals the "territoriality" of patents, which only protect the claims of the inventor in the country in which the patent is granted. Thus, inventors who wish to protect their inventions overseas will have to contend with some interesting differences between the patent systems of the jurisdictions in which they desire protection. The difficulties raised by Article 53 of the European Patent Convention and the European Commission's attempts to persuade European Parliament to agree on a Council Directive on the Legal Protection of Biotechnological Inventions are discussed. A decision of the European Patent Office on the patentability of the "Harvard mouse," and the relationship between the assessment of risks to the environment and the granting of patents are also examined.

*

This chapter traverses some recent European developments regarding the patentability of plants and animals in the context, *inter alia,* of opposition by special interest groups. It concludes that despite the arguments of opponents such as Greenpeace, the European Patent Office is correct to decide that it should not lightly impugn patents on organic inventions on the ground that they are contrary to morality. Attention is also drawn to the dangers for the Examining and Opposition Divisions of the European Patent Office of assuming what might be regarded as a regulatory role in relation to the environmental risks of inventions resulting from genetic engineering, particularly when there is little information about the risks of the technology. This is not to argue, however, that patent offices should ignore moral and environmental issues.

## Biodiversity and Patent Law

Although the patent system facilitates the dissemination of information about patentees' inventions, and thereby enhances the ability of others to build on research and development relating to biotechnological inventions, there are fears that the limited monopolies conferred by patents will foster a loss of genetic diversity by encouraging industry to focus on the production, promotion, and sale of particular genetically engineered plants and animals to the detriment of organisms that seem, at least in the present state of scientific understanding, to offer fewer genetic advantages (Reid et al. 1993). This is coupled with the concern that developing countries will be denied free access to modified plants and animals, even though their territory might have been the source, and sometimes the unique source, of the organic matter that has been transformed into patentable subject matter to the extent that it can no longer be regarded as a product of nature—as opposed to a novel and nonobvious product of humans (Stone, this volume).

At another level, genetic diversity may be enhanced by the patent system, not only because it creates a financial incentive to produce new genotypes, but also because of the deposit in "banks" of samples of the biological material necessary to satisfy patent disclosure requirements. In these depositories, scientists and technologists may in the future find genetic material that no longer exists elsewhere. It is against this background that the current position with regard to the patentability of genetically engineered organisms in Europe will be examined.

In addition to points of comparative interest that may cast light on the situation regarding patentability in the United States, European developments are of direct interest to American inventors due to what intellectual property lawyers describe as the territoriality of patents. Individuals, universities, or corporations obtaining patents in the United States cannot use them to prevent others abroad from freely making or using their invention. If inventors wish to protect their inventions overseas, they will have to apply for patents in the jurisdictions in which they desire protection. When they do so, they may encounter some surprising differences between national and supranational systems.

## The Nature and Scope of Patent Rights

When inventors receive patents, they do not acquire ownership. In general, however, they can require others to pay a royalty or obtain a licence if they wish to make or use the patented inventions. Thus, holders of patents on biotechnological inventions, as in the case of other inventions, are able to achieve what is in a sense negative control over their invention. Although there are some limitations in Europe on the types of biotechnological invention that may be patented, it is clear that, especially at the microbiological level, genetically modified organisms may be patented and a degree of monopoly obtained. This is not as surprising as it might first seem when it is remembered that we have no difficulty in regarding plants and animals as goods for the purposes of sale of goods legislation and contract law.

Two North American inventions serve to illustrate the types of biological invention that may be patented. The world's first patent on a genetically engineered organism was granted to the General Electric Company on a *pseudomonas* bacterium genetically engineered by Anand Chakrabarty so as to possess the capacity to consume oil.[1] If applied in sufficient numbers, these bacteria could, for example, be used to consume oil slicks. The organism developed by Chakrabarty was considered patentable because it was not simply a product of nature, but was altered sufficiently to be regarded as a human product—that is, involving an inventive step, and being novel and capable of industrial application. It is also important to note that the discovery of natural, although hitherto unknown, organisms does not render the newly discovered organisms patentable. They remain nothing more than products of nature unless they have, for instance, been specially purified. The second invention, the so-called Harvard onco mouse or myc mouse, is a mouse that has, for research purposes, been genetically modified in such as way as to be particularly susceptible to cancer.[2]

### The Chakrabarty and Harvard Applications

It is interesting to compare the progress of the relevant patent applications in the United States and Europe. Chakrabarty's patent application was granted by the U.K. patent office at the lowest level of examination and decisionmaking without any opposition or public controversy. By way of contrast, the same application provoked immense public and legal debate in the United States, and the application was fought through the U.S. patent system up to the U.S. Supreme Court before it was ultimately granted.[3] Shortly afterward, a similar application in the United States was also appealed all the way to the Supreme Court before being granted (Cripps 1980, 100 et seq.).

Since the late 1970s and early 1980s, when these applications were under discussion, controversy about the granting of patents on genetically engineered organisms appears to have diminished in the United States and increased in Europe. Although it is not always evident in the outcomes ultimately arrived at by the decisionmaking bodies, differences in reasoning can be traced, *inter alia*, to the existence of Article 53(a) of the European Patent Convention (1973).

Despite the fact that the European patents on multicellular plants had, in technical terms, paved the way for the Harvard mouse, the latter's arrival at the door of the European Patent Office stirred a controversy that has still not subsided. Although the mouse became the subject of a European patent in October 1991, having initially been rejected by the Examining Division of the European Patent Office,[4] it would not be unfair to say that it met with significantly less opposition in the United States when a patent was applied for in that jurisdiction and granted in 1988—almost exactly a decade after the legal battles over Chakrabarty's application.

## Controversies over Europe's Patent Convention: Contrary to Morality or Public Order

Article 53(a) of the European Patent Convention expressly excludes from patent protection any inventions that would be contrary to *ordre public* or morality if published or exploited. It also stipulates that the mere prohibition by law or regulation in some or all of the participating states does not cause an invention to fall foul of Article 53(a). When dealing with Harvard's application for the onco mouse, the Technical Board of Appeal clearly favored a patent on technical grounds, but it referred the case back to the Examining Division for a review of the following issues in relation to the morality clause in Article 53(a):

- Would it be better to perform cancer tests of this kind on nonanimal models?
- The purpose of the invented mouse was not to improve particular features, but to produce tumors in the test animals.
- Animals were regarded in the application as objects.
- Descendants of the transgenic animals might escape into the environment and spread malignant foreign genes through mating.
- Was evolution being drastically interfered with?

The Examining Division rightly concluded, on consideration of the moral issues that had been referred back to it, that the patent system was a less than ideal arena in which to resolve such problems (see Armitage and Davis 1994; compare to Beyleveld and Brownsword 1993).[5] The division stated that the *ordre public* and morality clause was designed to exclude from protection inventions likely to induce riot or public disorder, or to lead to criminal or other generally offensive behavior. They suggested that a new type of letter bomb is an example of an invention that would be excluded from patent protection by Article 53(a), the European Patent Office having adopted the view that inventions will only be contrary to morality if the "public in general" finds them "abhorrent." This is, of course, a strict test that will not be met by many inventions—a view that has been affirmed by the European Patent Office in the *Greenpeace UK v. Plant Genetic Systems NV* proceedings.[6] The Opposition Division in *Greenpeace* emphasized the undesirability of expecting technically trained patent assessors to rule on complex moral questions, remembering also that separate regulatory systems are in place with regard to genetic engineering and that the denial of a patent does not prevent an invention from being exploited.

The Examining Division in the Harvard mouse case decided that advantages in terms of cancer research outweighed the suffering to which the animals would be submitted as well as any risk to the environment. The division, however, preferred to avoid detailed discussion of the environmental issues, largely on suspects, for the reason that was subsequently given in the *Greenpeace* case, in which the Technical Board of Appeal pointed out that there was little scientific evidence available on the subject of the environmental risks of genetically engineered organisms.[7]

In the Harvard mouse case, the Examining Division stressed that their conclusions related exclusively to the onco mouse and did not create a general precedent for the patenting of animals. Indeed, a subsequent application by the pharmaceutical company Upjohn for a mouse genetically engineered to suit the purposes of research into hair growth was refused on the basis that the limited usefulness to humans did not justify the suffering of animals involved in the invention.

Patents granted on engineered plants have also been opposed under Article 53(a) based on the assertion that plants and animals are part of "the common heritage of humankind"—a concept that received a degree of implicit support, even if rejected as an overarching emphasis, in the Convention on Biological Diversity (CBD; appendix, this volume).[8] More specifically, the argument that has been raised by environmental groups opposing patents granted on genetically engineered plants is that since plants and animals are part of our common heritage, they should be freely available for use by all, unencumbered by intellectual property rights.[9]

Perhaps not surprisingly, this argument did not convince the European Opposition Division, although the terms of the division's rejection were unfortunate. The division decided that if a plant or animal were part of the common heritage, it could not be novel; therefore, since a patent would not be granted on what was novel, the patented plants under opposition could not have been part of our common heritage or they would never have qualified for patent protection. There is a degree of question begging here that, in its failure to address the significance of the alteration of natural products, unnecessarily trivializes the albeit doubtful "common heritage" argument.

The problems are aggravated by the wording of Article 53(b) of the European Patent Convention, which prohibits the granting of patents on varieties of plants and animals others than at the microbiological level.[10] This was arguably intended to deny patent protection to organic inventions other than microbiological ones. But the European Patent Office, when confronted with applications involving multicellular genetically engineered animals and plants, has on occasion granted patents if the applications in question contain claims not to a variety but to a group taxonomically larger than a "variety" (Christie 1989).[11] One might perhaps be forgiven for thinking that this interpretation diminished the effect of Article 53(b) in that it permitted claims in respect of groups taxonomically higher than varieties and, thus, to more than a variety—although it is difficult to see how such a claim did not, in practice, also encompass a variety.

The confusion surrounding Article 53(b) and claims involving varieties has been heightened by the decision of the Technical Board of Appeal of the European Patent Office in the *Greenpeace* case.[12] The patent under dispute in that case related to a plant genetically engineered to be resistant to herbicides. The board concluded that some of the claims in the patent were not allowable because they encompassed plant varieties and were thereby excluded from patentability by the first part of Article 53(b). Unfortunately, the Enlarged Board of Appeal, the European Patent Office's highest judicial authority, refused to give a ruling[13] when the president of the European Patent Office requested it in order to clarify what appeared to be a conflict between the decision of the Technical Board of Appeal in *Greenpeace* and the approach adopted by the same board in earlier cases, such as *Onco Mouse* and *Ciba-Geigy*.[14] The president asked whether "a claim which relates to plants or animals but wherein specific plant or animal varieties are not individually claimed contravene[s] the prohibition on patenting in Article 53(b) EPC, if it embraces plant or animal varieties?" (Roberts 1996). By refusing to entertain the issue, the enlarged board, which only has jurisdiction when there has been a conflict, in essence was arguing that there was no conflict because, unlike the technical boards in the earlier cases, the *Greenpeace* board had refused a patent on the ground that a plant having a stably maintained transgene necessarily was a variety, not that the claim merely covered varieties. This distinction is unsatisfactory not only because the enlarged board cravenly evaded the central issue, but also because one must question whether a plant (or an animal) having a stably maintained transgene is necessarily a variety.

Oppositional groups in Europe have also argued that patents cannot extend to cover progeny because, once a patent is granted on the parent plant or animal, the progeny cannot be novel and nonobvious as required by patent law. The Opposition Division within the European system has adopted a pragmatic approach to such arguments, no doubt bearing in mind economic factors and the need for Europe to compete in wider markets. It has held that the descendants of genetically engineered organisms ought not to be viewed differently from their parents since they would not exist if the initial technical manipulation had not been performed; if the parents are considered novel, the descendants are novel.[15]

Dolly, the cloned sheep, recently tottered onto the world stage as the subject of patent and trademark applications (the former under the Patent Cooperation Treaty of 1970). She may stumble at the domestic British equivalents of both Articles 53(a) and 53(b). It is unlikely that she will be regarded as anything but novel, because even cloning involves recombination of mitochondrial DNA, but her publication or exploitation as an invention may be regarded by the patent authorities as offensive or immoral. Or, by analogy with the decision of the Technical Board of

Appeal in the *Greenpeace* case, the claims relating to her may be excluded from patentability as encompassing an animal variety, depending on whether or not she is regarded as the product of microbiological process.

## The Human Genome Project

It is important not to lose sight of the fact that biodiversity includes human genetic diversity, and in this and other regards, the Human Genome Organization (HUGO) is of special interest. The human genome project, coordinated by HUGO, involves an unusual degree of collaboration between institutions such as the U.S. National Institutes of Health and the British Medical Research Council, which have previously acted as competitors in this and other fields. But the collaboration is not limited to the United States and the United Kingdom. Scientists in several countries are steadily unraveling the secrets of the human genetic code, identifying the functions of more and more gene segments. This increased understanding may well revolutionize medical science. Even when viewed as steps on the pathway to potential cancer cures, these scientific insights may be regarded as somewhat of a mixed blessing, as demographers and social scientists chart with growing concern the increase in the world's population. Difficulties also lie ahead with regard to the use of genetic information in employment and insurance decisions. Yet the deciphering of the code should not be stopped. Nor would it be easy to halt the scientific momentum completely, even if the will to do so existed. Against this background, the withdrawal by the National Institutes of Health and the British Medical Research Council of patent applications that they had independently filed on human gene segments of unknown function is greatly to be welcomed. It foreshadowed the recent and wise insertion into the proposed directive on the protection of biological inventions of a new provision, Article 3(3), which would require that "As regards a human gene or a partial sequence of a human gene, the identification of the function associated with the gene or sequence and the application giving rise to the patent must be sufficiently precise and identified" (European Commission 1995). Even if these isolated and purified segments could properly be regarded as inventions as opposed to discoveries, the lack of knowledge as to their function undermines patentability on the basis of industrial applicability, although it can be argued that the newly identified segments have some utility as test probes.

On April 4, 1997, 20 European nations signed the Convention on Human Rights and Biomedicine in Orviedo, Spain.[16] The convention, which was instigated by the Council of Europe, forbids the use of recombinant DNA techniques on humans other than for medical purposes in such a way as to amount to a ban, *inter alia*, on the cloning of humans in the signatory countries. It expressly prohibits the production of human embryos exclusively for research and the use of *in vitro* fertilization techniques in order to choose the sex of children, and it also proclaims a right not to be discriminated against on the basis of one's genotype. Signature by Britain was expected, but a general election campaign intervened; the matter is yet to be considered when a new government comes into power.

## The Legal Protection of Biotechnological Inventions

In states that are signatories to the European Patent Convention, patent applications may be made via a country's national patent office or under the European system to the European Patent Office. In the latter case, a European patent becomes a bundle of single patents coming under the jurisdiction of relevant individual states and the enforcement of such patents is a matter for national courts. The European Commission's (1988) initial proposal for a council directive on biotechnological inventions sought, among other things, to standardize divergent national interpretations regarding biotechnological patents. That proposal was not adopted by the European Parliament, in spite of the fact that a common position was reached, but the issues surrounding biotechnological inventions are to return to the parliament in the form of a new proposal (European Commission 1995), which has the same objectives as the earlier document. In the introduction, the commission notes that its new proposal is compatible with the CBD, in particular with Article 16(5), which provides that "The Contracting Parties, recognizing that patents and other intellectual property rights may have an influence on the implementation of this Convention, shall cooperate in this regard subject to national legislation and international law in order to ensure that such rights are supportive of and do not run counter to its objectives."

Although the commission points out that an invention is not defined as such in European patent laws (Cripps forthcoming), the distinction between unpatentable discoveries, on the one hand, and inventions, on the other, is emphasized by the commission in relation to the human body and its elements in their natural state. According to proposed Article 3(1), these are not to be regarded as patentable inventions. Article 3(2) of the draft, however, illustrates the limits of the commission's concept of the natural state. It declares that "the subject of an invention capable of industrial application which relates to an element isolated from the human body or otherwise produced by means of a technical process shall be patentable, even if the structure of that element is identical to that of a natural element" (European Commission 1995). Methods of human treatment involving germ-line therapy would be excluded from patentability if draft Article 9 were implemented, but successful claims to such methods would, in any event, be precluded by Article 52(4) of the European Patent Convention.

If implemented, the proposed directive would not significantly alter the situation in relation to the word "variety" in Article 53(b) of the European Patent Convention. The prohibition on the patenting of plant and animal "varieties" as such does not exist in the United States and Japan, despite the overlap between patent and plant variety systems in those countries, and it is interesting that it has been included in the proposed directive. It remains to be seen how it will fare when it comes before the European Parliament, in the wake of industry criticism of the Technical Board of Appeal's decision in the *Greenpeace* case.

The proposed directive also contains a controversial provision for a farmers' privilege, which would enable farmers to resow seed saved from first-plant crops. There is a parallel farmers' privilege available under the system of plant variety rights. Normally a purchaser of seed or other propagating material is not free to multiply it for use as propagation material, and industry has been reluctant to see a farmers' use privilege introduced into patent law in addition to its existing place in plant variety legislation. The new proposed directive would also create a "farm animal" exemption, which would permit farmers to breed from patented stock in order to replenish their herds.

## Plant Patents and Varieties

Plant varieties protectable under the system of plant variety rights may, unlike those for which patents are claimed, be obvious (see UPOV 1961; Byrne 1992; Straus 1987).[17] They may, for example, be created by traditional breeding techniques as long as there has been no prior commercialization. There is no separate animal variety rights system. And whereas patent protection covers variants of the core invention if they incorporate the inventive concept, under the plant variety system an applicant can only achieve protection for the new variety itself. Applicants can obtain protection under both the plant variety and patent systems, where appropriate, although that practice was previously precluded for the member states of the International Convention for the Protection of New Varieties of Plants (UPOV).

Article 14 of the proposed directive on the legal protection of biotechnological inventions would, if adopted, introduce a system of compulsory cross licensing, where a breeder could not acquire or exploit a variety right without infringing on a prior patent, and vice versa. Applicants would have to demonstrate that they have applied unsuccessfully to the holder of the patent or the plant variety right for a contractual license and that the exploitation of the plant variety or invention is dictated by the public interest and constitutes significant technical progress. It is not surprising that the proposed dependency licenses are less than popular with industry. They would, for example, relate to the situation in which a person has bred a new plant variety and obtained a plant variety right for it, but cannot exploit it without infringing on a prior patent. The very existence of such a situation highlights the difficulties identified earlier with regard to the prohibition, in Article 53(b) of the European Patent Convention, on the patenting of plant and animal varieties, as well as the way in which that part of Article 53 has been interpreted.

As a separate point, it should also be noted that there is a Council regulation on European Community plant variety rights (European Union 1994). This enables member states of the European Community to obtain plant variety right protection throughout the region by means of a single application to a central European Community office. One of the most interesting features of this regulation is that it implements the full provisions of the UPOV for the European Community before any member state of the community has done so (Ardley and Hoptroff 1996). These provisions include a number of extensions of plant breeders' rights with respect to the use of propagating material from protected varieties. The regulation does, however, make some concessions to the farmers' right to save seed for their own use.

## The TRIPS Agreement

The completion of the General Agreement on Tariffs and Trade's Uruguay Round in 1994 led to the creation of the World Trade Organization (WTO). WTO will, as part of its role, administer the Agreement on Trade-Related Aspects of Intellectual Property Rights (TRIPS) of 1994 (Cornish 1996, paras. 1–23, 1–28, 3–22, 7–44). The TRIPS agreement, which binds members of the WTO, was designed to increase protections for the holders of intellectual property rights. The European Commission has stated in the introductory notes to its Proposal for a Council Directive on the Legal Protection of Biotechnological Inventions (European Commission 1995) that the proposed directive, in addition to being compatible with the CBD, is compatible with Articles 27 and 30 of TRIPS. Article 27 of TRIPS relates to patentable subject matter, and Article 30 refers to exceptions to the rights conferred by a patent. It is not surprising that Article 31 of TRIPS is not among those listed as being compatible with the proposed directive, since it places severe restrictions on the granting of compulsory licenses. Given the antipathy displayed in TRIPS toward restrictions on patent holders' rights, it will be interesting to see what becomes of the European Commission's proposals for compulsory cross licensing and farmers' use exemptions when the relevant provisions come before the European Parliament.

## Access to Deposited Samples

Applicants must provide samples of the biological materials for which they seek patents. These samples will be deposited in approved culture collections. An important question bearing on biodiversity and arising in this context is that of third-party access to such deposits, not least because the biological materials may be reproduced, in large quantities if desired, from the samples.

Under U.S. and Japanese law, public availability of deposit samples is postponed until after applicants have obtained enforceable rights; European law requires availability to be coterminous with the first stage of publication of the European patent application—in other words, before the granting of a patent. The European Patent Convention limits access at this early stage to independent specialists acting for third parties. Such expert advisers may report to the third parties, but must not transfer any sample material to them. Once applicants obtain enforceable rights, third parties are permitted direct access to the samples. Article 15 of the new draft directive is designed to ensure that when an application is refused or withdrawn, only third parties' experts would continue to have access to samples. Such access is possible for 20 years (the length of a European patent) from the application date. In cases where a patent is granted, third parties may have direct access to samples.

Provision is also made in Article 17 of the proposed directive for assisting patent holders by placing the burden of proof on defendants in cases where there is a patent on a process that makes use of deposited material in order to create a new product, and where any person or persons, including third parties' experts, have had access to deposited samples.

### Contained Use and Deliberate Release

Although, strictly speaking, they are beyond the scope of this chapter, it is important to draw attention, even if only in passing, to the "contained use" and "deliberate release" directives of the European Community (Cripps 1992). Between them, they control the health, safety, and environmental aspects of laboratory work on genetically engineered organisms, as well as their deliberate release to the environment.[18] A scare over inadequate containment procedures in a laboratory in Birmingham, England, focused British attention on these directives and their implementing legislation. The laboratory was involved in work requiring a high degree of containment of a common cold virus—albeit disabled—carrying cancerous material.

In Britain, Section 109 of the Environmental Protection Act of 1990 requires those working with genetically modified viruses to use what is known as BATNIEC, an acronym that stands for the "best available techniques not involving excessive costs," for the purposes of keeping the organisms under control and preventing damage to the environment. This is a compromise between safety needs and environmental factors, on the one hand, and economic forces, on the other.

Discussion of environmental risk leads to thoughts of insurance, and it may be worth nothing that attempts to identify a firm in the London insurance market that would provide coverage for researchers or corporations seeking to protect themselves against losses that might be caused by work on or with genetically modified organisms have failed. Several insurance companies expressed the view that if such cover were available, the premiums would be excessively high, even for corporations.

## The CBD and the "Darwin" Initiative

The Council of the European Union has approved the CBD and, in so doing, has placed heavy emphasis on intellectual property rights. It has stated that compliance with intellectual property rights is an important facet of the implementation of policies for technology transfer and coinvestment. Moreover, transfers of technology and access to biotechnology under the convention will be carried out in compliance with the principles and rules of intellectual property law—in particular, with the multilateral and bilateral agreements signed or negotiated by the contracting parties for the convention.[19]

The British government has launched its own response to the Rio convention in the form of the "Darwin" initiative. This reminds government departments of the need to consider ways of putting the provisions of the convention into effect, and the Department of the Environment has been involved in arranging meetings between industry, voluntary bodies, and local communities in an attempt to further the convention's goal of sustainable development (Chartered Institute 1993).

## The Stockholm Environment Institute

In 1993 the Stockholm Environment Institute (SEI) established a Biotechnology Advisory Commission (BAC) with an international membership. This interdisciplinary committee is mainly comprised of eminent scientists, but also includes lawyers and economists. The SEI encourages initiatives involving the application of biotechnology to meet the requirements of developing countries in relation to sustainable production, waste treatment techniques, and associated matters. It established the BAC in order to create a commission that would, on request, advise governmental bodies on proposed introductions to the environment of genetically engineered organisms. In addition to establishing and administering the BAC, the SEI has now launched a specific initiative on biological diversity in an attempt to assist developing countries with access to and transfer of biotechnological materials and techniques (SEI 1994a, 1994b).

## Conclusion

Intellectual property systems provide incentives for innovation. They are driven by economic forces, including the forces of competition. Thus, patents represent rewards for investment in the incentive activity that they are designed to encourage. In deciding which inventions should receive patent protection, with all the advantages that confers, patent examiners should not close their eyes to the moral and environmental aspects of the claims that they are called on to assess—even though they are unlikely to be the best judges of those aspects and are understandably reluctant to intervene if only some uses of an invention are offensive. Although intellectual property systems have implications for biodiversity, the granting or withholding of patents will not constitute an adequate response to the social, moral, and environmental dimensions of the task of preserving the world's natural resources. We must not ignore the wider implications of biotechnology, but we run the risk of trivializing the problems involved in protecting biodiversity if we regard the "controls" or limitations in patent legislation as effective regulatory mechanisms.

## Notes

1 Patent no. 1,436,537 (United Kingdom). See Sagoff, this volume.
2 See the decision of the Technical Board of Appeal of the European Patent Office in *Harvard/Onco Mouse*, T19/90, *Official Journal of the European Patent Office*, OJ EPO 12/1990, 476; [1990] 7 *European Patent Office Reports*, EPOR 501. Note also the decision of the Examining Division after remittal by the Technical Board, OJ EPO 10/1992, 588.
3 *Diamond v. Chakrabarty*, 447 U.S. 303 (1980). See Sagoff, this volume.
4 OJ EPO 11/1989, 451; [1990] 1 EPOR 4.
5 OJ EPO 10/1992, 588. See also Cripps (1979) with regard, *inter alia*, to the assessment by patent offices of nontechnical questions.
6 *Greenpeace UK v. Plant Genetic Systems NV*. 1993. *International Review of Industrial Property and Copyright Law* (IIC). 24:618, at 621.
7 *Greenpeace UK v. Plant Genetic Systems NV*. 1997. *International Review of Industrial Property and Copyright Law* (IIC). 28:75, at 80.
8 See, for example, Articles 8, 9, and 18 of the convention (appendix, this volume). Note also Straus (1993).
9 See the opposition proceedings against Lubrizol's patent EP-B1-122 791 as discussed by Jaenichen and Schrell (1993). *Lubrizol/Hybrid Plants*, T320/87.
10 As we shall see, most countries also operate systems of plant, though not animal, variety rights that confer intellectual property rights distinct from those under the patent system (see also Stone, this volume).
11 See also *Lubrizol/Hybrid Plants*, T320/87.
12 *Greenpeace UK v. Plant Genetic Systems NV*. 1997. *International Review of Industrial Property and Copyright Law*. 28:75.
13 Enlarged Board of Appeal of the European Patent Office, G03/95, 27 November 1995.
14 *Ciba-Geigy/Propagating Material* [1984] OJ EPO 112 (T. 49/83).
15 See the decision of the Opposition Division of the European Patent Office in *Greenpeace UK v. Plant Genetic Systems NV*. 1993 (IIC). 24:618, at 619. The patent in question relates to a plant that was genetically engineered to be resistant to herbicides.
16 The United States, Japan, Canada, and the Vatican have reserved the right to become signatories; President Clinton has already banned the use of federal funds for cloning in the United States.
17 These are governed by the International Convention for the Protection of New Varieties of Plants (UPOV) and under corresponding legislation in UPOV's member states.
18 For discussion of a range of issues relating to the release of genetically engineered organisms to the environment, see the Royal Commission on Environmental Pollution (1989) and the United States (1988, 1986).
19 See Official Journal, L 309/1; L 309. Annex C. 13 December 1993.

# References

Ardley, J., and C. Hoptroff. 1996. Protecting plant 'invention': The role of plant-variety rights and patents. *Trends in Biotechnology* 14:67.

Armitage, E., and I. Davis. 1994. *Patents and Morality in Perspective.* London: Common Law Institute of Intellectual Property.

Beyleveld, D., and R. Brownsword. 1993. *Mice, Morality, and Patents.* London: Common Law Institute of Intellectual Property.

Byrne, N. 1992. *Commentary on the Substantive Law of the 1990 UPOV Convention for the Protection of Plant Varieties.* London: Queen Mary and Westfield College.

Chartered Institute of Patent Agents. 1993. *Briefing Paper on the United Nations Convention on Biological Diversity.* London: Chartered Institute of Patent Agents.

Christi, A. 1989. Patents for plant innovation. *European Intellectual Property Review* 11:394.

Cornish, W. 1996. *Intellectual Property.* 3d ed. London: Sweet and Maxwell.

Cripps, Y. 1979. Genetic engineering: A problem for the patent office? *New Zealand Law Journal* 11:232, 236, 464.

———. 1980. *Controlling Technology: Genetic Engineering and the Law.* New York: Praeger.

———. 1992. Evidence to the House of Lords Select Committee on Science and Technology. *Regulation of the United Kingdom Biotechnology Industry and Global Competitiveness.* 1992–93 sess., HL paper 80–I.

———. Forthcoming. Recombinant DNA technology: A patent case in the House of Lords. *Cambridge Law Journal.*

European Commission. 1988. *Proposal for a Council Directive on the Legal Protection of Biotechnological Inventions.* COM(88) 496 final.

———. 1995. *Proposal for a Council Directive on the Legal Protection of Biotechnological Inventions.* COM(95) 661 final.

European Patent Convention (Convention on the Grant of European Patents). 1973. *International Legal Materials* 15:5.

European Union. 1994. Council regulation 2100/94, Official Journal, 1.9.94, No L227/1.

Jaenichen, H., and A. Schrell. 1993. The European Patent Office's recent decisions on patenting plants. *European Intellectual Property Review* 12:466.

Reid, W.V., S.A. Laird, C.A. Meyer, R. Gámez, A. Sittenfeld, D.H. Janzen, M.A. Gollin, and C. Juma, eds. 1993. *Biodiversity Prospecting: Using Genetic Resources for Sustainable Development.* Washington, D.C.: World Resources Institute.

Roberts, T. 1996. Patentability of plant and animal varieties. *European Intellectual Property Review* 3:D–90.

Royal Commission on Environmental Pollution. 1989. *The Release of Genetically Engineered Organisms to the Environment.* 13th rep. London: Her Majesty's Stationery Office.

SEI (Stockholm Environment Institute). 1994a. *A Clearing-House Mechanism to Promote and Facilitate Technical and Scientific Co-operation under the UN Convention on Biological Diversity.* Stockholm: Stockholm Environment Institute.

———. 1994b. *The Sustainable Use of Genetic Material of Indigenous Plants and Animals: An Appraisal of Co-ordinated Arrangements for the Conservation of Genetic Material, Material and Technology Transfer, and Benefit Sharing.* Stockholm: Stockholm Environment Institute.

Straus, J. 1987. The relationship between plant variety protection and patent protection for biotechnological inventions from an international viewpoint. *International Review of Intellectual Property and Copyright Law* 18:723.

———. 1993. The Rio biodiversity convention and intellectual property. *International Review of Intellectual Property and Copyright Law* 24:602.

United States. Office of Technology Assessment. 1988. *New Developments in Biotechnology: Field Testing Engineered Organisms—Genetic and Ecological Issues.* Washington, D.C.: Government Printing Office.

———. White House Office of Science and Technology Policy. 1986. *Co-ordinated Framework for the Regulation of Biotechnology. Federal Register* 51 (123) (26 June): 23301–50.

UPOV (Convention for the Protection of New Varieties of Plants). 1961 and 1991 revision. *Plant Variety Protection Laws and Treaties.* UPOV publication no. 651(E).

# Animals as Inventions: Biotechnology and Intellectual Property Rights

by Mark Sagoff

This chapter distinguishes between two ways of justifying intellectual property rights: first, to promote useful inventions (the instrumental argument); and second, to respect the natural right of a person to own what is the result of his or her own labor (the moral right). The chapter contends that whatever the merits of the instrumental (and with it, the legal) argument, genetic engineers cannot claim to have created recombinant species in the way necessary to ground a "natural" property right. This is because engineered organisms remain essentially creations of nature to which biologists have not added new knowledge; there is no "invention." Religious leaders are correct in labeling as hubris the Faustian claims biologists may make that they understand how the genome works, for example, in the same way that Thomas Edison understood how the phonograph worked, and thus can claim natural property rights to grass and other products of nature.

*

John Locke's (1967) *First Treatise of Government*, which languishes in the shadow of his more famous *Second Treatise*, refutes Sir Robert Filmer (1947), who had argued for the divine right of kings. Filmer reasoned that just as God, by creating Adam, owned him absolutely, so Adam owned his children, and so princes, who inherit this right from Adam, hold absolute authority over their subjects. Locke replied that while ownership is, indeed, a consequence of authorship, God alone is able "to frame and make a living Creature, fashion the parts, and moulld and suit them to their uses." Humans, starting with Adam, beget children, but do not design and, therefore, do not own them. Locke wrote:

> What Father [in] a Thousand, when he begets a Child, thinks farther [than] the satisfying his present Appetite? God in his infinite Wisdom has put strong desires of Copulation into the Constitution of Men, thereby to continue the race of Mankind, which he doth most commonly without the intention, and often against the Consent and Will of the Begetter. And indeed those who desire and design Children, are but the occasions of their being, and when they

design and wish to beget them, do little more towards their making, than *Ducalion* and his Wife in the Fable did towards the making of Mankind, by throwing Pebbles over their Heads.

Today, the question is whether genetic engineers can claim an intellectual property right or patent in the organisms they produce. Although this chapter refers to cases and controversies in patent law, it offers a philosophical rather than a legal argument answer. Patent law stands on constitutional and statutory foundations; this chapter shall not be concerned with these, except insofar as they clarify the moral intuitions and philosophical arguments that inform our conception of intellectual property rights. These moral intuitions and philosophical arguments properly exert a gravitational force on the decisions patent examiners and judges may make in relation to hard cases, particularly in the area of biotechnology.

This chapter will explicate the concept of invention in relation to the products of biotechnology and genetic engineering. Whether these products should be patentable may turn, in part, on how we construe the act of creation—whether we believe biotechnologists design these creatures or serve as "but the occasions of their being." To the extent that ethical beliefs and analyses play a role in guiding legal decisions, reservations concerning the patentability of genetically engineered organisms may stem from an uncertainty about how much their producers, as it were, author or design them, and thus have contributed to the world's store not just of useful things but also of useful knowledge.

From a moral point of view, technologists may claim intellectual property rights in organisms, insofar as they design them, for example, as Edison designed the lightbulb, using ideas that are not found in nature but that are their own. If engineers—like Ducalion and his wife—simply produce without designing organisms, they may be able to claim the process, but not the product, as intellectual property. The difference lies in whether the inventor contributes to knowledge or simply borrows the design from nature that makes his or her invention work.

## Process and Product in Patent Law

"Justice gives every man a title to the product of his honest industry," wrote Locke (1967) in reply to Filmer.[1] If invention is an example of industry, this principle would suggest a moral right of creators to their ideas.[2]

Edison received a patent for the lightbulb because he contributed the novel idea that constituted the design that made it work, in this instance, the principle of the glowing carbon filament. He was thus able to satisfy the "description requirement" of patent law, which demands that an inventor set forth the new knowledge or novel idea his invention represents in a written "description of the invention . . . in such full, clear, concise, and exact terms as to enable any person skilled in the art . . . to make and use the same" (U.S. Patent Office 1995).

Similarly, King Gillette improved the lot of humankind by conceiving of a design that solved the problem of keeping a very thin and flexible blade rigid enough to shave whiskers. His patent specification describes his solution, namely, to "secure [the] blade to a holder" so that "it receives a degree of rigidity sufficient to make it practically operative" (U.S. Patent No. 775,134).

What made the lightbulb and the disposable razor intellectual property is that their inventors not only produced these objects, but also designed them. To put this distinction in Aristotelian language, the inventor can claim intellectual property rights if he or she provides not just the efficient, but also the formal cause of the mechanisms that he or she creates. For example, the electromagnetic spectrum and the principles that underlie it belong to nature not art. Accordingly, in 1853, *O'Reilly v. Morse* (15 How. 62, 112–29) held that Samuel Morse could patent the telegraphic instruments that he had invented, but not—as he wanted—the use of electromagnetic waves to send and receive signals. Morse claimed as intellectual property "the use of the motive power of the electric or galvanic current, which I call electro-magnetism, however developed for making or printing intelligible characters . . . at any distance." But the Supreme Court held that Morse had invented only a particular instrument to take advantage of electromagnetism, and that he could not prohibit others from harnessing the same natural materials or forces by other means.

The distinction between what one invents or designs and what one merely produces or discovers, underlies the refusal of the Patent Office and the courts to allow patents on products of nature. In a leading 1928 case, *General Electric Co. v. De Forest Radio Co.* (28 F.2d 641, 2[3d Cir.], *cert. denied*, 49 S. Ct. 180[1929]), an appeals court held that General Electric could not patent pure tungsten, but only its method for purifying it, since tungsten is not an invention but a "product of nature." In its pure form, tungsten possesses the ductility and tensile strength needed for constructing filaments for incandescent lamps and radio vacuum tubes. William Coolidge, working for General Electric, had developed a process for removing the impurities, chiefly oxygen, that are always combined with tungsten in nature. A lower court, under the impression that the metal Coolidge had produced was a new alloy, had upheld a patent on it. In invalidating the patent, the appeals court observed that Coolidge had not invented an alloy, but had purified an elemental mineral. Even though tungsten in a pure form is not found in nature but has to be produced by art, the appeals court concluded that Coolidge "cannot have a patent for it because a patent cannot be awarded for a discovery or for a product of nature or for a chemical element."

This and other leading cases illustrate a fundamental limitation on the scope of intellectual property rights. These rights can extend only to what someone designs—what he or she devises and constructs, as it were, from simpler materials by reason of a plan or principle that the person invents. Intellectual property does not include objects, such as natural forces and materials, that the individual does not design, but, rather, simply uses or manipulates by applying a novel instrument or process. The instrument (the telegraph) or the process (for purifying tungsten) is patentable, but not the object of nature (the airwaves or the metal), which may have been changed, but is not designed by the inventor. In these matters, as the court in *Monsanto Co. v. Rohm and Hass Co.* said: "It is basic to the grant of a patent that the scope of a patent should not exceed the scope of invention."[3]

In the past, the Patent Office and the courts have generally observed Jeremy Bentham's (1825) principle that a patent must be confined "to the precise point in which the originality of the invention consists." The "precise point [of] originality" may extend to the processes or instruments that are adequate to produce a result, but not necessarily to the result itself. Accordingly, the Patent Office—up until recently, at least—has not let the reach of intellectual property rights exceed what lies within the inventor's intellectual grasp.

## Patenting Organisms and Other Biological Objects

The practice of the Patent Office changed dramatically after a 1980 decision, *Diamond v. Chakrabarty* (447 U.S. 303), in which the Supreme Court held by a five to four majority that Anand Chakrabarty, a biologist, could patent a novel organism because "his discovery is not nature's handiwork, but his own."[4] The "discovery" in question was a bacterium that Chakrabarty had hybridized by inserting plasmids from other bacteria. The Court determined that the hybridized microorganism could be patented because it was not found in nature, but was "the result of human ingenuity and research" (447 U.S. 310).

This decision essentially reversed a series of court rulings that had rejected the idea that hybrids were inventions. In 1948, for example, the Supreme Court in *Funk Brothers Seed Co. v. Kalo Inoculant Co.* (333 U.S. 127) had held invalid a patent on a mixture of bacteria that did not occur together in nature. The Court stated that the mere combination of bacterial strains found separately in nature did not constitute "an invention or discovery within the meaning of the patent statutes." According to the Court, "a product must be more than new and useful to be patented; it must also satisfy the requirements of invention." The Court regarded the repackaging of genetic information taken from nature as a commercial, not a scientific advance. "Even though it may have been the product of skill, it certainly was not the product of invention" (333 U.S. 131–32).

The Supreme Court in the *Chakrabarty* case emphasized that the bacteria it had to consider differed from those in *Funk Brothers* because they had been genetically changed and were not naturally occurring strains. Yet the manipulation, in this instance, was no more than the recombination or hybridization of genetic materials found in nature, very much like what breeders had done for centuries. One might argue that there is no more design or invention in bringing naturally occurring bacteria together in new combinations—the situation in *Funk Brothers*—than in bringing plasmids together in bacteria. In either instance, the inventor manipulates a design already found in nature; he or she does not introduce a new design of his or her own. To delete a gene, to change the example, might be analogized to removing an impurity from a metal. The organism—although altered slightly from its naturally occurring state and therefore more useful—may result from the application of a novel process to natural materials, but the element of design resides in nature.

On the strength of the *Chakrabarty* decision, however, the Patent Office began to routinely award patents on hybridized and recombinant organisms—and very broad patents at that. The current high-water mark in this flood of patents came with the application of Agracetus of Middletown, Wisconsin, which received in 1992 U.S. patent no. 5,159,135, covering "all cotton seeds and plants which contain a recombinant gene construction [i.e., are genetically engineered]." By using a well-known process to introduce foreign genes into a cotton plant, Agracetus gained the right to exclude others from introducing any genes into that plant without its consent. "All transgenic cotton products . . . will have to be commercially licensed through us before they can enter the marketplace," a vice president of Agracetus said (*Pittsburgh Post-Gazette* 1995).

The Patent Office apparently reasoned that anything that is not a product of nature must, therefore, result from invention or design. This is perhaps not an unwarranted inference from *Chakrabarty,* but it does confuse invention with the production of anything novel and useful. The Patent Office went further to establish as intellectual property the genome of any organism—Agracetus also received a patent for soy—that a company managed to alter using novel technical means. In several instances, when a company seeking to patent an organism had used fairly well-known, if laborious, methods for manipulating its genome, the Patent Office still allowed a patent on that genome as the material for further genetic manipulation. The Patent Office, in other words, failed to distinguish manipulating or changing a genome from designing or inventing it.

Competitors were slow to challenge the Agracetus patent, perhaps believing that their time and interest would be better served by acquiring similar intellectual property rights to broccoli, bananas, and a host of other organisms whose genomes they could alter using the same technique that Agracetus applied. Eventually, challenges mounted up, including a major and successful suit by the U.S. Department of Agriculture. The Patent Office is reviewing the Agracetus patent and it is unlikely to be sustained.

The confusion over the extent of patents to living organisms began in 1988, when the creators of a transgenic mouse sought and received a patent not for their method of producing the creature, which was well known, but for the organism itself. Their procedure involved microinjecting genetic material directly into a fertilized egg that had been removed from an animal. The microinjected egg was then reinserted into the host female, who brought it to term.

The patentee, Harvard University, realized that the same trial-and-error laboratory procedure by which they injected an oncogene into a mouse egg could be used to inject the same gene (or some other gene) into virtually any eukaryotic animal; accordingly, it sought and received a patent broad enough to comprise the entire mammalian kingdom. The patent specifies that "the invention features a transgenic non-human eukaryotic animal." It did not apply simply to the process of microinjecting genes into eggs; indeed, the process, being well known or obvious beforehand, might not have been patentable. Instead, Harvard received a patent to the product—that is, any nonhuman mammal—not simply a mouse into which an oncogene has been inserted.[5]

Harvard did not design a mouse in the same sense that Edison "designed" the incandescent bulb. Edison dreamed up the principle that made the bulb work— the idea of running an electric current to heat a filament in a vacuum. In contrast, Harvard did not know, much less construct, the design by which animals work or even by which they may suffer from cancer. Had Harvard designed the genes it introduced, for example, by constructing them from amino acids according to a plan it had devised, then it would have contributed new knowledge to the world's store. By injecting genes into a mouse egg, however, it produced a new mouse but not new knowledge.

Critics may argue that the mythical characters Ducalion and his wife, by throwing pebbles behind them, managed to produce animals, but they neither designed nor constructed those organisms from simpler materials. They were the mere occasions of their being. Today, they could have applied for and received a patent on those organisms. In the area of biotechnology, the Patent Office seems to have substituted the concept of production for that of invention.

The difference between producing and inventing an organism is central to the question of whether intellectual property rights properly extend to novel organisms produced by genetic engineering. Karl Marx (1995) wrote that philosophers have "only tried to understand the world in various ways. The point, however, is to change it." Too often, genetic engineers may believe that by changing the world— that is, by altering genes—they understand it. What is crucial from the perspective of the moral underpinnings of patent law is that patentees contribute useful knowledge, not just that they make useful products—a principle that should exert a gravitational force on Patent Office decisions, even though it is an ethical and philosophical, rather than a legal argument.

## The Moral Basis of Intellectual Property Rights

Human beings have owned animals as personal property since prehistoric times; that right is not in contention. The question is whether people may rightly claim to own animals as intellectual property as well. Harvard, of course, did not file for a patent to protect its liberty to use, sell, or to exclude others from using the particular animals it raised. Rather, it sought to limit the liberty of others to make a similar mouse and, indeed, to produce any "non-human eukaryotic animal" carrying a similar oncogene.

The same moral principle, however, applies to the protection of intellectual and personal property, namely, the right of each person to the product of his or her honest industry. Plainly, an intellectual object—for example, a poem, a design, or a computer program—can result from a person's honest and creative effort. It would seem to follow, therefore, that if personal property rights have a moral basis, then intellectual property rights must have that basis as well.

Commentators often argue, however, that intellectual property rights have no moral basis, but are simply legal tools that governments create to promote social utility. Summarizing this common wisdom, two commentators write: "More than other forms of property ownership, intellectual property rights exist at the pleasure of the government and have been treated more as a means to social policy rather than the end of policy itself"(Evenson and Putnam 1987).[6]

Why do these commentators suppose that intellectual property rights—as distinct from rights in personal property—have no intrinsic or natural justification, and so exist at the discretion of the government? To be sure, patent law depends on statutory and constitutional foundations—but ethical intuitions and philosophical arguments may, and should, still inform the law.

A good way to tease out these intuitions is to note, first, the difference between things and ideas. The use of a thing—that is, personal property—by one person tends to exclude its being used at the same time by another. In contrast, intellectual property—that is, ideas—can be used by everyone at once. "Thus, in the absence of government sanction, ideas have the character of public goods" (Evenson and Putnam 1987).

Suppose, then, that ideas have the character of public goods; why should that affect their status as intellectual property? Consider the following argument. If one is legitimately to own personal property one creates, one must acquire all the ingredients honestly, for example, by purchasing them, or by appropriating unowned materials from the commons. With intellectual property, it is different. One generally avails oneself freely of the ideas of others, and on the basis of these borrowings—for which one apparently makes no account, and to which one may add little more than a wrinkle—one may patent the resulting invention. One might argue by this analogy that patenting is theft. One takes advantage of the ideas of others, but pays nothing for them, in creating one's own invention.

"A person who relies on human intellectual history and makes a small modification to produce something of great value," Edwin Hettinger (1989) has written, "should no more receive what the market will bear than should the last person needed to lift a car receive full credit for lifting it." He continues: "Given this vital dependence of a person's thoughts on the ideas of those who came before her, intellectual products are fundamentally social products." Hettinger adds that even if "a laborer's labor naturally entitles her to" her contribution, "[s]eparating out the individual contribution of the inventor, writer, or manager from this social/historical component is no easy task." He relies on this thought to conclude that rights to intellectual property turn not on moral arguments, but "on considerations of social utility."[7]

The government, however, can rely on a well-established principle of law to allow individuals to use the ideas of others in their intellectual endeavors without compromising their ability to patent the result: the concept of implicit compensation familiar in "takings" jurisprudence. Under state legislation, a local zoning authority can require, for example, that no house be built taller than three stories or that no estate be subdivided to smaller than half-acre lots. Everyone agrees that such requirements, which "take" certain property rights, remain consistent with the protection of property, as required by the Fifth Amendment, because "the restrictions imposed by the general legislation upon the rights of others serve as compensation for the property taken" (Epstein 1985).

In licensing everyone to use whatever knowledge he or she comes by honestly to produce more knowledge, whatever its source, the government creates a restrictive covenant, as it were, which keeps me from excluding you from using my ideas and keeps you from excluding me from using yours, in the common enterprise of creating new knowledge. In this respect, the government does not grant rights to intellectual property, but takes them. This "takings" is justified not necessarily because it serves the public interest, whatever that is, but because, insofar as a person contributes to the common stock of knowledge—here the analogy to a crew lifting a car is fairly apt—that person receives compensation from his or her efforts by being able to take equitable advantage of the efforts of others.

Inventors may refuse this *quid pro quo*, of course, by keeping their ideas as trade secrets. Since these inventors refuse to pay into the common fund of knowledge, they cannot justly draw from it; therefore, society will regard their inventions as if they were social or collective products to which the inventors have no intellectual property rights. Thus, if someone comes up with the same invention independently, reverse engineers it, or otherwise acquires the secret by honest means, then he or she can use and possibly patent it. Trade secrets are not protected as intellectual property, though a state may protect them from fraudulent disclosure, as when an employee, in breach of contract, sells the secret to a competitor.

If individuals have a similar moral claim to intellectual property as they do to personal property, one might wonder why the right to intellectual property should last for only a limited time, in practice for 20 years from the time of application, while rights to personal property have no such limit. The reason, from a moral point of view, has to do with the likelihood that others would create the same invention independently within some such period. Such an inventor, in theory, would have the same moral right to his or her invention, as a product of honest industry, and so should be equally free to license its use. The publication of the design of the invention, necessary to enter it into the common store of knowledge, however, effectively usurps the opportunity of others to create it independently. To make up for this usurpation of opportunity, the state may set a time limit to exclusive rights to the use of intellectual property.[8]

Inventors, to settle their accounts with the past—with their intellectual suppliers—must add significantly to the knowledge available to future inventors. In the United States, an inventor will not receive a patent, therefore, unless the knowledge that he or she provides is sufficiently new, nonobvious, and useful; the inventor must also describe the invention in a way that enables people in the field to duplicate it. This additional requirement serves two functions. First, it ensures that the patentee understands the design of his or her invention well enough to tell others not just how to produce it, but also how it works. Second, by putting this knowledge into the public till, the patentee can then draw from that till. Once the inventor has paid this *quid pro quo*, he or she may use what lies in the common store of knowledge to pursue a private enterprise. This would include making a product for sale, which is not a social or collective project, like lifting a car or promoting useful knowledge, but a private commercial activity.

To repeat: if an invention embodies a novel idea or piece of knowledge, and if the inventor discloses that knowledge to the world, then he or she commits no trespass on the intellectual property rights of others, even though he or she may only think their ideas a little farther. Inventors settle their accounts for intellectual materials by disclosing the new knowledge they create, as surely as they settle their accounts for physical materials by paying for them.

This means, of course, that if an individual adds nothing to science or knowledge, but merely applies conventional techniques—in ways obvious to those practiced in the art—to accomplish some task, then he or she has no patentable intellectual property in the result, however socially useful it may be. Such a person may have proprietary rights to any physical object created in this way, but he or she could not justly exclude others from accomplishing the same task, or creating the same kind of object, in similar ways. In the absence of new knowledge, no one can claim intellectual property rights in an object, however valuable, for there can be no right to a patent, where there is a want of invention. Where there is invention, however, then the patentee has a moral right to own the product of his or her honest intellectual industry. This is the case, at least, if the patentee describes the new knowledge in a way that permits others to acquire and apply it. In this way, the patentee not only shows that he or she has new knowledge to impart, but also, by imparting it, he or she pays compensation for the ideas of others that he or she may have borrowed.

## Intellectual Property versus Breeders' Rights

Plant and animal breeders have practiced their art for thousands of years and pursued it systematically for centuries. Over the past 100 years, in the light of Mendelian genetic theory, breeders have given us the crops we eat, the flowers we enjoy, the livestock we raise, and the pets we keep in our homes. They have not been able to protect their work, however, under the standard or utility patent act. Earlier this century, Luther Burbank decried this situation:

> I despair of anything being done at present to secure to the plant breeder any adequate returns for his enormous outlays of energy and money. A man can patent a mousetrap or copyright a nasty song, but if he gives the world a new fruit that will add millions to the value of the earth's annual harvests he will be fortunate if he is rewarded by so much as having his name connected with the result. . . . I would hesitate to advise a young man, no matter how gifted or devoted, to adopt plant breeding as a life work until America takes some action to protect his unquestioned rights to benefit from his achievements. (U.S. House 1930a)

Burbank—who created the shasta daisy, the Burbank potato, and scores of other important varieties—did fairly well, but never received the riches he deserved because he could not patent the varieties he created. Edison is reputed to have told him: "The things you have created for the American people are worth far more than what I have created for them. The law protects my creations but gives no protection to yours" (Diener 1953).

We can explain this situation. Edison could claim an intellectual property right to his inventions—and therefore a "utility" patent in them—since they arose from new knowledge that he discovered and disclosed, thus enabling others to intellectually reproduce his own inventions and to make many more besides. Burbank, a superb breeder for whose work we cannot be too grateful, did not provide new ideas or knowledge that would enable others to build further varieties. Rather, he depended entirely on his skill in conventional techniques of breeding. Thus, the daisy for which Burbank is justly famous can be reproduced biologically, but not intellectually. Burbank gave us many wonderful plants, but he left the science of plant breeding essentially unchanged.

The problem in patent law for plant and animal breeders arises in the require-ment that, to qualify for a standard or utility patent, an invention must be non-obvious, and the inventor must describe it and "the manner and process of making and using it, in such full, clear, concise and exact terms as to enable any person skilled in the art to which it pertains . . . to make and use the same" (U.S. Patent Of-fice 1995, Sec. 12). The "specification" must describe the invention so that others can understand precisely how it works and, therefore, how to build it from basic materials. In a leading case, in 1996, the Supreme Court defined the test as follows:

> The scope and content of the prior art are to be determined; differences be-tween the prior art and the claims at issue are to be ascertained; and the level of ordinary skill in the pertinent art resolved. Against this background, the obviousness or nonobviousness of the subject matter is determined (*Graham v. John Deere Co.* [383 U.S. 1.17])

In legal terms, the "nonobviousness" and "enablement" clauses presented insur-mountable obstacles to conventional breeders who wished patent protection for the new varieties they created. A breeder could do little more to describe a new variety than to refer to its ancestor plants and to the methods—such as hybridization, cross-pollination, backcrossing, and selective breeding— that he or she used to produce it. This would tell others little that they did not already know. If a breeder could enable others to build a plant up from basic materials—or even tell precisely how a plant differs genetically from its forebears—he or she might add to new knowledge. In 1930, however, officials concluded that no one could even differenti-ate new plant varieties on the basis of a written description (U.S. House 1930b).

Confronted with the inapplicability of intellectual property law to new varieties of plants and animals, Congress enacted the Plant Patent Act of 1930 and the Plant Variety Protection Act of 1970, which protect new varieties respectively against unauthorized asexual and sexual reproduction. In enacting these statutes, Congress sought to "afford agriculture, so far as practicable, the same opportunities to par-ticipate in the benefits of the patent system as has been given industry" (U.S. House 1970). Congress recognized that plant breeders could not claim to design the new varieties they created; indeed, they could hardly claim that the methods they used were anything but obvious, nor could they describe the genome of the plants they produced. Nevertheless, Congress wished to give an adequate economic incentive to breeders and to reward their efforts, even if those efforts did not produce new knowledge and, therefore, did not result in the creation of intellectual property.

These laws greatly weaken the nonobviousness and enablement tests; in fact, the 1970 statute does away with them entirely, mandating only that seeds be stored in a public repository (see U.S. Patent Office 1995, Sec. 103, 112). This may represent good social policy, insofar as it encourages plant breeders to continue their useful endeavors. It may also be fair, since Congress, by giving breeders rights over the use of their plants for reproduction, may recognize the rights of creators to control their personal property, including the seeds and tubers of the plants they produced. The genetic information contained in these seeds and tubers, which might be considered intellectual property, is the work not of the breeder, however, who could hardly specify what it is, but that of nature. For this reason, the Plant Variety Protection Act does not give a breeder intellectual property rights to this information. Rather, the statute provides that anyone who develops the same variety independently does not infringe the patent (Czarnetzy 1988).

All this changed dramatically in April 1987, when the U.S. Patent Office (1987, 1077) said that it "now considers nonnaturally occurring non-human multicellular living organisms, including animals, to be patentable subject matter." This ruling apparently overturned the traditional view that plants and animals developed by ordinary breeding techniques did not constitute intellectual property. The intellectual contribution of breeders, after all, may be as great as, or even greater than, that of genetic engineers with respect to the organisms they create. Breeders have to keep dozens of traits in mind and navigate among them, while engineers make a far more surgical strike on the genome of the organism. Moreover, plant breeding and artificial selection in general typically change the genome of target animals far more broadly than the more surgical technique employed by genetic engineers, which is usually to microinject many fewer foreign genes into fertilized eggs. Experts have testified, indeed, that "centuries of selective breeding have altered domestic animals far more than the next several decades of transgenic modifications are likely to alter them" (Adler 1988).[9]

It is unsurprising, therefore, that conventional plant breeders immediately saw an opening to claim their products as intellectual property—even though these "inventions" may exhibit no more new knowledge than those of Burbank. At first, however, Patent Office officials were unwilling to draw this inference. Commenting on the 1987 ruling, one official said:

> This won't affect our policy that products found in nature, such as farm animals produced by natural breeding, are not considered patentable. . . . For an animal to be covered under patent statutes, they [sic] have to be somehow created by man. (New York Times 1987)

Events have shown this statement to be whistling in the wind. "Under a 1985 decision of the United States Patent Office, Ex Parte Hibberd," an article in the Nebraska Law Review notes, "a plant breeder may obtain a utility patent on a newly developed plant variety" (Hamilton 1994). The traditional requirement that the patentee describe the design of his or her invention had been dropped when Chakrabarty was allowed, instead, to deposit a sample of his microorganism. Depositing samples became a way to satisfy what had been a requirement that patentees explain how their inventions work.

Organisms produced by conventional breeding techniques now routinely receive utility patents, as do products of genetic engineering. For example, several herbicide-tolerant crops, such as DuPont's sulfonylurea-tolerant soybeans, were developed through conventional breeding (Rotman and Farley 1995). It seems to no longer matter to the Patent Office whether new knowledge is involved; all that seems to matter is useful innovation. On the principle that production is as good as invention, Ducalion and his wife might now patent the novel varieties they produced, if we assume these organisms are useful, save that law prohibits the patenting of human beings.

## The Religious Rejoinder

Several religious thinkers have argued that life, even in its manifestation in genetic engineering, is no ordinary commodity. Rather, its creation involves a power so sublime that it cannot be understood or constrained within the conceptual framework of property law. The language of equity, property, justice, and so forth, according to this view, simply cannot express the moral issues that genetic engineering raises. We must use different conceptual categories, religious leaders argue, to discuss the regulation of biotechnology—categories that we still have to agree on and, indeed, may have yet to discover.

Karen Lebacqz (1983), a theologian, makes this point eloquently in relation to the book of Job. Job, whose faith is tested by God, suffers all kinds of torments, none of which he deserves; for example, he unfairly loses all of his possessions. He and his friends know that a just God would not visit this kind of suffering on an innocent man. While Job's friends assume that he is not innocent, Job accuses God of injustice. He asks God to justify His ways to humankind.

God replies to Job, in one of the most famous passages in the Bible, that His creative power—the power to make living animals from clay—has nothing to do with justice. This kind of power can hardly be understood, constrained, or regulated by the rights individuals may have within a Lockean conception of an appropriate political economy. God puts this thought in the following characteristic way:

> Where were you when I laid the foundations of the earth? . . . Have you ever given order to the morning? . . . Have you journeyed to the spring of the sea? . . . Who cuts a channel for the torrents of rain? . . . From whose womb comes the ice? (Book of Job 1991)

God then ridicules Job for applying concepts of property and justice to Him. God says: "Who has a claim against me that I must pay? Everything under heaven belongs to me." Lebacqz (1983) observes:

Job has approached God with the language of justice. God responds with the language of creation. God's response changes the very rubrics of the conversation. As the story is told, God in essence declares: "I am beyond your categories."... The rebuke constitutes a challenge to the paradigm accepted in common by Job and his friends.

Religious critics of biotechnology issue the same challenge to those who discuss biotechnology in the context of patents and property rights. These critics say that this context trivializes the issue: it treats the most momentous technological event in history as if it were simply a matter of benefits and costs. The patenting of animal life ignores the meaning of our new relationship with nature to concentrate, instead, on who will make money from it. Where the profits go, so religious leaders testify, is not the major issue confronting humankind. The more important question is whether we shall replace God as the author of life.

In testimony to Congress, religious leaders have emphasized the thought—as stated in the Book of Job—that "in God's hand is the life of every living thing" (U.S. House 1987). What is problematic about this conclusion is that it no longer appears to be true. In Job's time, it was easy to have faith in God's control over nature. After Charles Darwin, the creative authority passed to evolutionary processes, notably, random mutation and natural selection. Advances in DNA research have now given technology powers once reserved to nature and nature's God. Today, we might say that the life of every living thing lies in the hands of Calgene, Cambridge BioScience, and Genentech.

How do theologians respond to this? They suggest that as long as the genetic code remained a mystery, life maintained an independence from technology, which was not a bad thing. In the days before genetic engineering, it was evident that life was sacred and had a noble purpose (if one subscribed to a religious faith), or that life was random and had no direction (if one took Darwinian materialism seriously); in either case, however, humankind was not responsible for the ends of creation or evolution. Now, for better or worse, this responsibility falls into our hands.

The religious concern essentially may be a fear that we are not up to exercising this kind of power. Theologians observe that no consensus exists about the values we should serve—the ends to which we should direct nature. Indeed, no one discusses values—only property and profit, benefits and risks. It seems as if safety and the preferences that consumers are willing to pay to satisfy are all that we care about. We have no nobler ends, no better causes, no loftier purposes than those.

When the religious objection is stated this way, it sounds familiar; it decries our love of "the things of this world" and deplores our lack of attention to our "spiritual estate." What makes the contemporary jeremiad different is the extent of humankind's power over the things of this world and our new ability to include among them many of the things that were thought to be spiritual. The contemporary religious critique, in effect, concerns our ability to override or obliterate the distinctions between life and matter, nature and technology, and humanity and God on which our spiritual estate seemed to depend—distinctions that helped us to think that human beings belong to nature, not just the other way around. The Reverend Wesley Grandberg-Michaelson, in congressional testimony, made this point:

> Observing recent innovations of biotechnology, such as combining genetic characteristics of cows with pigs, inserting bovine growth material into salmon to create superfish, and uniting the phosphorescence of fireflies with tobacco plants, one gets the clear feeling that creation's inherent structures and boundaries are of little intrinsic worth. (U.S. House 1987)

An analogy may help us get to the quick of the problem. Parents sometimes joke that they know why their small children misbehave. "Little Johnny is terrified," the parent will observe. "He has discovered that he lives in a family where a two year old is in command." Genetic engineering puts us all in the position of the two year old—terrified at the prospect of our own power. And our moral vision is no better than that of the child who whines: "I want it. Give it to me now."

Theologians may concede that genetic engineering works and gives us vast new powers over the forms life takes. They may concede that we can exercise this power profitably to satisfy wants and preferences of the sort that are expressed in competitive consumer markets. The businesslike course, then, may be to offer patents for these life-forms; the benefits will exceed the costs. Nevertheless, it is possible to ask—as religious thinkers do—is this profitable course right?

How would one answer this question? Theologians have not provided a way. What they have done, essentially, is to ask for a moratorium on patenting animals until we can discuss the matter in terms more appropriate than those of market analysis. This seems to be the upshot of the Statement of Religious Leaders Against Animal Patenting, which asserts, rather blandly, that:

> The gift of life from God, in all its forms and species, should not be regarded solely as if it were a chemical product subject to genetic alteration and patentable for economic benefit. Moral, social, and spiritual issues deserve far more consideration before binding decisions are made in this area. (U.S. House 1987)

This statement and others like it call for a discussion of moral and religious values that are at stake in the patenting of animals. Yet these statements, while calling for more time, do not describe a conceptual and normative framework that would be appropriate for regulating biotechnology. Nor do they show that an ethical or spiritual approach must be incompatible with the framework of property rights, justice, and economic utility that now dominates the discourse over patents and similar policies. The difficulty, as religious leaders present it, comes in reconciling our reverence for life with our new ability to manipulate it. The central problem may lie not so much in the direct consequences, but on how we describe, understand, or interpret our actions. As Lebacqz (1983) writes: "The question is not so much what we *do* but how we *think* about it."

Until we learn how to "think about it," we are left with a dilemma. If we bring to bear no concepts other than those of property and utility, we will abandon many of the distinctions on which our cultural, moral, and religious identities depend. If we manage to maintain these distinctions and regulate technology to protect them, however, we may have to sacrifice many of the economic benefits of genetic engineering. What is the proper course? And who should have the power and responsibility to determine it?

## Breeders' Rights for Biotechnology: A Way Out?

If this analysis of the religious objection to patenting life is at all correct, then there seems a clear way to reconcile these objections with the needs of the biotechnology industry. Religious leaders are concerned with dispelling the idea that human beings create life in the same way that God created Adam out of clay, that is, by producing creatures out of basic and simple materials using a design humans themselves invented. Edison created the phonograph and incandescent bulb by virtue of his own design: He "laid the foundations" of these instruments in the same knowledgeable way that God laid the foundations of the earth. That is why Edison had a moral as well as legal right to his inventions as intellectual property. Religious leaders deny that genetic engineers—any more than conventional plant breeders—design organisms from scratch, and accordingly, that these organisms should be considered intellectual property of anyone but God.

As Dr. Richard B. Land of the Southern Baptist Convention testified:

> Humans and animals are pre-owned beings. We belong to the creator God. The PTO's [Patent and Trademark Office] decision to grant patents on animal or human genetic information represents a usurpation of the ownership rights of the Sovereign of the Universe. (quoted in Rubenstein 1995)

The biotechnology industry is also concerned with dispelling the idea that it is "playing God." It cares less about its pretensions to have designed life than its legal access to a particular set of rights to a monopoly position on certain products. Whether or not these rights carry with them a pretension to intellectual property in the moral sense is nothing that concerns industry; from its perspective, what matters is only the legal effect, not the ethical meaning or basis, of the law.

There seems to be room for compromise between the concerns of the industry and those of religious leaders. Indeed, each group has sought to accommodate the other. Activist Jeremy Rifkin, characterizing the position of a coalition of 80 religious leaders that he led to Washington on May 18, 1995, emphasized that they had "no problem" with process patents or even with protecting biotech products with the sort of marketing exclusivity conferred on "orphan" drugs (Rhein 1995). Rabbi David Saperstein, who heads the Religious Action Center for Reform Judaism, decried the idea that scientists would "arrogate to themselves the ownership of the life they are creating." He observed that there are "ways through contract law and licensing procedures to protect the economic investment these people make" (Rhein 1995). Like Rifkin, Saperstein objected to the Faustian hubris of claims that organisms are intellectual property, not to the legal regimes, *per se*, that are intended to reward and encourage effort and investment in biotechnology.

On the industry side, spokespersons have been anxious to assure their clerical critics that they neither want to "play God" nor claim to be the authors of life. What industry wants, they argue, is not to upstage the Creator, but to enjoy a legal regime that protects and encourages investment. Biotechnology Industry Organization President Carl Feldbaum emphasizes this point:

> A patent on a gene does not confer ownership of that gene to the patent holder. It only provides temporary legal protections against attempts by other parties to commercialize the patent holder's discovery or invention. This is a critical distinction because no one, in our view, can or should own life itself. (Rhein 1995)

It is not hard to see the outlines of a solution that can embrace both the views of religious leaders and the needs of the biotechnology industry. Novel organisms might be covered by a new patent statute that, like the old plant protection acts, eliminates the "description" and "enablement" requirements, thus recognizing that those who produce these organisms do not design them and, therefore, cannot claim to own them as intellectual property. Such a statute would accept the idea that only God has a claim to intellectual property in living things—at least until some scientist designs a new organism from basic inanimate materials.

After recognizing the bedrock conviction religious people share that only God can create—in the sense of design—life, the statute might not differ further from utility patents in the terms, conditions, and rights for the patenting of organisms. This approach would extend to today's needs and circumstances the compromise Congress wisely enacted in the plant protection acts of 1930 and 1970. These statutes intended to protect and encourage the efforts of plant breeders without implying that they had created intellectual property or truly invented new life.

A compromise along these lines would bring patent law into harmony with the important distinction between production and invention, which Locke drew at the beginning of this essay. To claim novel organisms as intellectual property is to confuse the way God and humans create life, as did Sir Robert Filmer. The difference is that God knew what He was doing in a way that plant breeders and even genetic engineers do not. To cross strains of plants or microinject genes into embryos is not to invent, design, or construct living things. Perhaps scientists someday will create living things from scratch, but the work of genetic engineers today more closely resembles what "Ducalion and his Wife in the Fable did towards the making of Mankind, by throwing Pebbles over their Heads."

## Notes

1 Locke's principal treatment of property rights is found in the *Second Treatise*, Chapter 5. For an excellent analysis of the Lockean argument, see Becker, 1977. Chap. 4 in *Property Rights*. London: Routledge and Kegan Paul.

   Alan Goldman (1987. Ethical issues in proprietary restrictions on research results. *Science, Technology, and Human Values* 12:23) understands the underlying premise to state "a person's right to the fruits of his or her labor, perhaps considered as an extension of his or her activity and therefore of him or herself." For a further discussion, see Nozick, R. 1974. *Anarchy, State, and Utopia*. New York: Basic Books.

2 Commentators generally agree that the basis of patent law consists in, first, the natural entitlement of inventors to the fruits of their labor, and second, the enhancement of social welfare or utility that follows from the patent protection. See, for example, United States. Office of Technology Assessment. 1989. *New Developments in Biotechnology: Patenting Life —Special Report*. OTA-BA-370. Washington, D.C.: Government Printing Office.

3 *Monsanto Co. v. Rohm and Hass Co.*, 312 F. Supp. 778, 790 (E.D. Pa. 1970), *aff'd* 456 F. 2d 592 (3d Cir.), *cert. denied* 407 U.S. 934 (1972).

4 This was not the first patent to be placed on an organism. In 1873, the United States granted patent no. 141,072 to Louis Pasteur, for a "yeast, free from organic germs of disease, as an article of manufacture."

5 Philip Leder and his co-worker, Timothy Stewart, introduced an activated *c-myc* oncogene into an early mouse embryo to create a "transgenic" mouse that carries the gene in both its somatic and sex cells. On this basis, Harvard applied for a patent written broadly to cover not only any such mouse, but also any nonhuman eukaryotic animal with the oncogene. See Transgenic Non-Human Mammals, patent no. 4,736,866 (Issued 12 April 1988).

6 These authors believe that the phrasing of the relevant passage from the U.S. Constitution supports this view. The Constitution states (Article 1, § 8): "Congress shall have the power . . . to promote the progress of science and the useful arts, by securing for limited times for authors and inventors the exclusive right to their respective writings and discoveries."

   The use of the word *secure*, however, suggests that the framers of the Constitution understood the rights in question to be natural or inherent, and therefore, that Congress has the power to recognize and legalize them. The Declaration of Independence, after declaring that persons have "inalienable rights," states that "to secure those rights, governments are instituted among men." James Madison also wrote: "The copyright of authors has been solemnly adjudged in Great Britain to be a right of common law. The right to useful inventions seems with equal reason to belong to the inventors. The public good fully coincides in both cases with the claims of individuals." See Madison, J. 1982. *The Federalist Papers: No. 43*. New York: Bantam Books.

7 Hettinger also argues that a person has no natural right to the market value of his or her intellectual

property—the same argument would apply to personal property, of course—because this value is created not by his or her labor, but by the demand translated through a socially constructed market. The question arises, however, whether someone who owns something is free to sell it in an open and fair market. If a person is not free to sell what he or she owns, what does owning it mean?

8 With respect to copyright on a literary or artistic work, of course, it is different, since the chances that someone would create the same work independently are very small. Accordingly, the state guarantees copyright for the lifetime of the artist and, if renewed, for 50 years more.

9 Reid Adler cites expert testimony to the effect that the new biotechnologies do not depart radically from historical practices. See p. 18 n. 99.

## References

Adler, R.G. 1988. Controlling the applications of biotechnology: A critical analysis of the proposed moratorium on animal patenting. *Harvard Journal of Law and Technology* 1:1–61, 20 n. 126.

Bentham, J. 1825. *The Rationale of Reward*. Quoted and cited in Greenfield, M.S. 1992. Recombinant DNA technology: A science struggling with patent law. *Stanford Law Review* 44:1082.

Book of Job. 1991. Chap. 38 in *The New Revised Standard Bible*. New York: Oxford University Press.

Czarnetzy, J.M. 1988. Altering nature's blueprints for profits: Patenting animals. *Virginia Law Review* 74:1327–62.

Diener, R. 1953. Patents for biological specimens and products. *Journal of the Patent and Trademark Office Society* 35:289.

Epstein, R. 1985. *Takings: Private Property and the Power of Eminent Domain*. Cambridge: Harvard University Press.

Evenson, R.E., and J.D. Putnam. 1987. Institutional change in intellectual property rights. *American Journal of Agricultural Economics* 6:404.

Filmer, R. 1947. Patriarcha: Or the natural power of kings. In *Two Treatises of Government*, edited by T. Cook. 1680. Reprint, New York: Hafner.

Hamilton, N.D. 1994. Why own the farm if you can own the farmer (and the crop)? *Nebraska Law Review* 73:91.

Hettinger, E.C. 1989. Justifying intellectual property. *Philosophy and Public Affairs* 18:31–52.

Lebacqz, K. 1983. The ghosts are on the wall: A parable for manipulating life. In *the Manipulation of Life*, edited by R. Esbjornson. San Francisco: Harper and Row.

Locke, J. 1967. *Two Treatises of Government*. Edited by P. Laslett. 1698. Reprint, Cambridge: Cambridge University Press.

Marx, K. 1995. *The Collected Works of Karl Marx*. New York: International Publishers.

*New York Times*. 1987. New animals will be patented. 17 April.

*Pittsburgh Post-Gazette*. 1995. Reported remarks of Russell Smestadm, vice president of Agracetus, a subsidiary of W.R. Grace and Co. of Boca Raton, Florida. 3 September.

Rhein, R. 1995. Gene patent crusade moving from church to court. *Biotechnology Newswatch* 5 (June): 1.

Rotman, D., and P. Farley. 1995. Transgenic crops head to market. *Chemical Week* 157:25.

Rubenstein, B. 1995. Genetic patent pits clients, religion, government. *Corporate Legal Times* (September): 1.

U.S. House 1930a. *Rep. Purcell quoting from a telegram received from the widow of Luther Burbank*. H.R. Rep. No. 1129. 71st Cong., 2d sess.

———. 1930b. Committee on Patents. *Memorandum of Patent Commissioner Robertson. Hearings on H.R. 11372*. 71st Cong., 2d sess.

———. 1970. *Hearing before the Subcommittee on Departmental Operations of the Committee on Agriculture*. 91st Cong., 2d sess.

———. 1987. Subcommittee on Courts, Civil Liberties, and the Administration of Justice of the Committee of the Judiciary. *Appendix: Testimony by Reverend Wesley Grandberg-Michaelson.* 100th Cong., 1st sess.

U.S. Patent No. 775,134 (issued 15 November 1904). For litigation, see *Gillette Safety Razor Co. v. Clark Blade and Razor Co.,* 187 F. 149 (C.C.D.N.J. 1991) *aff'd* 195 F. 421 (3d Cir. 1912).

U.S. Patent Office. 1987. *U.S. Patent and Trademark Office Gazette.*

———. 1995. Vol. 35, *Patents and Related Materials, U.S. Code.*

# The Convention on Biological Diversity:
# A Polemic

by Lakshman D. Guruswamy

The Convention on Biological Diversity fails to address the problems it was meant to remedy. It declined to institutionalize the common responsibility of humanity to protect biodiversity, rejected the extension of state responsibility for damage to the global commons, and effectively spurned the concept of sustainable development.

\*

The U.N. Convention on Biological Diversity (CBD; appendix, this volume), ceremoniously signed at the Earth Summit,[1] is a much-heralded chapter in international legislation dealing with biodiversity (Reid, this volume). Although envisioned as a comprehensive law that would remedy the fragmented and regional approach to biodiversity protection evident in earlier treaty making (Hendricks, this volume), the CBD has belied these expectations, and may halt the advance of international environmental law on three fronts.[2]

First, it rejects the concept of sustainable development by prioritizing economic growth over environmental protection, allowing international resources earmarked for the protection of biodiversity to be expended on economic growth that could destroy biodiversity. Second, it denies state responsibility for damage to the global commons. Finally, it repudiates the idea that the plant, animal, insect, and genetic resources of the world (our biodiversity) are the common heritage of humankind, and that it is the responsibility of the community of nations to protect this heritage.

Before embarking on this brief exegesis, it is important to point out that the CBD does not belong to the "soft" law genus of Agenda 21 and the Rio declaration, or to the "nonlegally binding" category of the forestry principles, but behooves to be judged on its own terms: as a species of "hard" treaty law.[3] If the distinction between nonlegally binding principles, hortatory declarations, and obligatory laws is to mean anything, hard treaty law protecting biodiversity—as distinct from soft law or nonlegally binding instruments—ought to embody legally recognizable duties, obligations, and rights pertaining to the protection of biodiversity.[4] Typically, hard legal obligations can be distinguished from soft law or nonlegally binding aspirations by asking whether the provisions create binding legal obligations as opposed to aspirational exhortations. It is precisely because they were aware of these distinctions that the lawmakers at Rio created three different types of instruments.

To say this is not to diminish or detract from other social forces that can act more effectively and efficiently than law to solve problems, or even avoid those problems *ex ante*. Increasing knowledge and awareness about the dangers of extinction; educational campaigns; appeals to ethics, equity, and morals; economic incentives; and market mechanisms could preserve biodiversity without being institutionalized as law. Additionally, there may also be societal needs for establishing legal rights, duties, and obligations to complement, facilitate, empower, or harness the use of social mechanisms and techniques.

Once it is determined by lawmakers that they want to create hard treaty laws, it is essential that they carefully craft the substance, content, and form of such rights and duties. International laws, like all laws, reflect the minds and intentions of their lawmakers. The precision, extent, and force of legal prescriptions in treaties are the product of human design, not accident. The weight of a legal obligation, duty, or right will depend on the extent to which it commands or demands actions, and cures the perceived mischief by advancing a remedy. In assessing the CBD, therefore, it would be useful to identify the main concerns and problems about biodiversity that it was meant to address, and then inquire if the remedies offered by the CBD help to solve these problems.

### The Import of the Convention

The factors that influence and mold the mind-set and intentions of lawmakers will continue to be the subject of perennial inquiry. There can be little debate, however, that two critical questions had faced the international community when forging the CBD. First, what problems did they confront? And second, how willing were they to take legal measures to address these problems with measures, to the extent possible, that were as binding and enforceable in the international arena as statutes are under municipal law? While the CBD answered these questions by articulating the nature of the challenge caused by vanishing biodiversity, it set its face against remedying those problems, and even turned the clock back on numerous developments in international environmental policy and law.

To begin with, the CBD rejects the concept of sustainable development—the very *grundnorm* of the Earth Summit. Sustainable development has not been authoritatively defined, but its key attributes are identifiable. In essence, it calls for economic growth that can relieve the great poverty of the less-developed countries (LDCs), based on policies that sustain and expand the environmental resource base (WCED 1987, 1). Consequently, sustainable development becomes environmentally sensitive development that meets the needs of the present generation without compromising the ability of future generations to meet their own needs (8).

Sustainable development, therefore, gives parity of status to economic growth and environmental protection. It rejects economic development and growth that is not environmentally sensitive or that destroys the resource base. It is a new concept precisely because it embraces both development and environmental protection. But the CBD states, both in its Preamble and in critical articles dealing with the financing of the CBD that "economic and social development and poverty eradication are the first and overriding priorities of developing countries" (Preamble; see Article 20[4]). By any analysis, the elevation of development and the diminishment of biodiversity is clear. Not only is biodiversity subjected to the preeminence of development, but also what constitutes development is subjectively to be determined by developing countries. By diminishing environmental protection, the CBD effectively disowns sustainable development (Pallemaerts 1996, 630–85).

This diminishing of biodiversity and accentuation of development is confirmed by the financial provisions. To enable LDCs to implement the CBD, developed countries agree both to pay the "full incremental costs" of such implementation (Article 20[2]) and to transfer technology to LDCs (Article 16). An examination of the commitments of developing countries, in exchange for this transfer of money and technology, is revealing. The CBD lucidly states that "economic and social development and eradication of poverty are the first and overriding priorities of the developing country Parties" (Article 20[4]). Having made the overriding principle clear, it then develops the implementing structure. The institutional structures as well as the "policy, strategy, programme priorities and eligibility criteria relating to" access to those transferred resources and technologies will be determined by the Conference of the Parties to the CBD (Article 21[1]).

Where does this leave us? In the absence of an explicit commitment to protect biodiversity, any resources transferred under the CBD could be used by a small minority of zealous developing countries to advance their own concept of economic and social development. If, for example, they decide that road building, "reclamation" for beach development and marinas, or even the cutting down of tropical forests is necessary for economic and social development, they would be acting within the powers and privileges granted to them.

A somewhat foreboding omen of the future direction of the CBD is offered by its treatment of tropical rain forests. It is estimated that tropical forests are home to at least 50 percent of plant and insect diversity (WCED 1987, 151; Myers 1988, 35; Wilson 1992, 259). Yet all references to tropical forests were systematically, and deliberately, excised from the CBD.[5] As its first flaw, then, the CBD clearly undermines the original environmental emphasis of sustainability within the concept of *sustainable development*.

Second, the CBD tilts against an emerging and developing pattern of regional customary and treaty law that, in the last 50 or so years, has sought to establish the common responsibility of humankind to protect biodiversity.[6] Many involved in the development of international environmental law hoped that the CBD would consolidate these endeavors, and provide an instrument that dealt comprehensively, globally, and more specifically with the nature of the obligation to protect biodiversity. Instead, the CBD contains no substantive protection obligation.

Although the collective obligation to protect biodiversity was seen by the United Nations Environmental Programme, the World Conservation Union, and numerous other nongovernmental organizations as constituting the foundations of the new treaty, the CBD rejects such an obligation, instead proclaiming that states have the "sovereign right to exploit their own resources pursuant to their own environmental policies" (Article 3). In a similar vein, the CBD rejects the principle that biodiversity is the natural heritage of humankind.[7] Accepting biodiversity as our common natural heritage gives rise to the corollary obligation that we protect and preserve such a heritage. Instead, the CBD settles for an effete and legally nonbinding recitation that biodiversity is the common "concern" of humankind. Furthermore, the attenuated affirmation that "biological diversity is a common concern of humankind" is found only in the Preamble, even though it ranked as a fundamental principle throughout the drafting process.[8]

Even when the CBD attempts to protect biological diversity by *in situ* conservation, *ex situ* conservation, and sustainable use in Articles 6-14, it makes sure that every obligation assumed (except those related to research, training, education, and public awareness) yield to the caveat: "as far as possible and as appropriate." Moreover, Article 7, which deals with the key elements of identification and monitoring, allows each contracting party to do such identification. This contrasts with earlier expectations and drafts that provided for the establishment of Global Lists of Biogeographic Areas of Particular Importance for the Conservation of Biological Diversity and of Species Threatened with Extinction on a Global Scale to be internationally, not nationally, determined.[9]

Any obligations to protect the common heritage of humankind need not fall disproportionately on the poor and deprived. Given the enormous disparities of wealth among nations, equity, fairness (Rawls 1971, 103), and efficiency require that discharging the burden of protection should fall differentially and more heavily on the richer nations. Biological diversity is a public good that is of critical importance to all humanity, and therefore, it ought to be protected by the entire international community. In the absence of a system of international government that can act to protect public goods for collective benefit, other mechanisms should be found. One fecund suggestion is to give areas of biodiversity a designated value, and pay the owner country an interest or financial allotment for the conservation or preservation of such areas (Sedjo 1988). The burden of such payments should be proportionately heavier for the richer nations.

It is also clear that the duty to preserve huge extents of forestland, marshes, or coral reefs rich in biological diversity could entail daunting opportunity costs to

LDCs. For example, an obligation to protect rain forests placed on LDCs is tantamount to denying them the right to cut down and develop such forests in order to provide land, housing, and food to their desperately poor populace. Accordingly, it becomes necessary not only to affirm the responsibility of the entire community of nations, including the biologically rich, to protect biodiversity, but also to establish measures and mechanisms to ameliorate the costs borne by LDCs that may be biologically rich yet economically poor.

This is not at all the same as the "burden sharing" referred to in Article 21(1) of the CBD. The present arrangements deny the legal responsibility of the community of nations to protect and preserve biodiversity, while asserting that any commitment by developing countries to protect biodiversity will depend on the extent to which they are bankrolled by developed countries (Article 20[4]). Unfortunately, by denying legal responsibility, the CBD also declines to impose any tangible duties or responsibilities on the community of nations to protect biodiversity. This is a serious defect, and the second flaw of the CBD.

Third, the challenge facing the CBD was to extend state responsibility for extraterritorial harm to damage caused to the global commons.[10] The global commons may include the critical habitats or homes of life-forms physically located within the territorial jurisdiction of nation states. But such an extension of state responsibility was roundly rejected by the CBD, and its application has been strictly confined to extraterritorial damage. The Fifth Revised Draft Convention on Biological Diversity had asserted the principle that states are responsible "for the conservation and sustainable use of their biological resources."[11] While a weaker formulation of that principle is retained in the Preamble (para. 5), it is effectively emasculated by the assertion that states have a sovereign right over their biological resources (para. 4).

What emerges is a deeply flawed convention that fails, at its core, to live up to expectations. On the contrary, it very nearly interdicts the obligation to protect biodiversity, fails to institutionalize the principle of differentiated responsibility, and rejects sustainable development. The conclusion that the CBD flounders in holding the ring between the global need for biological diversity and the sovereign right of states to control and develop their own resources is a somber one.

**The Way Forward**

Laws have a strong and important psychological dimension. The significance of international law in regulating vanishing biodiversity was immensely boosted by the publicity attending the ceremonial signing of the CBD at the Earth Summit, possibly the greatest gathering of world leaders in history. The Earth Summit placed biological diversity on the international agenda, gave the protection of biodiversity a world stage, and introduced a new term to the international political lexicon. The symbolic importance of the CBD—and the stimulus that it has given to thought, scholarship, discussion, and action on the protection of biodiversity—is undeniable. It is worth noting, however, that the psychological and symbolic importance of the CBD gives rise to incipient dangers. By extolling this convention despite its fundamental flaws, a false sense of achievement may be created. Indeed, such tactics may anesthetize the public, and lull national and international decisionmakers into a spurious sense of security. By offering and acclaiming specious remedies, we may be placing ourselves in the precarious predicament of thwarting genuine ones. In the final analysis, we may be encouraging the belief that biodiversity is safe, when it is actually in peril.

Assuming that the Earth Summit was an impressive launching pad for international law and the CBD, a powerful launch will not put a rocket into orbit unless it is possessed of its own power. The somber truth is that the CBD lacks such power. This deeply flawed instrument may fail in its mission, and in the process, damage not only itself (the CBD), but also the mission of protecting biodiversity. What are the implications of such a depressing conclusion, and how ought the international community react to the biodiversity outcome of Rio? It is necessary to note at this juncture that the United States has not ratified the CBD and may not do so, given the mood of the present Congress.

A way out of the present impasse is called for that recognizes the reality both of U.S. suspicions about the CBD, and the fact that the United States can and should play a role in protecting global biodiversity. This chapter does not presume to chart the path of global environmental diplomacy dealing with biodiversity, but it does advocate a strategy for the United States to play a key role in making the best of a bad convention. The strategy advanced below may be interwoven into Brent Hendricks's (this volume) imaginative suggestions for "transforming" and "reinventing" the CBD.

First, the primary aim of any strategy must remain a "World Forestry Convention." Such a convention should protect old forests, particularly tropical ones, which are home to up to 50 percent of the plant and insect biological diversity of the world. The Clinton administration should use Senate approval as the mechanism for reaching accord on such a convention. Though the Bush administration rejected any attempts to bring old growth areas in the United States under the protective umbrella of a World Forestry Convention, the Clinton administration appears ready to do so. Using the new U.S. policy as a bargaining chip for a larger international commitment toward preserving biodiversity may also have the additional benefit of convincing a hostile U.S. Congress that this is not another "sellout" to corrupt Third World countries.

There are hopeful signs emanating from the Conference of the Parties required by the CBD. At their first meeting, they issued a statement to the Commission on Sustainable Development that emphasized the importance of the conservation, management, and sustainable use of forests for achieving the objectives of the CBD.[12] This was followed at the second Conference of the Parties, in 1995, by the adoption of an even more commendable statement on forests and biological diversity, demonstrating the crucial role of forests in maintaining global biodiversity.[13] This kind of soft hortatory statement is a long way from the hard obligations required by an international forest convention, but it is a development that should be seized on by the United States. The United States should abandon its meaningless silence on this issue, and conduct a strong diplomatic and political campaign for a World Forestry Convention. Such a move may well be the only prudent way of securing the approval of the U.S. Senate for the CBD. Doing so may, of course, invoke the ire of a minority of developing countries that are aggressively and relentlessly destroying biodiversity, as well as their misdirected environmentalist appeasers, who seem politically coded toward capitulating to ecological aggression. But, that is a risk worth taking.

Second, it is impossible to devise any strategy without a global financial commitment to sustainable development. A commitment to sustainable development will embrace precisely the kind of financial and technological transfers that the United States has eschewed until now. This is not an easy task for the Clinton administration, which has signed the CBD but faces a hostile Congress. The administration would need to demonstrate that the real bucks invested in protecting biodiversity will result in real gains. To do so, it must establish that resource transfers will be directed toward the protection of biodiversity, not some enterprise within the "developmental" discretion of the recipient nation. Given the confusing state of the financial arrangements under the CBD, this can only be accomplished if the United States directs its full diplomatic efforts toward transforming the procedures under the CBD, through separate protocols, and reforming the manner of its implementation. In short, it is time to reach out for a new design that addresses the infirmities and fundamental failures of the current CBD.

## Notes

1   The Earth Summit spawned two "hard" law treaties: the U.N. Convention on Climate Change and the Convention on Biological Diversity. It also gave rise to two "soft" law instruments: the Rio Declaration on Environment and Development, and Agenda 21; and one "nonlegally binding" document, the Non-legally Binding Authoritative Statement of Principles for a Global Consensus on the Management, Conservation, and Sustainable Development of All Types of Forests. (See United Nations Conference on Environment and Development. 1992. *Adoption of Agreements on Environment and Development.* Resolution 1, A/CONF.151/6 [vol. 1], 4–13 June, reprinted in 31 I.L.M. 881 [1992]; Earth Summit Press Summaries. 1992. 3–14 June; *New York Times.* 1992. 15 June, 5.)
2   This critique does not deal with questions affecting industry and trade, such as intellectual property rights to nature, technology transfers, and the relationship between biotechnology and biodiversity. It does not endorse or support the arguments advanced by the Bush administration for not signing the CBD. The present criticisms are directed to the deficiencies of the CBD from an environmental perspective.

3   For the limited purpose of this piece, the main difference between hard and soft laws lies in the way that norms, expectations, or consensus among the parties is expressed or articulated. Soft laws often express themselves as political statements, values, and exhortations as distinct from binding rules and clear standards that are justiciable. (See Palmer, G. 1992. New ways to make international law. *American Journal of International Law* 86:259, 269; Gruchalla-Wesierski, T. 1984. A framework for understanding "soft law." *McGill Law Journal* 30:37, 44–45, 52–55.) The very fact that such a distinction is recognized and acted on is sufficiently strong ground for asserting that hard law is distinguishable from soft law and nonlegally binding documents.

4   Making this point does not in any way deny that it is possible to find within instruments of soft law—such as, for example, Agenda 21 and the Rio Declaration on Environment and Development—a few provisions that restate, codify, or express existing rules or principles of customary international law.

5   The World Conservation Union drafts attempted to protect tropical and rain forests by including the principle, carried right through to the fifth revised draft of the convention, that states are responsible "for the conservation and sustainable use of their biological resources" (Article 3[2][a] of the Fifth Revised Draft Convention on Biological Diversity, UNEP/bio.div/N7–INC.5/2, 20 February 1992). This principle was excised from the CBD and replaced, instead, by one that asserts the sovereign right of states to exploit their own resources, subject to the duty not to cause extraterritorial harm (Article 3 of the final document).

6   For example, the Declaration of the United Nations Conference on the Human Environment, UN DOC A/CONF. 48/14, Preamble, Principles 1, 2, 4, and 5, Stockholm, 16 June 1972; Convention for the Protection of the World Cultural and Natural Heritage, especially Articles 4 and 6, Paris, 16 November 1972; the Berne Convention on the Conservation of European Wildlife and Natural Habitats, Preamble, Articles 3, 4, 5–9, and 13, Berne, 19 September 1979; the Apia Convention on the Conservation of Nature in the South Pacific, 12 June 1976; the ASEAN Convention on the Conservation of Nature and Natural Resources, Kuala Lampur, 9 July 1985; the Protocol Concerning Mediterranean Specially Protected Areas, Geneva, 3 April 1982; and the Protocol on the Conservation of Common Natural Resources, Khartoum, 24 January 1982.

7   The natural heritage of humankind is to be distinguished from the common heritage of humankind (CHH) that has been applied to the deep seabed and the ocean floor beyond the limits of national jurisdiction (United Nations Convention on the Law of the Sea, Articles 133, 136, and 156–69, A/CONF.62/122, 10 December 1982, reprinted in 21 I.L.M. 1261 [1982]), and the outer space regime, respectively (Agreement Governing the Activities of States on the Moon and Other Celestial Bodies, Article 11[1], G.A.Res.68, 34 UN GAOR Supp. [No. 46] at 77, UN Doc. A/RES/34/68, 5 December 1979, reprinted in 18 I.L.M. 1434 [1979]). At its core, the CHH involves the inclusive enjoyment and sharing of the products of the common heritage, and its thrust remains redistribution, not conservation. The essential feature of CHH, whether based on *res communes* or *res publica*, is the entitlement of the entire international community to exploit the seabed and share the fruits of exploitation. CHH is not a conservationist principle because it is directed toward maximizing resource exploitation and economic returns. Moreover, it is so suffused in traditional nonconservationist resource economics as to render it constitutionally incapable of nurturing a regime of sustainable development.

The attempt in the Food and Agriculture Organization (FAO) to secure an International Undertaking on Plant Genetic Resources was based on genetic resources being accepted as a heritage of humankind. If the FAO did attempt to bring plant genetic resources within that rubric, their attempt did not improve the conservationist credentials of CHH. The FAO undertaking involves the repudiation of property rights from all germ plasm—both natural and improved—including genetically engineered plants, seed, and tissue culture. But it is clear that the thrust of the agreement is exploration and utilization, rather than conservation. By contrast, the natural heritage of humankind refers to the biological necessities of the world, essential for the existence and development of all humankind, that may fall within the national jurisdiction of states.

8   Including the Fifth Revised Draft Convention on Biological Diversity, UNEP/bio.div/N7– INC.5/2, 20 February 1992.

9   Ibid. Articles 15 and 25, note 8.

10  The principle of state responsibility for extraterritorial harm has been accepted as international law since at least the Stockholm Declaration of 1972. It has subsequently been affirmed in numerous other treaties and instruments, for example, the United Nations Convention on the Law of the Sea (1982), the Rio Declaration on Environment and Development (1992), and the World Charter for Nature (1982).

11  UNEP/bio.div/N7–INC. 5/2, 20 February 1992. Article 3(2)(a), note 8.

12  Statement from the Conference of the Parties to the Convention on Biological Diversity to the Commission on Sustainable Development at Its Third Session, Article 15, UNEP/CBD/COP/1/17, 28 February 1995.

13  Statement on Biological Diversity and Forests from the Conference on Biological Diversity to the Intergovernmental Panel on Forests, UNEP/CBD/COP/2/19, 30 November 1995.

## References

Myers, N. 1988. Tropical forests and their species: Going, going . . . ? In *Biodiversity*, edited by E.O. Wilson. Washington, D.C.: National Academy Press.

Pallemaerts, J. 1996. International environmental law in the age of sustainable development: A critical assessment of the UNCED process. *Journal of Law and Commerce* 15:623.

Rawls, J. 1971. *A Theory of Justice*. Cambridge: Harvard University Press.

Sedjo, R. 1988. Property rights and the protection of plant genetic resources. In *Seeds and Sovereignty: The Use and Control of Plant Genetic Resources*, edited by J.R. Kloppenburg Jr. Durham, N.C.: Duke University Press.

(WCED) World Commission on Environment and Development. 1987. *Our Common Future*. Oxford: Oxford University Press.

Wilson, E.O. 1992. *The Diversity of Life*. Cambridge, Mass.: Belknap Press of Harvard University Press.

# Transformative Possibilities: Reinventing the Convention on Biological Diversity

by Brent Hendricks

When originally contemplating a worldwide agreement on biodiversity, the international community imagined an "umbrella" convention, one which would coordinate or perhaps consolidate the multitude of global and regional treaties in the field. Due to politics and impracticality, this effort was abandoned in favor of the current Convention on Biological Diversity (CBD)—a loose framework convention much maligned for its lack of substance. Rather than further indict the CBD, this chapter attempts to reinvent it in order to realize its earlier umbrella aspirations. To do so, this chapter considers the fundamental commitments of the "framework" approach to lawmaking: an open political process and an institutional capacity for self-revision. Using these as constitutive principles for transformation, recommendations are offered to bolster specific institutions within the CBD, so as to resurrect that treaty into an operational, if limited, umbrella convention.

*

The international legal community originally conceived of the CBD (appendix, this volume) as an "umbrella" convention—one which might consolidate the existing panoply of global and regional treaties under a single administrative structure.[1] The hope was "to rationalize current activities" in the field of biodiversity, eliminating jurisdictional overlap and filling perceived gaps (Burhenne-Guilmin and Casey-Lefkowitz 1993). Very soon, however, the utopian plan for an umbrella convention proved both legally difficult and politically unattainable. In the end, the contracting parties to the CBD adopted a fairly loose "framework" treaty, one which failed to provide the CBD with even a *primary* coordinating role vis-à-vis the prior treaties. Rather than malign the hollowness of the CBD, however, this chapter offers suggestions as to how current institutions within the CBD might function to resurrect the treaty into an operational, if limited, umbrella convention.[2]

Without action to further endow the CBD, the present institutionalization of cacophony will continue. Even a cursory glance at table 1, appended to this chapter, reveals evidence of such confusion—an incoherence of created duties leading to a misapplication of scarce resources. Concerning marine species, for example, at least 13 treaties contend for attention in a world already distracted by a growing

TABLE 1  International Conventions

| International Conventions | Focus of Protection | Mechanism | Duties Created |
|---|---|---|---|
| Ramsar Convention (Convention on wetlands of international importance, especially as waterfowl habitat) (1971) | Wetlands habitat | Listing of protected wetlands<br>Funding for developing countries | Preservation and conservation |
| World Heritage Convention (UNESCO Convention concerning the protection of the world cultural and natural heritage) (1972) | Cultural/natural heritage (habitat) | Listing of world heritage sites<br>Listing of world heritage sites in danger<br>Funding for developing countries | Preservation and conservation |
| Bonn Convention (Convention on the conservation of migratory species of wild animals) (1979) | Migratory species | Listing of endangered species (appendix 1)<br>Listing of species with unfavorable conservation status (appendix 2) | Appendix 1: Conservation and prohibition of takings<br>Appendix 2: Duty to enter into cooperative agreements with "range" states |
| CITES (Convention on International Trade in Endangered Species of Fauna and Flora) (1973) | Trade in endangered species | Listing of species threatened with extinction (appendix 1)<br><br>Listing of species who may become threatened with extinction without trade restrictions (appendix 2)<br>Listing of species by individual parties who regulate within own jurisdiction and need international cooperation to control (appendix 3) | Appendix 1: No trade allowed unless approved by scientific authority<br>Import and export permits required<br>Appendix 2: Trade allowed but still must seek approval by scientific authority<br>Export permit required<br>Appendix 3: Trade allowed without approval by scientific authority<br>Export permit required<br>Certificate of origin presented at country of import |

TABLE 1 (continued)

| International Conventions | Focus of Protection | Mechanism | Duties Created |
|---|---|---|---|
| Antarctic Treaty System | | | |
| 1) Protocol on environmental protection to the Antarctic treaty (Madrid Protocol) (1991) Note: Antarctic treaty signed by contracting parties, 1959 | All Antarctic fauna (without derogating rights and obligations under the International Convention for the Regulation of Whaling [ICRW]; and all Antarctic flora | Environmental impact assessment | Stages of prior assessment with varying degrees of scrutiny |
| | | Permit for taking or harmful interference of fauna and flora | Takings or interference only for scientific or similar purposes |
| | | Listing of specially protected species | Takings or interference only for compelling scientific or similar purposes |
| | | Designation of specially protected areas | Restricted entry by humans |
| | | Designation of specially managed areas | Development of management plans to coordinate access and protection |
| (2) Convention on the Conservation of Antarctic Marine Living Resources (CCAMLR) (1980) | Antarctic marine ecosystem (without derogating rights and obligations under the ICRW and the CCAS) | Harvesting quotas including identification of protected species | Duty to conserve (based on "rational use") |
| (3) Convention for the Conservation of Antarctic Seals (CCAS) (1972) | Seals within the Antarctic treaty area | Regulation of kill limits, hunting seasons, hunting methods, and protected species | Duty to provide information concerning satisfaction of all sealing regulations |
| | | Permit system | Duty to implement domestic sealing permit system, "as appropriate" |
| Law of the Sea Convention (1982) | Marine environment | Creation of exclusive economic zone (EEZ) (200-nautical-mile jurisdictional zone) | EEZ: State must ensure conservation of living resources (art. 61 [2]) |

TABLE 1 (continued)

| International Conventions | Focus of Protection | Mechanism | Duties Created |
|---|---|---|---|
| | | Species regimes: • Migratory species (MS) | • State must seek to coordinate conservation measures with other states for MS found in EEZ (art. 63) |
| | | • Highly migratory species (HMS) | State must seek to coordinate conservation measures with other states for HMS found in EEZ (art. 64) |
| | | • Marine mammals (MM) | State may strictly regulate the exploitation of MM found in EEZ (art. 65) |
| | | • Anadromous species (AA) (species, like salmon, which originate in rivers and migrate through EEZ to high seas) | State in whose rivers AA originate has primary responsibility for ensuring conservation; taking on high seas discouraged (art. 66) |
| | | • Catadromous species (CS) (species, like eels, which originate on high seas and migrate to EEZ) | State in whose EEZ CS spend greater part of life cycle has responsibility for management; taking on high seas prohibited (art. 67) |

*Note:* Highlighting disarray, the above table outlines the basic duties created by the major global conventions concerning biodiversity. On top of this scheme rest the numerous regional treaties, which only compound an already confused regulatory structure. The proliferation of duties, creating overlap as well as gaps, fostered the original attempt to develop an umbrella convention. Unfortunately, the end result is the present CBD, which on its face, possesses no umbrella powers.

number of environmental problems (Birnie and Boyle 1992). Without fostering a hierarchical centralization of authority (practically an impossibility, even if the goal were desirable), a reinvigorated CBD could more equitably weave together the present structures of dissonance.

## The Convention as a Framework Treaty

The impetus for such transformation lies in the conceptual underpinnings of the framework convention. As a relatively new method of international lawmaking, the framework treaty has provided the model for a number of major international environmental agreements, including the Vienna Convention for the Protection of the Ozone Layer (1985; hereinafter Ozone Convention) and its accompanying Montreal Protocol on Substances That Deplete the Ozone Layer (1987; hereinafter Montreal Protocol), the Basel Convention on the Control of Transboundary Movements of Hazardous Wastes (1989), and the United Nations Framework Convention on Climate Change (1992).

Also known as the convention-protocol approach, the framework convention attempts to create a flexible system for decisionmaking so that parties may adapt to new scientific and/or socioeconomic information as such data becomes available. Though each example differs in design and application, the defining attribute of a framework approach exists in its two-step process of law creation. Initially, the parties agree to a relatively vague convention that generally imposes few substantive duties; in other words, parties feel politically compelled to sign a document with apparently insubstantial downside risks and an abundance of public relations benefits. In the next step, after the document has come into force, the parties meet regularly to set more specific requirements—which in its most formal manifestation, leads to separate and binding protocols on particular subjects related to the original convention.[3] At this formal level of more specific lawmaking, the framework approach shifts to a "majoritarian" process, striving for consensus, but in its absence, allowing for varying degrees of majority vote in the development of protocols.[4]

With the CBD, we have a typically vague framework convention, but also one that typically creates the institutional structure for further action. For our purposes, the most significant institutions include the Conference of the Parties (COP), the Secretariat, and the Subsidiary Body on Scientific, Technical, and Technological Advice (SBSTTA). The COP exists as the "legislative" organization of the convention, meeting regularly and voting on protocols, amendments, and other administrative matters. The Secretariat, functioning as the administrative agency, carries out actions designated by the CBD and any further tasks delegated by the COP. On technical matters, the SBSTTA, providing legitimacy and a sound basis for decision, compiles the required scientific data on which the COP (and the Secretariat, if so delegated) can take action. Yet, and this is the most important point, these institutions exist only as the vehicles of change. Propelling any meaningful transformation is the political philosophy behind the CBD—a philosophy that provides the signposts by which to measure any attempts to reform the CBD into a more dynamic umbrella convention.

The power of a framework convention thus lies in its two fundamental philosophical commitments: to open up the political process and to allow for the political structures created by that process to reinvent themselves. In this sense, the framework convention embodies the "law as conflict" description of legal structures—one that sees international law not as a formal set of rules, rights, and obligations, but as an ongoing political discourse about what needs to be done.[5] To accommodate the first commitment requires both a transparency of operation and the widest possible range of concerned voices. The CBD, for example, demonstrates these qualities in its COP, where parties meet to agree (or disagree) on matters of the highest import. At these gatherings, a range of nations, from the weakest to the most powerful, confer in an open forum that candidly reflects the vigorous political process of consensus building—replete with the frustrating complexity of competing coalitions and interests. Any agreements reached by such a process—either of the formal legal variety or otherwise—do not arise mystically out of state practice or from a "closed" meeting of a small number of powerful players. Rather, a text emerges that democratically (and honestly) reflects the political nature of the discussion. Of course, such procedures have their own potential pitfalls, including administrative inefficiency and "lowest common denominator agreements"—the latter occurring when the parties at the COP level fail to transcend the vague language of the original convention (Susskind 1994). Problems such as these exist as real impediments to a forward-moving legal regime, and are dealt with in the specific suggestions at the conclusion of this chapter.

The second principle informing the law as conflict approach is the commitment to institutional reinvention. This, of course, speaks to the overt premise of any framework convention—which seeks a fundamental flexibility in the range of powers to amend the original convention, adopt protocols, and conduct everyday business. This opportunity for reinvention is especially available in the CBD, as evidenced by the power of the COP to amend the convention with only a two-thirds majority. Such a capacity for revision compares favorably to the three-fourths majority required by the similar body in the Ozone Convention. Perhaps of greater significance, the everyday business of the COP may take on a dynamic process that informally alters the formal regime set up by the convention or its future protocols. As Gehring (1990) has shown with regard to the ozone regime, the transparently political process of a COP may at times ignore, circumvent, and even contradict the formal rules established in the parent documents.[6] As a framework convention, therefore, the CBD possesses a powerful transformative capacity that allows for rapid and dramatic self-revision. In this way, the CBD truly takes on the character of an ongoing political discourse—parties at an "open" forum reach agreements that they can readily alter with further discussion. The question remains as to how this "conversation" might prove fruitful with regard to the recasting of the CBD as an umbrella convention.

## Reinventing the Convention

As presently constituted, the CBD has no "strong" umbrella powers as it neither subsumes nor supersedes any of the global or regional treaties concerning biodiversity. More significantly, the contracting parties even refused to endow it with the "weak" umbrella powers of leadership and coordination vis-à-vis other treaties. The CBD, however, does mandate that its COP

> [c]ontact, through the Secretariat, the executive bodies of conventions dealing with matters covered by this Convention with a view to establishing appropriate forms of cooperation with them (Article 23 [4][h]).

In this way, the CBD only creates a duty to contact and cooperate, still far removed from even the weak umbrella powers of coordination and leadership. Yet with the coming into force of the treaty, the COP recently moved to expand these powers. At the second meeting of the COP in November 1995, the parties made the following informal decision:

> [The COP requests the Secretariat] to *coordinate* with the [s]ecretariats of relevant biodiversity-related conventions with a view to:
> (a) Facilitating exchange of information and experience;
> (b) Exploring the possibility of recommending procedures for harmonizing, to the extent desirable and practicable, the reporting requirements of Parties under those instruments and conventions;
> (c) Exploring the possibility of *coordinating their respective programmes of work*;
> (d) Consulting on how such conventions and other international instruments can contribute to the implementation of the provisions of the Convention on Biological Diversity.[7] (emphasis added)

As such, the COP—informally, but quite appropriately—acted to endow the Secretariat with a limited coordination and leadership authority, an authority not bestowed by the original convention. Clearly, such action displays the flexibility of the framework approach to lawmaking, exhibiting the capacity of the COP to create new and informal obligations for its own Secretariat. On the other hand, and this looms as the most difficult problem, little weight resides behind the CBD's self-proclaimed weak umbrella powers as participation by the different treaties remains voluntary.

Employing a latent mechanism in the treaty itself, however, the Secretariat might overcome this obstacle and thus reinvent those duties into more substantive controls. The key lies in Article 22(1) of the CBD, which provides that

> ... the provisions of this Convention shall not affect the rights and obligations of any Contracting Party deriving from any existing international agreement, *except where the exercise of those rights and obligations would cause a serious damage or threat to biological diversity.* (emphasis added)

In other words, the CBD trumps all other treaties dealing with biodiversity whenever the exercise of the rights and duties created by such treaties "would cause a serious damage or threat to biological diversity."[8] A broad interpretation of "serious damage" or "threat" would, therefore, provide the CBD with preeminence over all conflicting treaty provisions, and at the same time, imbue its own Secretariat with some real leverage in coordinating "their respective programmes of work" with other secretariats. In fact, given the magnitude of the biodiversity crisis, a liberal interpretation of the phrase seems easily justified: any waste of the world's scarce financial and human resources in the conflicting or duplicative application of duties under competing treaties constitutes a "serious harm or threat to biological diversity." To contend otherwise, so the argument would go, suggests a failure to grasp the enormity of the problem.

The ease in instituting this interpretation again reflects the flexibility of the CBD. For example, the exercise of this power requires only the tacit interpretation by the Secretariat, coupled with an appropriate signal to the other secretariats of its broadly perceived authority. In communicating such a wide translation of its powers, either expressly or otherwise, the Secretariat can rely on the further interpretive powers of the COP. The trump card held by the COP comes into play given any open dispute between the various negotiators in developing specific cooperative arrangements—a scenario in which the COP would be asked to choose between its own Secretariat and that of a foreign treaty or treaties. Given the significant reliance on its own Secretariat for a wide range of responsibilities, coupled with the COP's individual stake in strengthening the treaty, it seems improbable that the latter would rule against its own administrative agency. Therefore, in the rare instance of an open dispute, the COP would tend to interpret what constitutes a "serious damage or threat" quite broadly in the particular case. The CBD, thus, contains a simple though dynamic interpretive mechanism by which to convert the Secretariat's newly created weak umbrella authority, as provided by the COP, into much-expanded powers of coordination and leadership.

These powers, though widened, still remain limited to the weak umbrella variety and, therefore, do not point to an unwieldy centralization of authority. Instead, bolstered by the COP, the Secretariat should be seen as a "mediator" between the multitude of institutional players involved in the protection of biodiversity. In fact, the arrangements it facilitates should often arise out of multiparty negotiations, with the Secretariat functioning as an arbiter between the secretariats of other treaties on a specific subject of overlapping interest. For example, on the topic of protecting migratory birds in Europe and North Africa, the Secretariat might invite the secretariats of the Bonn Convention (1979), the Ramsar Convention (United Nations 1971), the World Heritage Convention (United Nations 1972), the Berne Convention (1979), and the African Convention (United Nations 1968), as well as a representative from the European Union (articulating the EEC Birds Directive).[9] The broader implied powers of the Secretariat should help to ensure attendance. In addition, in arbitrating the most effective and efficient arrangement, the Secretariat's reserved umbrella powers should also help to coerce an agreement given resistant or recalcitrant parties.

In the end, the resilience found in this framework treaty offers an indirect opportunity to achieve the original purpose of the CBD. While not creating the option of a strong umbrella treaty, the CBD does allow for its reinvention into a promising weak umbrella convention. Wielding the powers of cooperation and leadership, the CBD may therefore still accomplish much in the area of streamlining the international legal institutions of biodiversity. Its ability to do so resides in the continuing adherence to the twin commitments of the framework concept: the open political process and the transformative power of its political structures.

## Implementing the Transformation

*1. Enhancing Nongovernmental Organization (NGO) Access.* Best characterizing the commitment to an open political process is the necessity to allow the greatest possible NGO access to the three major institutions of the CBD—the Secretariat, the COP, and the SBSTTA. Today's NGOs, from Greenpeace to the World Conservation Union (IUCN), have clearly established themselves as important political players in the development of international environmental agreements, and in the monitoring of both institutions and sovereign nations. Their information-gathering capacity, as well as their ability to publicize their views, makes them powerful actors on the international stage. As an acknowledgment of this reality, the current trend in international environmental agreements is to recognize this political power in a limited fashion. In the CBD, for example, qualified NGOs receive "observer" status at the COP—unless one-third of the parties present at the meeting object to their appearance (Article 23[5]). Though the CBD allows the usual rights to speak and distribute information—as similarly conferred by the Ramsar, CITES, and Bonn regimes—NGOs may still receive further powers if conferred by the COP. For instance, the COP may provide qualified NGOs with standing to bring formal grievances under the dispute resolution machinery of any future protocol. Presently, only parties are privy to the dispute resolution process of the CBD, but the COP possesses the capacity to alter this arrangement for any subsequent protocol. Of course, any provision for NGO standing would run contrary to the traditional notion of sovereignty as the organizing principle of international law. Again, however, if we re-envision international law as an ongoing political discussion, then NGOs deserve the status commensurate with their political power. Indeed, the very functioning of the discussion requires their inclusion.

Unlike the COP, the SBSTTA has no mandate to include NGOs as observers. This seems especially infelicitous given the high level of scientific expertise frequently displayed by a number of NGOs. Reports by the IUCN and the World Resources Institute, for example, receive great respect throughout the international community. Such expertise could significantly help the SBSTTA in its task of compiling and analyzing data. Fortunately, the COP still has the ability to alter this inequity through Article 25(3) of the CBD, which states that "[t]he functions, terms of reference, organization and operation of this body may be further elaborated by the Conference of the Parties."

Similarly, the Secretariat has no overt duty to include NGOs in performing its functions under the convention, unless the phrase "[t]o coordinate with other relevant international *bodies*" (emphasis added) in Article 24(1)(d) is interpreted as such. Regardless, the Secretariat will no doubt openly receive the reports and advice of the most significant NGOs, as the Secretariat itself must "prepare reports on the execution of its functions under this Convention and present them to the Conference of the Parties" (Article 24[c]). In addition, NGOs may prove particularly useful in helping to arrange and facilitate the types of negotiations envisioned between multiple secretariats of competing conventions. Here, as a disinterested party, a well-placed NGO may play an additional "mediator" role between the various secretariats and the CBD Secretariat itself. Therefore, as well as contributing through its usual advisory capacity, an NGO may also help liaison between the Secretariat and the other parties whenever the Secretariat assumes more dominant characteristics. Generally, such a situation would occur when negotiations become intractable and the Secretariat feels compelled to flex its umbrella powers in order to spur further discussion.

In the final analysis, by enhancing NGO access, and thus adhering to the commitment of openness in each of its primary structures, the CBD further enables itself to meet the task of streamlining the international regime for biodiversity.

2. *Strengthening the Secretariat.* The commitment to transformative power, reified in the convention's COP, also allows the COP to transform the Secretariat at any moment into a strong, activist organization. At the most basic level, this means creating a well-funded and fully staffed Secretariat. Of all the steps toward reinventing the CBD as an umbrella convention, the empowering of the Secretariat looms as the most crucial. In considering the rationale behind this appraisal, two categories of

analysis emerge: the effect on the functioning of the COP and the effect on the negotiation of cooperative arrangements between the various treaty secretariats.

For the COP, a strong Secretariat would help to ameliorate any inefficiencies caused by the former's open political process. At a general level, without the help of a sophisticated Secretariat, the COP may find itself debating inconsequential questions *ad nauseam*. A dynamic Secretariat, however, could carefully screen any unnecessary stumbling blocks, leaving only the most significantly troublesome issues for the COP. Concerning our specific issue—the coordination of arrangements with prior biological diversity treaties—a strong Secretariat could present only the most elusive problems to the COP, resolving all others through its own competent administrative means. Similarly, a powerful Secretariat also could help the COP to sidestep what negotiators call lowest common denominator agreements (Susskind 1994). Such agreements occur whenever the COP fails to transcend the vague language of the convention, offering little more than the reiteration of platitudes. For our purposes, if a lack of perceived power or simply a lack of administrative capability causes the Secretariat to negotiate ineffective arrangements among the competing secretariats, then the COP can only embrace such agreements or ask for a continued attempt to improve them. Nonetheless, the result is the continuation of the status quo, a confused medley of treaties played inarticulately against the backdrop of an empty convention.

Interestingly, a dynamic Secretariat also contributes to the dynamism of the CBD as a whole. For example, the COP can only reinvent the convention to the extent that it possesses adequate information on which to do so. Only a proficient Secretariat can provide such information, properly focusing the COP as to the necessity of a new protocol or the advisability of significant changes to organizational structures. Only a strong Secretariat, for instance—after negotiating arrangements between the various secretariats—can offer the COP sufficiently detailed information on which to assess the productiveness of the terms. In short, much of the generative dynamism of the CBD rests with the information-providing capabilities of a credible, influential Secretariat.

As to the actual effect of an empowered Secretariat on the negotiations of arrangements between competing secretariats, we might recall the importance given earlier to the perception of authority in transforming the Secretariat's nascent umbrella faculties into more substantive powers of coordination and leadership. The leverage provided by a liberal interpretation of the Article 22(1) phrase, "serious damage or threat to biological diversity," only materializes to the extent raised by the Secretariat and backed by the COP. The clearest signal to the other secretariats that the CBD Secretariat indeed possesses such authority is to endow it with the greatest possible financial and political support. Only through the channeling of resources and prestige to the Secretariat can that body emerge with the necessary stature for the complete transformation of the CBD into a weak umbrella treaty.

In addition, only a powerful and proficient Secretariat could employ a supplemental leverage available to it under the CBD: issue linkage.[10] For our purposes, the term is used specifically to mean the option potentially available to the Secretariat to invoke its power, if delegated by the contracting parties, to facilitate the transfer of financial resources to "existing financial institutions."[11] In developing a more coherent framework for biodiversity protection, the Secretariat, therefore, possesses the ultimate "carrot"—financial remuneration—by which to induce concessions from the secretariats of other treaties. In this sense, "existing financial institutions" means the specific financial mechanisms set up under other biodiversity treaties, such as the World Heritage Fund of the World Heritage Convention.[12]

One simple way forward exists by which the contracting parties may create this incentive, also reflecting the institutional flexibility of the framework approach. At the current time, financial resources for the conservation and sustainable use of biodiversity must be channeled through the newly restructured Global Environment Facility (GEF). Finding a method to employ issue linkage through the GEF appears a difficult means of acquiring financial rewards for other biological diversity treaties, as the COP cannot recommend specific projects to its own financial mechanism. Instead, rather than proceeding through the GEF, the Secretariat may soon have another, more direct source of funding. At the first and second meetings of the parties, the COP requested that the Secretariat study the availability of financial resources "additional" to those provided through the restructured GEF.[13] These new resources could, and should, be made available to the Secretariat for the purposes of issue linkage. In fact, the use of funds for this objective—to support the "existing financial institutions" of other biodiversity treaties—may provide the COP with the best return on its dollar as compared to individual projects for individual countries.[14]

In an individual or group setting, then, the Secretariat can essentially "bribe" another secretariat both to come to the negotiating table and to "trade" away any impediments to streamlining the global biodiversity regime. To illustrate the point, let us consider the prior example of protecting migratory species of birds in Europe and North Africa. In this instance, assume that the secretariat of the World Heritage Convention has long desired to protect a site in a region presently included on its "List of World Heritage in Danger." Without the money to do so in the World Heritage Fund, however, the project has lain dormant. In order to finance the same project through the CBD, the World Heritage Secretariat may be willing to allow input into its listing process, permitting the CBD Secretariat to coordinate the protection provided by that convention with the overlapping protection provided by the listing processes of the Ramsar and Bonn Conventions, respectively. Through issue linkage, then, the secretariats of other conventions may cede a degree of their own sovereignty in order to gain funding for specific projects.

## Conclusion

Perhaps we hear the first strains of "a new stream" of international lawmaking, a stream of dialogue flowing from abstract questions about the nature of law toward the effective development of innovative solutions (Kennedy 1988). Thus, in experiencing international law as a "conversation" rather than a rigid set of rules, we can better appreciate the component voices underpinning the framework treaty: the open political process and the transformative potential of institutions. This recognition, in turn, amplifies the tenor of very practical umbrella suggestions, such as enhanced NGO access and an empowered Secretariat. Over time, the divergent interests of multiple treaties may speak together on the problem, brought closer by a revitalized Convention on Biological Diversity.

## Notes

1   In this context, an umbrella treaty is seen to encompass either of the following two varieties. A "strong" umbrella treaty would simply absorb or supersede prior biological diversity treaties in whole or in part, supplanting any organizational structures with its own. A "weak" umbrella treaty would retain all or most of the structures of the prior treaties, adopting a leadership or coordinating role for the new convention.

2   For a more theoretical examination of this thesis, see Hendricks, 1996. Postmodern possibility and the convention on biological diversity. *New York University Environmental Law Journal* 4:1.

3   By *formal*, I mean agreements that, before they become fully binding, remain subject to ratification or other official approval by the state party. In this sense, the protocol exists as the most formal because, on coming into force, it emerges as an independent and binding agreement under international law for all those who have signed and approved it (see CBD, Article 34). Another formal mechanism, amendments to the original convention and to protocols, also invariably remains subject to "ratification, acceptance or approval," as in the present case of the CBD (see Article 29[4]). *Informal* agreements, meaning those that take effect without such official approval, include the adoption and amendment of annexes and appendices. These typically contain very particular information, often setting critical standards and requirements, yet take effect without the time-consuming and possibly insuperable procedures of ratification. In the CBD, annexes are restricted "to procedural, scientific, technical and administrative matters"—a category that, in fact, may encompass a great deal of the parties' work—and amendments or adoption of these have full effect on all the parties, except those filing written objections (CBD, Article 30). One further example of an informal agreement exists in the general decisions of the parties with regard to substantive matters, which again take effect for all parties at the time of voting (see Article 23[4])[i]).

4   In the movement from generalities to particulars, which typically parallels the movement from formal to informal agreements, the framework approach tends to reflect increased reliance on the majoritarian process. At the formal level, of course—involving amendments to protocols and the original convention—Rule 34 of the Ozone Convention prevents a party from being bound against its will. Still, in the CBD, the relatively low two-thirds majority necessary to pass either type of amendment does, in fact, put pressure on recalcitrant parties, much in the same fashion as it compels parties to sign onto a protocol. At the more informal and specific level, such as in the adoption and amendment of annexes, even more pressure comes to bear on reluctant parties as the burden shifts, and they must officially file a timely objection to the change passed by a two-thirds majority of their

peers (CBD, Article 30[2]). With regard to other informal decisions, a few holdouts at the first Conference of the Parties (COP) to the CBD managed to block a two-thirds majority voting standard—a delay repeated at the second Conference of the Parties—though in time, most probably a non-consensus approach will prevail (COP Report I, Annex III, Rules of Procedure, Rule 40; COP Report II, Agenda Item 3, 31–32). Regardless, the give-and-take at the level of everyday business, when coupled with a self-evident environmental problem, should tend to influence agreement among the parties. Thus, in the ideal version of the framework approach, in the movement from formal to informal lawmaking, we see an institutionalized turning of the screw toward majority-impelled decisionmaking.

5   The phrase comes from Martti Koskenniemi (1989), and suggests the general approach adopted by "postmodern" scholars in international law. For our purposes, without delving into an intricate intellectual debate, these scholars attempt to uncover the political and normative bias lurking behind all solutions, including seemingly objective rules and procedures. The relative (or rhetorical) nature of all legal answers, however, in no way precludes a struggle to find and implement them. As Koskenniemi asserts:

> The legitimacy of critical solutions does not lie in the intrinsic character of the solution but in the openness of the process of conversation and evaluation through which it has been chosen and in the way it accepts the possibility of revision—in the authenticity of the participant's will to agree (487).

Such a description, in fact, sounds like the very formula for decisionmaking within a framework treaty.

6   Lauding the dynamic power of the ozone regime, Gehring (1990) cites specific examples of the COP's informal lawmaking power, including "clarifications" of vague treaty obligations, such as extending the reductions of controlled substances beyond those stated in the Montreal Protocol at 47–48; the circumvention and disregarding of relevant convention and protocol provisions, for instance, amending the Montreal Protocol with less than the required two-thirds majority (48); and the creation of a new institution without the formal authority to do so, such as establishing the Interim Multilateral Fund (49). Of course, whether the COP of the CBD adopts such informal powers remains to be seen, but especially in the "clarifying" of vague treaty language, the COP possesses a great deal of latent transformative power.

7   COP Report II, Annex II, Decision II/14.

8   The CBD specifically exempts the "law of the sea" from this exception clause, providing the conventional and customary law of the sea with priority (Article 22 [2]). To this extent, we now have two dominant biodiversity conventions, the CBD and the Law of the Sea Convention. The exact nexus between the two remains a question for study, but the legal lever offered by Article 22(1) would not be available to influence, for example, the regional seas protocols, as those agreements trace their legal basis to the Law of the Sea Convention. In such circumstances, however, the CBD may still use the financial incentive of "issue linkage," discussed later in this chapter.

9   The first three treaties—the Bonn, Ramsar, and World Heritage Conventions—are global in scope and are included in table 1, International Conventions. On the other hand, the 1968 African Convention on the Conservation of Nature and Natural Resources (the African Convention) and the 1979 Berne Convention on the Conservation of European Wildlife and Natural Habitats (the Berne Convention), the latter prepared within the Council of Europe, are regional mechanisms. For those members of the European Union, a further regulatory scheme exists under the 1979 EEC Birds Directive (79/409/EEC) as well as the more general 1992 Council Directive on the Conservation of Natural Habitats and of Wild Fauna and Flora (92/43/EEC).

10  *Issue linkage* is a term used by negotiators to describe "trading across what appear to be completely

different realms" (Susskind 1994). It can help to bring reluctant parties to the negotiating table by promising consideration of their special "issues," as well as by providing a way out of an impasse by offering a trade of issues among parties.

11 See CBD, Article 21(4). This provision states that "[t]he Contracting Parties shall consider strengthening existing financial institutions to provide financial resources for the conservation and sustainable use of biological diversity."

12 The World Heritage Fund provides financial assistance to parties to carry out their respective obligations under that convention. Of course, not all biodiversity treaties have established such a formal mechanism, though all do contain institutional procedures through which to finance treaty functions. Whether this would make all biodiversity treaties eligible for assistance as "existing financial institutions" under CBD Article 21(4) remains a question of interpretation—but one easily settled by the COP. Without such an interpretation, the present treaties undoubtedly would create formal mechanisms for assistance, such as the World Heritage Fund, if money became available through the CBD.

13 COP Report I, Decision I/2 (7); COP Report II, Decision 11.6.9.

14 The use of funds in this way should be seen as a successful compromise for both developed and developing countries. The developed countries would receive the financial benefits of any grant, and the developed countries would feel better protected from wasteful spending, knowing that the funds would be channeled through a treaty mechanism designed to serve the common good.

To facilitate this compromise, the COP would probably have to alter the current eligibility criteria for funding to include the "existing financial institutions" of other biodiversity treaties. The present informal language of COP I appears to only allow funding for developing country parties: "[o]nly developing countries that are parties to the Convention are eligible to receive funding upon the entry into force of the Convention for them" (COP Report I, Annex I [I]). Again, in a "win-win" situation for competing coalitions, the flexibility of the CBD would allow another informal decision by the COP to include all biodiversity conventions.

## References

Basel Convention on the Control of Transboundary Movements of Hazardous Wastes. 1989. *International Legal Materials* 28:657.

Berne Convention (Convention on Conservation of European Wildlife and Natural Habitats). 1979. *Available in westlaw* 1979 WL 42275. 19 September.

Birnie, P.W., and A.E. Boyle. 1992. *International Law of the Environment.* Oxford: Clarendon.

Bonn Convention (Convention on the Conservation of Migratory Species of Wild Animals). 1979. *Available in westlaw* 1979 WL 37754. 23 June.

Burhenne-Guilmin, F., and S. Casey-Lefkowitz. 1993. The convention on biological diversity: A hardwon global achievement. *Yearbook of International Environmental Law* 3:43–59.

Gehring, T. 1990. International environmental regimes: Dynamic sectoral legal systems. *Yearbook of International Environmental Law* 1:35–56.

Kennedy, D. 1988. A new stream of international law scholarship. *Wisconsin International Law Journal* 7:1–89.

Koskenniemi, M. 1989. *From Apology to Utopia: The Structure of International Legal Argument.* Helsinki: Lakimiesliiton Kustannus.

Montreal Protocol on Substances That Deplete the Ozone Layer. 1987. *International Legal Materials* 26:1550.

Susskind, L.E. 1994. *Environmental Diplomacy: Negotiating More Effective Global Agreements.* New York: Oxford University Press.

United Nations. Treaty Series. 1968. African convention on the conservation of nature and natural resources (with list of protected species). 15 September. *Treaties and International Agreements Registered or Filed or Reported with the Secretariat of the United Nations*, vol. 1001, no. 3 (1969).

————. Treaty Series. 1971. Convention on wetlands of international importance, especially as waterfowl habitat. 2 February. *Treaties and International Agreements Registered or Filed or Reported with the Secretariat of the United Nations*, vol. 1976, no. 245.

————. Treaty Series. 1972. UNESCO convention concerning the protection of the world cultural and natural heritage. 16 November. *Treaties and International Agreements Registered or Filed or Reported with the Secretariat of the United Nations*, vol. 1037, no. 151.

United Nations Framework Convention on Climate Change. 1992. *International Legal Materials* 31:849.

Vienna Convention for the Protection of the Ozone Layer. 1985. *International Legal Materials* 26:15.

# Conclusion: How to Save the Biodiversity of Planet Earth

by Jeffrey A. McNeely and Lakshman D. Guruswamy

## Systemic Solutions or Small Victories?

This book has brought together a wide range of views from a variety of disciplines and we now attempt to reweave these strands of thinking into more cohesive conclusions. We also make specific, albeit selective, recommendations for converting these conclusions into practical action.

Three foundational premises require explanation. First, policies to conserve biodiversity must be seen as part of an overall effort to promote global welfare. Second, protecting global biodiversity calls for fundamental changes in the way people behave and relate to the environment. Third, the problem of endangered biodiversity, as we noted in the introduction, is part of a more comprehensive and interconnected web.

The need for global welfare arises from the fact that the majority of the world's terrestrial biodiversity is confined to forested areas in poor, tropical developing countries, which are under great pressure to utilize these areas for growing food, settling people, or harvesting timber and firewood. Without international assistance to improve the management of biological resources and provide hope for the peoples of these countries, the future of the world's animals, plants, fungi, and microorganisms looks dismal.

That future must include changes in human behavior. What needs to be learned at this juncture is that it is only from the organisms that are saved that people will be able to build the productive systems and ecological communities of the future. Building technical competence in nations around the world so that they can manage their own biodiversity for their own benefit, and thus ultimately for the benefit of the planet, remains one of the best investments that developed countries can make in our common future.

A world scheme for accomplishing the preservation of the maximum amount of biodiversity possible would be the most important single contribution that the people of our generation could make to the future. Prescriptions for a sustainable future based on principles of conservation biology often involve restricting access to resources, expecting people to forego material benefits, assigning values to resources that are elusive or difficult to measure, and requiring payment today for abstract future benefits. This is not an attractive recipe for most people, so other elements must be added to the mix.

It is apparent that public support is crucial to any successful conservation program; such support will need to be based on a sound ethical footing, good information, and economic benefits. Conservation biologists will need to build on science to demonstrate the benefits of conserving biodiversity to those who farm, fish, ranch, and log; to balance the attention given to the loss of biodiversity with a concern for sustainable use of harvestable species; and to build a broader constituency among business, the public, and academics (Ludwig, Hillborn, and Walters 1993).

Experience suggests, however, that the most popular public policies are those calling for modest changes in current practices to address immediate, proximate causes rather than imposing comprehensive changes in deeply embedded social behavior (Tobin 1990). Popular policies coincide with prevailing public opinion, and do not require people to change their lifestyles or cause them great inconvenience; they distribute material benefits to a majority or a politically significant and effectively organized minority. Furthermore, they provide more benefits than costs, thereby favoring policies with easily monetized values, such as goods traded in the marketplace or development that provides jobs; and they generate concentrated, immediate benefits while deferring and diffusing costs. The popularity of this approach is indicated by the budget deficits of many governments, as voters favor discounting future benefits to enjoy present ones.

It is clear that biodiversity is part of a more complex pattern of problems. Emerging global strategies to conserve biodiversity will necessarily involve social equity issues. Improved agricultural and forestry practices (for example, using cutover lands rather than mature forests for new enterprises) may halt, and eventually reverse, the activities that are leading to global warming and other drastic alterations of the earth's environment. Some strategies may limit overconsumption in industrialized countries to levels that the world can sustain, and build a strong public ethic in favor of conservation. As Rappaport (1993) says, "We are facing such a multiplicity of quandaries, dilemmas, crises, inequities, iniquities, dangers, and stresses ranging from substance abuse, homelessness, teenage pregnancy, and prevalence of stress disease among minorities to global warming to ozone depletion that they cannot all be named, much less studied." The traditional approach of seeking to understand systems by reducing them to components and analyzing the interactions between them might facilitate "problem solving," but cannot provide an adequate understanding of complex systems. Conserving biodiversity, in the view of many, therefore requires moving toward a comprehensive view that synthesizes contributions from numerous sectors (WRI, IUCN, and UNEP 1992).

The classic counter to this view was articulated by Lindblom (1959). He argued that a "rational-comprehensive" decisionmaking process that adopts a synoptic perception of a problem, collects all relevant information, and explores all relevant solutions after considering all relevant answers, is quite impossible. Such an approach—which is admirably marked by a clarity of objective, an explicitness of evaluation, a high degree of comprehensiveness of overview, and a quantification of values for mathematical analysis—is only possible when dealing with small-scale problems with a very limited number of variables. Lindblom suggested, therefore, that poor as it is, incremental politics ordinarily offers the best chance of producing beneficial political changes.

Lindblom cogently argued that precisely because everything is interconnected, environmental problems are beyond our capacity to control them in one unified policy. The very enormity of the interconnected environment makes it impossible to treat as a whole. Tactically defensible, or strategically defensive, points of intervention have to be found (Lindblom 1973), suggesting that a step-by-step approach will help to solve a problem better than a grand solution based on the necessarily incomplete analysis offered by comprehensive rationality.

In many cases, and for most people, it is simply too overwhelming to think concurrently about whole litanies of problems; the response is to sink into passive despair. Instead, building a series of "small wins" creates a sense of control, reduces frustration and anxiety, and fosters continued enthusiasm on the part of the public, conservation biologists, and politicians (Heinen and Low 1992). We endorse this view, appreciative of the fact that these small wins can be real victories only if they contribute to an overall strategy for conserving biodiversity. This requires a politically sophisticated approach involving multidisciplinary actions, such as those discussed in this book and reformulated below.

## The Sociopolitical Setting

When conservation was confined to endangered species or national parks that did not involve critical national or international interests, the stakes were sufficiently small to be left to conservation biologists and a few specialized agencies representing the "scientific" perspective. The issues arising were not important enough to merit attention on the international agenda. But this changed dramatically when "biodiversity" was merged with the concept of development during the debate and discourse leading to the Convention on Biological Diversity (CBD).

The convention may herald a new dawn for the protection of biodiversity. Signed by 157 nations at the United Nations Conference on Environment and Development (the Earth Summit, held in Rio de Janeiro in June 1992), it entered into force at the end of 1993, and over 170 nations had ratified it by the end of 1997. Despite its substantial shortcomings (Guruswamy, this volume), the CBD is a commitment by governments to conserve biological diversity, use biological resources sustainably, and promote equitable sharing of the benefits arising from the use of such resources. However, advancing from the aspirational commitments and non-obligatory exhortations to changes in behavior means facing a number of formidable political obstacles at both national and international levels (Mathews 1991; Sanchez and Juma 1994).

This development also presages a new political era in the protection of biodiversity. Issues discussed in this book, such as equitable sharing of benefits, intellectual property rights, sustainable development, access to genetic resources, and national sovereignty, are at the very center of modern conservation. Biologists are now sharing a larger and more important political stage with agricultural scientists, anthropologists, ethnobiologists, lawyers, economists, pharmaceutical firms, farmers, foresters, tourism agencies, industrialists, indigenous and traditional peoples, and many others. These competing groups claim resources, powers, and privileges through a political decisionmaking process in which biologists and conservationists have become inextricably embroiled. This political immersion is not something that they are accustomed to by training, or for which they have a predilection, but it is an unavoidable reality.

Many of the most important decisions affecting biodiversity—especially on issues of budgets, priorities, information, and resource management policies—are made by politicians and "nonconservation" sectors of government (ministries of finance, trade, defense, and so on); and perhaps even more crucial decisions are made by the private sector. Scientists, who tend to believe that action must follow logically from their findings, discover to their consternation that governments and industries often treat them as just another interest group. As a result, scientists are in danger of being little more than concerned bystanders when policies are formulated to address the problems of conserving biodiversity (Tobin 1990).

Most problems affecting biodiversity, in fact, reflect a conflict of interests. For example, biologists have found that the annual runs of adult salmon in the Columbia River basin in the United States have declined by 75 to 85 percent. Groups with an interest in the policy response to this observed decline include electric utilities; environmental advocates; the barge industry; recreational boaters; agricultural irrigators; logging and mining companies; the aluminum industry; government agencies; conservation biologists; and commercial, sports, and tribal fishing groups (Hyman and Wernstedt 1991; Sagoff, "On the Uses of Biodiversity," this volume).

Scientists may contend that reliable information is the basis for sound decisionmaking about managing the salmon. But most will admit that they have inadequate knowledge about the life cycle of the salmon, natural fluctuations in populations, relations with other variables in the ecosystem, and impacts of various fishing regimes—indeed, since many of these are hot scientific issues, no consensus is yet available. Thus, for the salmon, as for many other harvested resources, an increasingly knowledgeable and skeptical public is asking questions that cannot be answered by scientists with a suitable degree of certainty (Binkley 1992). Therefore, competing interest groups tend to fill the information vacuum with self-serving interpretations of the "truth," often based on incomplete knowledge or misinterpretations of the research data collected by scientists. But even if better knowledge were available, the policy environment is still highly volatile; because many mutually exclusive choices are possible, the long-term implications of management alternatives are difficult to predict, and different groups have different access to the political process. Political forces will ultimately decide what to do about the salmon (and other resources), but scientists are most likely to have their point of view heard when they are able to couch their arguments in terms that bureaucrats and politicians find convincing—and feel that their constituents will support.

## What Do We Know?

Set against the social and political background outlined above, the authors of this volume have reached a number of conclusions from their various disciplines. We have not attempted to forge consensus, but have tried, instead, to analyze the emerging diversity of views and perspectives within the organizational framework of the book.

Scientific Issues
First, the undeniable. Environmental change will continue, and even accelerate, because the amount of space on our planet and the natural resource base are fixed, but both energy consumption and the human population are expanding. It is inevitable that pressure on fixed resources will increase, and that various forms of ecosystem management are necessary, if only to support life on earth. Four main areas with rapidly evolving paradigms of thought are driving management: the perception of a rapidly changing world; the notion of spatial and temporal hierarchies; the resiliency of ecosystems; and the human dimension of management, including humans as agents of environmental change.

Second, the sobering awareness that scientific knowledge about many crucial issues of biodiversity is still inadequate. Ecologists have not yet developed a body of quantitative science to predict changes in species composition or explain the way in which species composition influences the functioning of ecosystems. Much research is still required to understand species composition in many parts of the world (Norton, this volume).

Increased effort is certainly required to document life on earth, but given the enormity of the task and the increasing rapidity of extinction, other alternatives—including *ex situ* methods such as cryopreservation (preservation by freezing)—must also be explored (Benford, this volume; Sagoff, "On the Uses of Biodiversity," this volume). If life is seen as a library of information carried by DNA, would one rather have the card catalog or the books themselves? Given the books, one can later catalog at leisure. This is especially the case since current levels of knowledge make us almost illiterate at trying to read this "library of life." Biotechnology is at the brink of major new discoveries that will enable scientists to read and use genetic "texts" in ways that are currently unimaginable, but they will not be able to work on texts that have been lost or destroyed by the current generation (Horsch and Fraley, this volume). We are limited by a lack of imagination, which is perhaps understandable at the early stages of a profound, dramatic revolution in biological technology, rather like the Wright brothers trying to envision a moon landing within three generations (Benford, this volume).

Third, while rates of extinction may be useful in drawing public attention to the problems of biodiversity, these rates should not be driving public policy because they are still too uncertain. Saving individual species may eventually play a less central role in biodiversity policy, and the species-protection policy is perhaps best seen as a temporary expedient. Rather, policy should be driven by the need to conserve all biodiversity, from genes to life zones, so that we can support sustainable development and the high quality of life on our planet (Lugo, this volume).

Fourth, in many cases, good ecosystem management will be identical to good species-level management. Nature needs to be managed on multiple scales, including the ecological system scale of time and space, as well as the short-term economic scale that seems to dominate current thinking. While attention to individual species may still be required in some cases, it is clear that managing large-scale systems for one or a few resources can lead to increased brittleness, making them susceptible to collapse or gradual degradation. Therefore, management actions need to be based on system-level characteristics and the dynamic processes that they represent (Ostrom, this volume).

Conditions on tropical islands presage what might become the environmental fate of the world's continents in the future. These islands typically have high population densities, exhibit highly fragmented landscapes, and have already experienced significant extinction events. On the other hand, their landscapes are now enriched by the addition of exotic plants and animals that form new combinations of communities and ecosystems.

Rehabilitation of degraded ecosystems will be an increasingly important activity, particularly in the proximity of urban areas, where damaged lands are so prevalent. Sometimes the first option for ecosystem management (that is, the use of native systems and natural succession) is either not available or not sufficient to accomplish the agreed management goals. For example, natural ecosystems may have low net productivity and, thus, limitations in situations where maximizing a net yield is necessary. When habitats are excessively damaged due to careless human activity, native systems and natural successions may not be effective for rehabilitation because native species may grow slowly and succession is arrested. Under these conditions, it may be necessary to import genetic material from other geographic areas to accelerate the healing process of the ecosystem (Lugo, this volume).

While human activity is not necessarily incompatible with the maintenance of biodiversity, some important components of biodiversity are most likely to prosper in areas that are remote from human influence, often where extreme environmental conditions prevail. In forging policy, however, we should keep in mind that some components remain inextricably associated with conditions selected by humans as in the case of protected areas such as forests or ecosystems that are managed directly by people (Ostrom, this volume).

Finally, successful policymaking requires continuous feedback from field-level resource management activities. This is accomplished by monitoring ecosystem structures and processes so that the results of previous management actions can be compared with expectations of the plans that led to the actions. Results from monitoring programs must be made available to planners, managers, policymakers, and scientists so that they can adjust plans, management actions, policies, and research programs. A loop called *adaptive management* is created between implementing field actions, monitoring the ecosystem, comparing the results against expectations, and adjusting future actions, with each reiteration of activity based on past experience. The adaptive management approach to protecting biological resources rests on a willingness and ability to react to new information as it becomes available. Policies need to be based on the best available science, protect both species and the ecological processes associated with them, and yield new information to support further policy actions (Lugo, this volume; Raven and McNeely, this volume).

The first of our conclusions on the relevance of technology may seem obvious, but it is worth reiterating: technology will never be a complete substitute for the free goods and services that humans derive from biodiversity. Even so, a critical component of strategies to deal with the loss of biodiversity will be continued investment in new, more efficient, and more sustainable technologies for food production. Agricultural technologies that improve productivity and reduce labor-intensive practices may be an essential factor in stabilizing the human population on our planet.

Second, biotechnology has important contributions to make. It can incorporate new insect-resistant genes into crop plants, thereby eliminating some of the need to consume petroleum resources and energy in manufacturing pesticides and packaging and in distributing the chemical product, as well as the need to dispose of the wastes generated in the process. If genetic improvements are able to improve pest control, overall yields can be expected to increase; these benefits may be possible without huge capital investments or ongoing resource consumption for manufacturing, since they will be built into the seed without significant additional costs of production. Herbicide-resistant crops are also expected to help increase productivity and minimize the environmental impacts of farming, enabling producers to get the most out of their land and livestock in the safest, most efficient manner possible. This can help the best agricultural lands to be more productive, thereby reducing pressure on marginal lands, which may be important for conserving biodiversity (Horsch and Fraley, this volume).

On the other hand, biotechnology may also work against the efforts to protect biodiversity. Genetic engineers are creating increasing proportions of their own materials, for example, through the computer-assisted design of molecules; and biotechnology may encourage the domestication of nature, leading to the replacement of wild habitats with bioindustrial systems of aquaculture, silviculture, and agriculture. New crops genetically engineered to grow on marginal lands may help meet subsistence and commercial needs, but they will also increase agricultural use of land that is now *de facto* protected from development (Reid, this volume).

Third, high-tech agricultural inputs—such as purchased fertilizer, herbicides, and genetic materials—are often substitutes for the labor of farmers, who once managed ecological and evolutionary processes to earn their living from the land. Services and products formerly provided by the agricultural ecosystem through the skills and wisdom maintained by culture and practiced by individuals have now become industrialized. While this has produced more food, at least in the short term, the result in many countries has been growing numbers of poor people who are unable to purchase the food that is thus produced (Jackson, this volume).

Fourth, while population growth has often been held responsible for the loss of biodiversity, the regions of the world with the lowest population growth—namely, the technology-intensive industrial countries—account for most of the damage to the global environment. This trend could, of course, easily be reversed in the future, since the rapidly growing populations of the tropical countries could greatly increase their impact on the world's resources, especially as their economies expand and they adopt more technology-intensive forms of production.

Fifth, it appears likely that biological diversity in agricultural systems was higher in earlier times, when large numbers of different cultivators had long-term stakes in the land they farmed and control over their own technology. These systems of land management were highly variable, following a range of different rules to take into account specific attributes of the physical systems within which they were found, cultural views of the world, and the economic and political relationships that existed in the setting.

Economic Issues

Economics provides numerous approaches to dealing with problems of conserving biodiversity. An essential element is establishing clear property rights. One can only buy or sell goods for which property rights are well defined, where the seller truly owns the goods and has the right to transfer the goods to others (Chichilnisky, this volume). Biodiversity, on the other hand, is a public good that is provided to everyone, rather like law and order or defense (Heal, this volume). Market economies, if left to themselves, typically underprovide public goods. Thus, property rights work well for bread as a private good, but much less well for a public good, such as genetic variation in wheat types. The full social benefits of tigers, rhinos, portfolios of germ plasm, marine resources of the global commons, and so forth, are public goods beyond appropriation by markets, even when the market value is fully enhanced by all the devices of the law (Stone, this volume).

Many biological assets, however, both in the wild and in agro-ecosystems, are potentially private goods, whose use is rival, whose control can be made exclusionary, and whose value is commercially marketable. For these goods, the ideal is to ensure that the resources are priced at true marginal cost, requiring the establishment and enforcement of property boundaries around resource areas, and deploying various measures to internalize externalities (Stone, this volume).

Conserving biodiversity often requires preventing land from being converted from uses that maintain biodiversity to uses that tend to simplify biological systems, even though the latter may generate higher economic returns. Therefore, incentives for conservation must increase the willingness to pay for the marginal hectare. It could be argued that investing in research capacity may increase the marginal value of biological resources by making their qualities more accessible (Sittenfeld and Lovejoy, this volume). Expanding the knowledge base should involve the full range of biological research, rather than being focused on genetic prospecting for pharmaceuticals, where the return on investment is likely to be relatively limited. Managing biodiversity requires complex governance systems (Ostrom, this volume). This is especially the case since some biological resources should not be subject to property rights, because of the broad public benefits they provide. It may not be appropriate for any one entity to own key aspects of complex biological systems (Heal, this volume; also Gupta, this volume).

While the idea of compensating farmers for conserving genetic diversity seems useful, the actual application of this is difficult; current technology does not enable germ plasm to be assigned to any particular site and most new varieties draw from a large number of sources of germ plasm (Toenniessen, this volume).

Institutional Issues

Most strategies to protect biological resources will emphasize the wise and sustainable use of resources, rather than the unrealistic protection of resources from any human use. Older agricultural systems often shared basic design principles, such as clearly defined boundaries; specific rules on how much, when, and how different products could be harvested; involvement of the affected people in collective choices; a system of monitoring the use of resources; sanctions on those who violated the operational rules; inexpensive local mechanisms for resolving conflict among appropriators; the rights of appropriators to devise their own institutions; and a way of organizing all of these activities in multiple layers of nested enterprises (Ostrom, this volume).

It is clear that in larger systems it is quite difficult to devise rules that are well matched to all aspects of the provision and appropriation of that system at any level of organization (Ostrom, this volume). Among long-enduring and self-governed systems of resources, therefore, smaller-scale organizations tend to be nested in ever larger ones. Thus, efforts to pass national legislation establishing a uniform and detailed set of rules for an entire country are likely to fail in many of the specific locations where biological resources are most at risk. Enabling users to manage their resources locally may be a more effective way of dealing with the immense variability in diversity from site to site. This is especially the case where the benefits that local users may obtain from carefully managing their resources remain high (augmented by future flows of benefits appropriately taken into account), while the costs of monitoring and sanctioning rule infractions at a local level tend to be relatively low (Gupta, this volume). A key aspect of all these proposals is the effort to enable institutions of multiple scales to more effectively blend local indigenous knowledge with scientific knowledge.

With incentives provided by intellectual property rights, private sector agricultural research will generate many future advances in crop improvement technology, but will not directly address the needs of the majority of the people in developing countries; experience suggests that the private sector is unlikely to be a major developer or supplier of improved seed to farmers with limited purchasing power. Mechanisms, therefore, need to be developed that will allow technology to flow from advanced laboratories and the private sector to the international system while maintaining incentives for further research. Such mechanisms require a strong, well-supported public sector agricultural research system committed to meeting the needs of those with limited purchasing power (Toenniessen, this volume).

Intellectual property rights issues in the field of biotechnology have created difficulties for scientists who suddenly find their professional interests for access to genetic resources addressed through international agreements rather than scientist-to-scientist exchanges. Trying to balance issues of intellectual property rights and biotechnology relative to germ plasm exchange and collection has become an important unresolved issue for breeders and inventors (Toenniessen, this volume).

In the future, much of what the international system needs from industrialized countries will be proprietary property in those countries. And the increased value of plant genetic resources is causing developing countries to rethink their policies concerning access to the natural biodiversity they control; restrictions already exist in some locations. Unlike developments in many other areas of technology, however, the genetic improvement of plants is a process in which each enhancement is based directly on preceding generations, and the process of adding value requires access to the plant material itself (Reid, this volume; Sagoff, "Animals as Inventions," this volume).

As a new approach to making use of biological resources, the kind of biodiversity prospecting being undertaken by INBio may offer tropical countries a means of integrating biodiversity into scientific, technological, and economic development and improving the quality of life for their people, while ensuring that the resource is protected and used sustainably. By generating resources for conservation purposes and promoting economic development, it helps provide returns to the source country. It requires an appropriate macropolicy framework, biodiversity inventories, biodiversity information management, business development, and access to appropriate technology. In return, it generates information about taxonomy, the distribution of organisms, ecology, natural history, markets, market needs, major actors, in-country capabilities and needs, and institutional goals; encourages formation of collaborative arrangements to achieve institutional and national strategies; and provides access to relevant technology (Sittenfeld and Lovejoy, this volume).

Moral Issues

The primary reasons to value species and ecosystems are found in their intrinsic qualities, not in the benefits they may confer on us. Nature possesses aesthetic, historical, and expressive qualities that are different from the utility or monetary benefits it provides to humans. We value nature because of what it means to us, not just for what it does for us economically (Norton, this volume; Sagoff, "On the Uses of Biodiversity," this volume). The principal reason for protecting biodiversity is not to ensure an enormous inventory of raw materials that could have economic use and applications, for example, in biotechnology; rather, the main reason to preserve biodiversity is because we value it for its own sake. In finding only material uses for biodiversity, we offer a utilitarian rationalization for a conclusion based on ethical grounds (Sagoff, "On the Uses of Biodiversity," this volume).

Legal Issues

Intellectual property rights in organisms can legitimately be claimed when they are constructed from simple materials using ideas that are not found in nature, but rather, are creations of the inventor. Simply producing organisms without designing them may justifiably lead to claiming a patent on the process, but not the product, as intellectual property. Thus, patents should not be valid for the *result* of a certain process, as that would tend to prevent others from producing the same product by any means, thereby discouraging the kind of progress that is necessary to promote human welfare (Sagoff, "Animals as Inventions," this volume).

Yet much of the debate about intellectual property rights may essentially be a side issue. While intellectual property systems may have significant implications for biodiversity, the granting or withholding of patents is unlikely to be an effective response to the social, moral, political, and environmental dimensions of the task of conserving biodiversity. Patent systems, after all, are driven by economic forces, including the need to compete with other jurisdictions, and those who administer intellectual property are unlikely to be appropriate assessors of these additional factors; indeed, they typically are reluctant to base their decisions on such factors, tending to narrowly define their decisions on the basis of legal precedent (Cripps, this volume). Thus, while developing biotechnology and the need to protect the world's genetic resources are crucial issues, the challenges of protecting biodiversity may be trivialized if patent legislation is considered an appropriate regulatory mechanism (Sagoff, "Animals as Inventions," this volume).

Although the patent system is designed to disseminate information about new inventions, thereby enhancing the ability of others to build on research and development, the limited monopolies conferred by patents on biotechnological inventions may foster a loss of genetic diversity by encouraging industry to focus on promoting particular genetically engineered organisms to the detriment of others that apparently offer fewer genetic or economic advantages. Developing countries are concerned that they will be denied free access to modified organisms, even though their territory may have been the source—even the sole source—of the organic matter that has been transformed into patentable subject matter under current legislation. On the other hand, genetic diversity may be enhanced by the incentives to produce new genotypes and the availability of samples of biological material in patent system depositories to satisfy the patent disclosure requirements.

Since no single approach is perfect, we need to rely on multiple strategies. The best beginning is to recognize the institutional and economic structure that underlies and shapes the problem. Many nations host biological assets that are either pure public goods or, as joint products with marketable goods, retain an appreciable layer of public good benefit, even after the domestic legal regime has gone as far as it can go to appropriate benefits and internalize costs through available legal instruments. While "stick" devices such as taxes, lawsuits, and trade sanctions do exist, their utility is almost certainly going to remain marginal and will be applied only to the most egregious situations (Stone, this volume).

Finally, increasing knowledge and awareness about the dangers of extinction, educational campaigns, appeals to ethics, equity, morals, economic incentives, and market mechanisms could preserve biodiversity without being institutionalized as law. Perhaps international law should be seen as an ongoing political discourse about what needs to be done, rather than a formal set of rules, rights, and obligations—indeed, obligations that most governments remain unlikely to accept (Hendricks, this volume). But there may also be felt societal needs for establishing legal rights, duties, and obligations to complement, facilitate, empower, or harness the use of social mechanisms and techniques (Guruswamy, this volume).

## Converting Conclusions into Actions

Based on these conclusions, and drawing from the wisdom of the authors of the preceding chapters, we identify six ways in which global biodiversity might be protected more effectively and efficiently. They are: education leading to knowledge and awareness; coordination of international resource management; market-based remedial actions; international financial and trade mechanisms; international cooperation in science; and implementation of the CBD.

1. Education Leading to Knowledge and Awareness

- Too little is spent on protecting biodiversity and too much is spent on destroying it. Much effort must be devoted to converting a public largely ignorant of this reality into one that recognizes that sound ecological development is also sound economic development.
- Efforts should be expanded to encourage national and international efforts to learn more about biodiversity, so that it can be managed—in all senses of that word, including conservation—more effectively.
- Excessive land conversion and habitat destruction are often encouraged by subsidies to ranching, logging, farming, and other activities that result in biodiversity loss. Governments should investigate ways to redirect these subsidies in order to foster the conservation of biodiversity.
- Human cultural diversity must also be taken seriously into account if biological diversity is to be preserved; the two are intimately connected. Biologists need to seek better ways of working with social scientists, and vice versa.

2. Coordination of International Resource Management

- The preservation of selected natural and seminatural areas is the major strategy that will result in the preservation of the greatest amount of biodiversity at the lowest cost. This strategy should include managing ecosystems everywhere for maintaining biological diversity, and to the extent possible, limiting further human-caused losses of biodiversity in relatively undisturbed natural areas.
- Because it remains an open question whether the key concepts of dynamic, ecosystem-level management—namely, ecosystem health and integrity—can be defined with sufficient clarity to guide biodiversity policy, the comprehensive

endangered species protection system developed in the United States should be maintained until the ecosystem management concepts have been defined in a scientifically acceptable and politically viable way. Thus, U.S. policy should be to protect species, even as scientists continue to search for a more precise measure of exactly what needs to be emphasized over the long term.

3. Market-Based Remedial Actions

• Market-based instruments such as a "tradable depletion rights" scheme would give tradable rights to deplete biodiversity. Unused depletion rights, resulting from a country having depleted less than its allocation, would have a market value and could be sold for hard currency on world markets. Such depletion rights could thus provide a clear economic incentive *not* to deplete.

• By making conservation more profitable to the managers of the land and resources, each of the various internalizing mechanisms can temper the pace of destruction. Much work needs to be done to refine and refit traditional mechanisms in the context of new technology; it may even be defensible to adapt patent or other intellectual property laws to provide countries of origin with an additional incentive to conserve especially promising land and habitats.

4. International Financial and Trade Mechanisms

• While the economic value associated with the transfer of genetic and biochemical materials is often exaggerated, effective mechanisms for dealing with these transfers and ensuring equity in the distribution of the benefits is an important measure requiring a range of actions. This is especially the case since mechanisms that may be useful for controlling genetic resources in the pharmaceutical industry, may not make sense for the use of similar resources in agriculture and forestry.

• Whereas it is theoretically possible to deal with the problem of overconsumption of public goods by punishing the offending nation with taxes, lawsuits, trade sanctions, and so forth, as both a practical matter and a matter of international law, the room for such measures is distinctly limited. A better approach is through transfer payments, and the case for a series of selective subsidies appears all the more important when one considers how much of the resource base lies in countries that are poor, debt-pressured, and operating under deep discount rates. Rich countries should subsidize the conservation of the earth's biological assets.

• Royalty payments from pharmaceutical agreements to tropical countries are far from assured, but the prospecting collaboration should include elements that provide direct contributions to conservation in return for the information and samples provided to industries for commercial development. Prospecting activities should seek to increase the value by moving beyond simple collection and distribution services.

- Consideration should be given to establishing an international fund that obtains its resources from a minuscule tax on the seed, pharmaceutical, and biotechnology industries. Because many economic benefits of biodiversity are available to all, fewer incentives exist for any one individual or nation to conserve the resource than would make sense from the standpoint of society as a whole. Thus, the conservation of biodiversity should draw on some form of intervention to correct for the market failures involved—for example, offsetting the substantial subsidies that promote economically and environmentally damaging resource exploitation in agriculture, forestry, mining, and so forth.

- Charges also should be imposed on the use of the global commons, such as taxes on ocean pollution, high seas fishing, and atmospheric carbon emissions. The funds raised could be placed into a global environmental fund that would underwrite the costs of environmental management in the global commons.

5. International Cooperation in Science

- Many organisms will be preserved only if they are brought into cultivation; protected in zoos, type culture centers, or similar facilities; deep frozen; or otherwise preserved in a living condition outside of their natural habitats. Such *ex situ* measures should, therefore, be seen as a crucial part of any biodiversity conservation strategy.

- Electronic data processing provides an essential tool for the efficient handling of information about biodiversity, and should be utilized fully in this area of human knowledge.

6. Implementation of the CBD

- A new forest protocol should be negotiated as an integral element of the implementation of the CBD. The use of primary forests for the production of timber should be avoided when disrupted communities provide alternatives that are less destructive of biodiversity.

- Rights under the CBD should be extended to local communities. Key elements of the convention, such as prior informed consent obtained by the collectors and benefit sharing on mutually agreed terms, should apply to private landowners, local communities, and indigenous peoples with territorial claims.

- Developed countries such as the United States should support the building of institutional capacity in developing countries faced with new obligations under the CBD to survey and protect species as well as control the transboundary flow of genetic resources. This will involve developing legal and technical expertise related to intellectual property rights and technology transfer, thereby harnessing the new opportunities to benefit from biodiversity as more information is obtained.

- Because of a range of economic factors, the supply of biologically rich areas is doomed to be suboptimal without a concerted effort by the world community to make conservation an attractive option to the rural people who have practical jurisdiction over the resource. The network of international agencies and nongovernmental organizations that are in the business of promoting conservation should also be strengthened.

Our recommendations reflect the resolve of the authors of this book to look for solutions to the problems of global biodiversity both within and beyond their own disciplines. Consequently, we have tried to direct our search toward the broader conceptual intersection of the horizontal natural sciences axis with the vertical sociopolitical pivot. We hope this has resulted in a fresh look at well-known themes from an explicitly multidisciplinary perspective.

## References

Binkley, C.S. 1992. Forestry after the end of nature. *Journal of Forestry* 90 (10): 33–37.

Heinen, J.T., and R.S. Low. 1992. Human behavioral ecology and environmental conservation. *Environmental Conservation* 19 (2):105–16.

Hyman, J.B., and K. Wernstedt. 1991. The role of biological and economical analyses in the listing of endangered species. *Resources* (summer): 5–9.

Lindblom, C.E. 1959. The science of muddling through. *Public Administration Review* 19:79–82.

———. 1973. Incrementalism and environmentalism. In *Managing the Environment*. Washington, D.C.: U.S. Government Printing Office.

Ludwig, D., R. Hillborn, and C. Walters. 1993. Uncertainty, resource exploitation, and conservation: Lessons from history. *Science* 260:17, 36.

Mathews, J.T. ed. 1991. *Preserving the Global Environment: The Challenge of Shared Leadership*. New York: W.W. Norton.

Rappaport, R.A. 1993. Distinguished lecture in general anthropology: The anthropology of trouble. *American Anthropologist* 95 (2):295–303.

Sanchez, V., and C. Juma. 1994. *Biodiplomacy: Genetic Resources and International Relations*. Nairobi: ACTS Press.

Tobin, R. 1990. *The Expendable Future: U.S. Politics and the Protection of Biological Diversity*. Durham, N.C.: Duke University Press.

WRI (World Resources Institute), IUCN (World Conservation Union), and UNEP (United Nations Environment Programme). 1992. *Global Biodiversity Strategy: Guidelines for Action to Save, Study, and Use Earth's Biotic Wealth Sustainably and Equitably*. Washington, D.C.: World Resources Institute, World Conservation Union, and United Nations Environment Programme.

# Appendix

# Convention on Biological Diversity

## Preamble

The Contracting Parties,

Conscious of the intrinsic value of biological diversity and of the ecological, genetic, social, economic, scientific, educational, cultural, recreational and aesthetic values of biological diversity and its components,

Conscious also of the importance of biological diversity for evolution and for maintaining life sustaining systems of the biosphere,

Affirming that the conservation of biological diversity is a common concern of humankind,

Reaffirming that States have sovereign rights over their own biological resources,

Reaffirming also that States are responsible for conserving their biological diversity and for using their biological resources in a sustainable manner,

Concerned that biological diversity is being significantly reduced by certain human activities,

Aware of the general lack of information and knowledge regarding biological diversity and of the urgent need to develop scientific, technical and institutional capacities to provide the basic understanding upon which to plan and implement appropriate measures,

Noting that it is vital to anticipate, prevent and attack the causes of significant reduction or loss of biological diversity at source,

Noting also that where there is a threat of significant reduction or loss of biological diversity, lack of full scientific certainty should not be used as a reason for postponing measures to avoid or minimize such a threat,

Noting further that the fundamental requirement for the conservation of biological diversity is the in-situ conservation of ecosystems and natural habitats and the maintenance and recovery of viable populations of species in their natural surroundings,

Noting further that ex-situ measures, preferably in the country of origin, also have an important role to play,

Recognizing the close and traditional dependence of many indigenous and local communities embodying traditional lifestyles on biological resources, and the desirability of sharing equitably benefits arising from the use of traditional knowledge, innovations and practices relevant to the conservation of biological diversity and the sustainable use of its components,

Recognizing also the vital role that women play in the conservation and sustainable use of biological diversity and affirming the need for the full participation of women at all levels of policy-making and implementation for biological diversity conservation,

Stressing the importance of, and the need to promote, international, regional and global cooperation among States and intergovernmental organizations and the non-governmental sector for the conservation of biological diversity and the sustainable use of its components,

Acknowledging that the provision of new and additional financial resources and appropriate access to relevant technologies can be expected to make a substantial difference in the world's ability to address the loss of biological diversity,

Acknowledging further that special provision is required to meet the needs of developing countries, including the provision of new and additional financial resources and appropriate access to relevant technologies,

Noting in this regard the special conditions of the least developed countries and small island States,

Acknowledging that substantial investments are required to conserve biological diversity and that there is the expectation of a broad range of environmental, economic and social benefits from those investments,

Recognizing that economic and social development and poverty eradication are the first and overriding priorities of developing countries,

Aware that conservation and sustainable use of biological diversity is of critical importance for meeting the food, health and other needs of the growing world population, for which purpose access to and sharing of both genetic resources and technologies are essential,

Noting that, ultimately, the conservation and sustainable use of biological diversity will strengthen friendly relations among States and contribute to peace for humankind,

Desiring to enhance and complement existing international arrangements for the conservation of biological diversity and sustainable use of its components, and

Determined to conserve and sustainably use biological diversity for the benefit of present and future generations,

Have agreed as follows:

*Article 1. Objectives*

The objectives of this Convention, to be pursued in accordance with its relevant provisions, are the conservation of biological diversity, the sustainable use of its components and the fair and equitable sharing of the benefits arising out of the utilization of genetic resources, including by appropriate access to genetic resources and by appropriate transfer of relevant technologies, taking into account all rights over those resources and to technologies, and by appropriate funding.

*Article 2. Use of Terms*

For the purposes of this Convention:

"Biological diversity" means the variability among living organisms from all sources including, inter alia, terrestrial, marine and other aquatic ecosystems and the ecological complexes of which they are part; this includes diversity within species, between species and of ecosystems.

"Biological resources" includes genetic resources, organisms or parts thereof, populations, or any other biotic component of ecosystems with actual or potential use or value for humanity.

"Biotechnology" means any technological application that uses biological systems, living organisms, or derivatives thereof, to make or modify products or processes for specific use.

"Country of origin of genetic resources" means the country which possesses those genetic resources in in-situ conditions.

"Country providing genetic resources" means the country supplying genetic resources collected from in-situ sources, including populations of both wild and domesticated species, or taken from ex-situ sources, which may or may not have originated in that country.

"Domesticated or cultivated species" means species in which the evolutionary process has been influenced by humans to meet their needs.

"Ecosystem" means a dynamic complex of plant, animal and micro-organism communities and their non-living environment interacting as a functional unit.

"Ex-situ conservation" means the conservation of components of biological diversity outside their natural habitats.

"Genetic material" means any material of plant, animal, microbial or other origin containing functional units of heredity.

"Genetic resources" means genetic material of actual or potential value.

"Habitat" means the place or type of site where an organism or population naturally occurs.

"In-situ conditions" means conditions where genetic resources exist within ecosystems and natural habitats, and, in the case of domesticated or cultivated species, in the surroundings where they have developed their distinctive properties.

"In-situ conservation" means the conservation of ecosystems and natural habitats and the maintenance and recovery of viable populations of species in their natural surroundings and, in the case of domesticated or cultivated species, in the surroundings where they have developed their distinctive properties.

"Protected area" means a geographically defined area which is designated or regulated and managed to achieve specific conservation objectives.

"Regional economic integration organization" means an organization constituted by sovereign States of a given region, to which its member States have transferred competence in respect of matters governed by this Convention and which has been duly authorized, in accordance with its internal procedures, to sign, ratify, accept, approve or accede to it.

"Sustainable use" means the use of components of biological diversity in a way and at a rate that does not lead to the long-term decline of biological diversity, thereby maintaining its potential to meet the needs and aspirations of present and future generations.

"Technology" includes biotechnology.

## Article 3. Principle

States have, in accordance with the Charter of the United Nations and the principles of international law, the sovereign right to exploit their own resources pursuant to their own environmental policies, and the responsibility to ensure that activities within their jurisdiction or control do not cause damage to the environment of other States or of areas beyond the limits of national jurisdiction.

*Article 4. Jurisdictional Scope*

Subject to the rights of other States, and except as otherwise expressly provided in this Convention, the provisions of this Convention apply, in relation to each Contracting Party:

(a) In the case of components of biological diversity, in areas within the limits of its national jurisdiction; and

(b) In the case of processes and activities, regardless of where their effects occur, carried out under its jurisdiction or control, within the area of its national jurisdiction or beyond the limits of national jurisdiction.

*Article 5. Cooperation*

Each Contracting Party shall, as far as possible and as appropriate, cooperate with other Contracting Parties, directly or, where appropriate, through competent international organizations, in respect of areas beyond national jurisdiction and on other matters of mutual interest, for the conservation and sustainable use of biological diversity.

*Article 6. General Measures for Conservation and Sustainable Use*

Each Contracting Party shall, in accordance with its particular conditions and capabilities:

(a) Develop national strategies, plans or programmes for the conservation and sustainable use of biological diversity or adapt for this purpose existing strategies, plans or programmes which shall reflect, inter alia, the measures set out in this Convention relevant to the Contracting Party concerned; and

(b) Integrate, as far as possible and as appropriate, the conservation and sustainable use of biological diversity into relevant sectoral or cross-sectoral plans, programmes and policies.

*Article 7. Identification and Monitoring*

Each Contracting Party shall, as far as possible and as appropriate, in particular for the purposes of Articles 8 to 10:

(a) Identify components of biological diversity important for its conservation and sustainable use having regard to the indicative list of categories set down in Annex I;

(b) Monitor, through sampling and other techniques, the components of biological diversity identified pursuant to subparagraph (a) above, paying particular attention to those requiring urgent conservation measures and those which offer the greatest potential for sustainable use;

(c) Identify processes and categories of activities which have or are likely to have significant adverse impacts on the conservation and sustainable use of biological diversity, and monitor their effects through sampling and other techniques; and

(d) Maintain and organize, by any mechanism data, derived from identification and monitoring activities pursuant to subparagraphs (a), (b) and (c) above.

*Article 8. In-situ Conservation*

Each Contracting Party shall, as far as possible and as appropriate:

(a) Establish a system of protected areas or areas where special measures need to be taken to conserve biological diversity;

(b) Develop, where necessary, guidelines for the selection, establishment and management of protected areas or areas where special measures need to be taken to conserve biological diversity;

(c) Regulate or manage biological resources important for the conservation of biological diversity whether within or outside protected areas, with a view to ensuring their conservation and sustainable use;

(d) Promote the protection of ecosystems, natural habitats and the maintenance of viable populations of species in natural surroundings;

(e) Promote environmentally sound and sustainable development in areas adjacent to protected areas with a view to furthering protection of these areas;

(f) Rehabilitate and restore degraded ecosystems and promote the recovery of threatened species, inter alia, through the development and implementation of plans or other management strategies;

(g) Establish or maintain means to regulate, manage or control the risks associated with the use and release of living modified organisms resulting from biotechnology which are likely to have adverse environmental impacts that could affect the conservation and sustainable use of biological diversity, taking also into account the risks to human health;

(h) Prevent the introduction of, control or eradicate those alien species which threaten ecosystems, habitats or species;

(i) Endeavor to provide the conditions needed for compatibility between present uses and the conservation of biological diversity and the sustainable use of its components;

(j) Subject to its national legislation, respect, preserve and maintain knowledge, innovations and practices of indigenous and local communities embodying traditional lifestyles relevant for the conservation and sustainable use of biological diversity and promote their wider application with the approval and involvement of the holders of such knowledge, innovations and practices and encourage the equitable sharing of the benefits arising from the utilization of such knowledge, innovations and practices;

(k) Develop or maintain necessary legislation and/or other regulatory provisions for the protection of threatened species and populations;

(l) Where a significant adverse effect on biological diversity has been determined pursuant to Article 7, regulate or manage the relevant processes and categories of activities; and

(m) Cooperate in providing financial and other support for in-situ conservation outlined in subparagraphs (a) to (l) above, particularly to developing countries.

*Article 9. Ex-situ Conservation*

Each Contracting Party shall, as far as possible and as appropriate, and predominantly for the purpose of complementing in-situ measures:

(a) Adopt measures for the ex-situ conservation of components of biological diversity, preferably in the country of origin of such components;

(b) Establish and maintain facilities for ex-situ conservation of and research on plants, animals and micro-organisms, preferably in the country of origin of genetic resources;

(c) Adopt measures for the recovery and rehabilitation of threatened species and for their reintroduction into their natural habitats under appropriate conditions;

(d) Regulate and manage collection of biological resources from natural habitats for ex-situ conservation purposes so as not to threaten ecosystems and in-situ populations of species, except where special temporary ex-situ measures are required under subparagraph (c) above; and

(e) Cooperate in providing financial and other support for ex-situ conservation outlined in subparagraphs (a) to (d) above and in the establishment and maintenance of ex-situ conservation facilities in developing countries.

*Article 10. Sustainable Use of Components of Biological Diversity*

Each Contracting Party shall, as far as possible and as appropriate:

(a) Integrate consideration of the conservation and sustainable use of biological resources into national decision-making;

(b) Adopt measures relating to the use of biological resources to avoid or minimize adverse impacts on biological diversity;

(c) Protect and encourage customary use of biological resources in accordance with traditional cultural practices that are compatible with conservation or sustainable use requirements;

(d) Support local populations to develop and implement remedial action in degraded areas where biological diversity has been reduced; and

(e) Encourage cooperation between its governmental authorities and its private sector in developing methods for sustainable use of biological resources.

*Article 11. Incentive Measures*

Each Contracting Party shall, as far as possible and as appropriate, adopt economically and socially sound measures that act as incentives for the conservation and sustainable use of components of biological diversity.

*Article 12. Research and Training*

The Contracting Parties, taking into account the special needs of developing countries, shall:

(a) Establish and maintain programmes for scientific and technical education and training in measures for the identification, conservation and sustainable use of biological diversity and its components and provide support for such education and training for the specific needs of developing countries;

(b) Promote and encourage research which contributes to the conservation and sustainable use of biological diversity, particularly in developing countries, inter alia, in accordance with decisions of the Conference of the Parties taken in consequence of recommendations of the Subsidiary Body on Scientific, Technical and Technological Advice; and

(c) In keeping with the provisions of Articles 16, 18 and 20, promote and cooperate in the use of scientific advances in biological diversity research in developing methods for conservation and sustainable use of biological resources.

*Article 13. Public Education and Awareness*

The Contracting Parties shall:

(a) Promote and encourage understanding of the importance of, and the measures required for, the conservation of biological diversity, as well as its propagation through media, and the inclusion of these topics in educational programmes; and

(b) Cooperate, as appropriate, with other States and international organizations in developing educational and public awareness programmes, with respect to conservation and sustainable use of biological diversity.

*Article 14. Impact Assessment and Minimizing Adverse Impacts*

1. Each Contracting Party, as far as possible and as appropriate, shall:

(a) Introduce appropriate procedures requiring environmental impact assessment of its proposed projects that are likely to have significant adverse effects on biological diversity with a view to avoiding or minimizing such effects and, where appropriate, allow for public participation in such procedures;

(b) Introduce appropriate arrangements to ensure that the environmental consequences of its programmes and policies that are likely to have significant adverse impacts on biological diversity are duly taken into account;

(c) Promote, on the basis of reciprocity, notification, exchange of information and consultation on activities under their jurisdiction or control which are likely to significantly affect adversely the biological diversity of other States or areas beyond the limits of national jurisdiction, by encouraging the conclusion of bilateral, regional or multilateral arrangements, as appropriate;

(d) In the case of imminent or grave danger or damage, originating under its jurisdiction or control, to biological diversity within the area under jurisdiction of other States or in areas beyond the limits of national jurisdiction, notify immediately the potentially affected States of such danger or damage, as well as initiate action to prevent or minimize such danger or damage; and

(e) Promote national arrangements for emergency responses to activities or events, whether caused naturally or otherwise, which present a grave and imminent danger to biological diversity and encourage international cooperation to supplement such national efforts and, where appropriate and agreed by the States or regional economic integration organizations concerned, to establish joint contingency plans.

2. The Conference of the Parties shall examine, on the basis of studies to be carried out, the issue of liability and redress, including restoration and compensation, for damage to biological diversity, except where such liability is a purely internal matter.

*Article 15. Access to Genetic Resources*

1. Recognizing the sovereign rights of States over their natural resources, the authority to determine access to genetic resources rests with the national governments and is subject to national legislation.

2. Each Contracting Party shall endeavour to create conditions to facilitate access to genetic resources for environmentally sound uses by other Contracting Parties and not to impose restrictions that run counter to the objectives of this Convention.

3. For the purpose of this Convention, the genetic resources being provided by a Contracting Party, as referred to in this Article and Articles 16 and 19, are only those that are provided by Contracting Parties that are countries of origin of such resources or by the Parties that have acquired the genetic resources in accordance with this Convention.

4. Access, where granted, shall be on mutually agreed terms and subject to the provisions of this Article.

5. Access to genetic resources shall be subject to prior informed consent of the Contracting Party providing such resources, unless otherwise determined by that Party.

6. Each Contracting Party shall endeavour to develop and carry out scientific research based on genetic resources provided by other Contracting Parties with the full participation of, and where possible in, such Contracting Parties.

7. Each Contracting Party shall take legislative, administrative or policy measures, as appropriate, and in accordance with Articles 16 and 19 and, where necessary, through the financial mechanism established by Articles 20 and 21 with the aim of sharing in a fair and equitable way the results of research and development and the benefits arising from the commercial and other utilization of genetic resources with the Contracting Party providing such resources. Such sharing shall be upon mutually agreed terms.

*Article 16. Access to and Transfer of Technology*

1. Each Contracting Party, recognizing that technology includes biotechnology, and that both access to and transfer of technology among Contracting Parties are essential elements for the attainment of the objectives of this Convention, undertakes subject to the provisions of this Article to provide and/or facilitate access for and transfer to other Contracting Parties of technologies that are relevant to the conservation and sustainable use of biological diversity or make use of genetic resources and do not cause significant damage to the environment.

2. Access to and transfer of technology referred to in paragraph 1 above to developing countries shall be provided and/or facilitated under fair and most favourable terms, including on concessional and preferential terms where mutually agreed, and, where necessary, in accordance with the financial mechanism established by Articles 20 and 21. In the case of technology subject to patents and other intellectual property rights, such access and transfer shall be provided on terms which recognize and are consistent with the adequate and effective protection of intellectual property rights. The application of this paragraph shall be consistent with paragraphs 3, 4 and 5 below.

3. Each Contracting Party shall take legislative, administrative or policy measures, as appropriate, with the aim that Contracting Parties, in particular those that are developing countries, which provide genetic resources are provided access to and transfer of technology which makes use of those resources, on mutually agreed terms, including technology protected by patents and other intellectual property rights, where necessary, through the provisions of Articles 20 and 21 and in accordance with international law and consistent with paragraphs 4 and 5 below.

4. Each Contracting Party shall take legislative, administrative or policy measures, as appropriate, with the aim that the private sector facilitates access to, joint development and transfer of technology referred to in paragraph 1 above for the benefit of both governmen-

tal institutions and the private sector of developing countries and in this regard shall abide by the obligations included in paragraphs 1, 2 and 3 above.

5. The Contracting Parties, recognizing that patents and other intellectual property rights may have an influence on the implementation of this Convention, shall cooperate in this regard subject to national legislation and international law in order to ensure that such rights are supportive of and do not run counter to its objectives.

### Article 17. Exchange of Information

1. The Contracting Parties shall facilitate the exchange of information, from all publicly available sources, relevant to the conservation and sustainable use of biological diversity, taking into account the special needs of developing countries.

2. Such exchange of information shall include exchange of results of technical, scientific and socio-economic research, as well as information on training and surveying programmes, specialized knowledge, indigenous and traditional knowledge as such and in combination with the technologies referred to in Article 16, paragraph 1. It shall also, where feasible, include repatriation of information.

### Article 18. Technical and Scientific Cooperation

1. The Contracting Parties shall promote international technical and scientific cooperation in the field of conservation and sustainable use of biological diversity, where necessary, through the appropriate international and national institutions.

2. Each Contracting Party shall promote technical and scientific cooperation with other Contracting Parties, in particular developing countries, in implementing this Convention, inter alia, through the development and implementation of national policies. In promoting such cooperation, special attention should be given to the development and strengthening of national capabilities, by means of human resources development and institution building.

3. The Conference of the Parties, at its first meeting, shall determine how to establish a clearing-house mechanism to promote and facilitate technical and scientific cooperation.

4. The Contracting Parties shall, in accordance with national legislation and policies, encourage and develop methods of cooperation for the development and use of technologies, including indigenous and traditional technologies, in pursuance of the objectives of this Convention. For this purpose, the Contracting Parties shall also promote cooperation in the training of personnel and exchange of experts.

5. The Contracting Parties shall, subject to mutual agreement, promote the establishment of joint research programmes and joint ventures for the development of technologies relevant to the objectives of this Convention.

### Article 19. Handling of Biotechnology and Distribution of its Benefits

1. Each Contracting Party shall take legislative, administrative or policy measures, as appropriate, to provide for the effective participation in biotechnological research activities by those Contracting Parties, especially developing countries, which provide the genetic resources for such research, and where feasible in such Contracting Parties.

2. Each Contracting Party shall take all practicable measures to promote and advance

priority access on a fair and equitable basis by Contracting Parties, especially developing countries, to the results and benefits arising from biotechnologies based upon genetic resources provided by those Contracting Parties. Such access shall be on mutually agreed terms.

3. The Parties shall consider the need for and modalities of a protocol setting out appropriate procedures, including, in particular, advance informed agreement, in the field of the safe transfer, handling and use of any living modified organism resulting from biotechnology that may have adverse effect on the conservation and sustainable use of biological diversity.

4. Each Contracting Party shall, directly or by requiring any natural or legal person under its jurisdiction providing the organisms referred to in paragraph 3 above, provide any available information about the use and safety regulations required by that Contracting Party in handling such organisms, as well as any available information on the potential adverse impact of the specific organisms concerned to the Contracting Party into which those organisms are to be introduced.

*Article 20. Financial Resources*

1. Each Contracting Party undertakes to provide, in accordance with its capabilities, financial support and incentives in respect of those national activities which are intended to achieve the objectives of this Convention, in accordance with its national plans, priorities and programmes.

2. The developed country Parties shall provide new and additional financial resources to enable developing country Parties to meet the agreed full incremental costs to them of implementing measures which fulfil the obligations of this Convention and to benefit from its provisions and which costs are agreed between a developing country Party and the institutional structure referred to in Article 21, in accordance with policy, strategy, programme priorities and eligibility criteria and an indicative list of incremental costs established by the Conference of the Parties. Other Parties, including countries undergoing the process of transition to a market economy, may voluntarily assume the obligations of the developed country Parties. For the purpose of this Article, the Conference of the Parties, shall at its first meeting establish a list of developed country Parties and other Parties which voluntarily assume the obligations of the developed country Parties. The Conference of the Parties shall periodically review and if necessary amend the list. Contributions from other countries and sources on a voluntary basis would also be encouraged. The implementation of these commitments shall take into account the need for adequacy, predictability and timely flow of funds and the importance of burden-sharing among the contributing Parties included in the list.

3. The developed country Parties may also provide, and developing country Parties avail themselves of, financial resources related to the implementation of this Convention through bilateral, regional and other multilateral channels.

4. The extent to which developing country Parties will effectively implement their commitments under this Convention will depend on the effective implementation by developed country Parties of their commitments under this Convention related to financial resources and transfer of technology and will take fully into account the fact that economic and social

development and eradication of poverty are the first and overriding priorities of the developing country Parties.

5. The Parties shall take full account of the specific needs and special situation of least developed countries in their actions with regard to funding and transfer of technology.

6. The Contracting Parties shall also take into consideration the special conditions resulting from the dependence on, distribution and location of, biological diversity within developing country Parties, in particular small island States.

7. Consideration shall also be given to the special situation of developing countries, including those that are most environmentally vulnerable, such as those with arid and semi-arid zones, coastal and mountainous areas.

*Article 21. Financial Mechanism*

1. There shall be a mechanism for the provision of financial resources to developing country Parties for purposes of this Convention on a grant or concessional basis the essential elements of which are described in this Article. The mechanism shall function under the authority and guidance of, and be accountable to, the Conference of the Parties for purposes of this Convention. The operations of the mechanism shall be carried out by such institutional structure as may be decided upon by the Conference of the Parties at its first meeting. For purposes of this Convention, the Conference of the Parties shall determine the policy, strategy, programme priorities and eligibility criteria relating to the access to and utilization of such resources. The contributions shall be such as to take into account the need for predictability, adequacy and timely flow of funds referred to in Article 20 in accordance with the amount of resources needed to be decided periodically by the Conference of the Parties and the importance of burden-sharing among the contributing Parties included in the list referred to in Article 20, paragraph 2. Voluntary contributions may also be made by the developed country Parties and by other countries and sources. The mechanism shall operate within a democratic and transparent system of governance.

2. Pursuant to the objectives of this Convention, the Conference of the Parties shall at its first meeting determine the policy, strategy and programme priorities, as well as detailed criteria and guidelines for eligibility for access to and utilization of the financial resources including monitoring and evaluation on a regular basis of such utilization. The Conference of the Parties shall decide on the arrangements to give effect to paragraph 1 above after consultation with the institutional structure entrusted with the operation of the financial mechanism.

3. The Conference of the Parties shall review the effectiveness of the mechanism established under this Article, including the criteria and guidelines referred to in paragraph 2 above, not less than two years after the entry into force of this Convention and thereafter on a regular basis. Based on such review, it shall take appropriate action to improve the effectiveness of the mechanism if necessary.

4. The Contracting Parties shall consider strengthening existing financial institutions to provide financial resources for the conservation and sustainable use of biological diversity.

*Article 22. Relationship with Other International Conventions*

1. The provisions of this Convention shall not affect the rights and obligations of any Contracting Party deriving from any existing international agreement, except where the ex-

ercise of those rights and obligations would cause a serious damage or threat to biological diversity.

2. Contracting Parties shall implement this Convention with respect to the marine environment consistently with the rights and obligations of States under the law of the sea.

*Article 23. Conference of the Parties*

1. A Conference of the Parties is hereby established. The first meeting of the Conference of the Parties shall be convened by the Executive Director of the United Nations Environment Programme not later than one year after the entry into force of this Convention. Thereafter, ordinary meetings of the Conference of the Parties shall be held at regular intervals to be determined by the Conference at its first meeting.

2. Extraordinary meetings of the Conference of the Parties shall be held at such other times as may be deemed necessary by the Conference, or at the written request of any Party, provided that, within six months of the request being communicated to them by the Secretariat, it is supported by at least one third of the Parties.

3. The Conference of the Parties shall by consensus agree upon and adopt rules of procedure for itself and for any subsidiary body it may establish, as well as financial rules governing the funding of the Secretariat. At each ordinary meeting, it shall adopt a budget for the financial period until the next ordinary meeting.

4. The Conference of the Parties shall keep under review the implementation of this Convention, and, for this purpose, shall:

(a) Establish the form and the intervals for transmitting the information to be submitted in accordance with Article 26 and consider such information as well as reports submitted by any subsidiary body;

(b) Review scientific, technical and technological advice on biological diversity provided in accordance with Article 25;

(c) Consider and adopt, as required, protocols in accordance with Article 28;

(d) Consider and adopt, as required, in accordance with Articles 29 and 30, amendments to this Convention and its annexes;

(e) Consider amendments to any protocol, as well as to any annexes thereto, and, if so decided, recommend their adoption to the parties to the protocol concerned;

(f) Consider and adopt, as required, in accordance with Article 30, additional annexes to this Convention;

(g) Establish such subsidiary bodies, particularly to provide scientific and technical advice, as are deemed necessary for the implementation of this Convention;

(h) Contact, through the Secretariat, the executive bodies of conventions dealing with matters covered by this Convention with a view to establishing appropriate forms of cooperation with them; and

(i) Consider and undertake any additional action that may be required for the achievement of the purposes of this Convention in the light of experience gained in its operation.

5. The United Nations, its specialized agencies and the International Atomic Energy Agency, as well as any State not Party to this Convention, may be represented as observers at meetings of the Conference of the Parties. Any other body or agency, whether governmental or non-governmental, qualified in fields relating to conservation and sustainable use

of biological diversity, which has informed the Secretariat of its wish to be represented as an observer at a meeting of the Conference of the Parties, may be admitted unless at least one third of the Parties present object. The admission and participation of observers shall be subject to the rules of procedure adopted by the Conference of the Parties.

*Article 24. Secretariat*

1. A Secretariat is hereby established. Its functions shall be:
(a) To arrange for and service meetings of the Conference of the Parties provided for in Article 23;
(b) To perform the functions assigned to it by any protocol;
(c) To prepare reports on the execution of its functions under this Convention and present them to the Conference of the Parties;
(d) To coordinate with other relevant international bodies and, in particular to enter into such administrative and contractual arrangements as may be required for the effective discharge of its functions; and
(e) To perform such other functions as may be determined by the Conference of the Parties.

2. At its first ordinary meeting, the Conference of the Parties shall designate the Secretariat from amongst those existing competent international organizations which have signified their willingness to carry out the Secretariat functions under this Convention.

*Article 25. Subsidiary Body on Scientific, Technical and Technological Advice*

1. A subsidiary body for the provision of scientific, technical and technological advice is hereby established to provide the Conference of the Parties and, as appropriate, its other subsidiary bodies with timely advice relating to the implementation of this Convention. This body shall be open to participation by all Parties and shall be multidisciplinary. It shall comprise government representatives competent in the relevant field of expertise. It shall report regularly to the Conference of the Parties on all aspects of its work.

2. Under the authority of and in accordance with guidelines laid down by the Conference of the Parties, and upon its request, this body shall:
(a) Provide scientific and technical assessments of the status of biological diversity;
(b) Prepare scientific and technical assessments of the effects of types of measures taken in accordance with the provisions of this Convention;
(c) Identify innovative, efficient and state-of-the-art technologies and know-how relating to the conservation and sustainable use of biological diversity and advise on the ways and means of promoting development and/or transferring such technologies;
(d) Provide advice on scientific programmes and international cooperation in research and development related to conservation and sustainable use of biological diversity; and
(e) Respond to scientific, technical, technological and methodological questions that the Conference of the Parties and its subsidiary bodies may put to the body.

3. The functions, terms of reference, organization and operation of this body may be further elaborated by the Conference of the Parties.

*Article 26. Reports*

Each Contracting Party shall, at intervals to be determined by the Conference of the Parties, present to the Conference of the Parties, reports on measures which it has taken for the implementation of the provisions of this Convention and their effectiveness in meeting the objectives of this Convention.

*Article 27. Settlement of Disputes*

1. In the event of a dispute between Contracting Parties concerning the interpretation or application of this Convention, the parties concerned shall seek solution by negotiation.

2. If the parties concerned cannot reach agreement by negotiation, they may jointly seek the good offices of, or request mediation by, a third party.

3. When ratifying, accepting, approving or acceding to this Convention, or at any time thereafter, a State or regional economic integration organization may declare in writing to the Depositary that for a dispute not resolved in accordance with paragraph 1 or paragraph 2 above, it accepts one or both of the following means of dispute settlement as compulsory:
(a) Arbitration in accordance with the procedure laid down in Part 1 of Annex II;
(b) Submission of the dispute to the International Court of Justice.

4. If the parties to the dispute have not, in accordance with paragraph 3 above, accepted the same or any procedure, the dispute shall be submitted to conciliation in accordance with Part 2 of Annex II unless the parties otherwise agree.

5. The provisions of this Article shall apply with respect to any protocol except as otherwise provided in the protocol concerned.

*Article 28. Adoption of Protocols*

1. The Contracting Parties shall cooperate in the formulation and adoption of protocols to this Convention.

2. Protocols shall be adopted at a meeting of the Conference of the Parties.

3. The text of any proposed protocol shall be communicated to the Contracting Parties by the Secretariat at least six months before such a meeting.

*Article 29. Amendment of the Convention or Protocols*

1. Amendments to this Convention may be proposed by any Contracting Party. Amendments to any protocol may be proposed by any Party to that protocol.

2. Amendments to this Convention shall be adopted at a meeting of the Conference of the Parties. Amendments to any protocol shall be adopted at a meeting of the Parties to the protocol in question. The text of any proposed amendment to this Convention or to any protocol, except as may otherwise be provided in such protocol, shall be communicated to the Parties to the instrument in question by the Secretariat at least six months before the meeting at which it is proposed for adoption. The Secretariat shall also communicate proposed amendments to the signatories to this Convention for information.

3. The Parties shall make every effort to reach agreement on any proposed amendment to this Convention or to any protocol by consensus. If all efforts at consensus have been exhausted, and no agreement reached, the amendment shall as a last resort be adopted by a two-third majority vote of the Parties to the instrument in question present and voting at

the meeting, and shall be submitted by the Depositary to all Parties for ratification, acceptance or approval.

4. Ratification, acceptance or approval of amendments shall be notified to the Depositary in writing. Amendments adopted in accordance with paragraph 3 above shall enter into force among Parties having accepted them on the ninetieth day after the deposit of instruments of ratification, acceptance or approval by at least two thirds of the Contracting Parties to this Convention or of the Parties to the protocol concerned, except as may otherwise be provided in such protocol. Thereafter the amendments shall enter into force for any other Party on the ninetieth day after that Party deposits its instrument of ratification, acceptance or approval of the amendments.

5. For the purposes of this Article, "Parties present and voting" means Parties present and casting an affirmative or negative vote.

*Article 30. Adoption and Amendment of Annexes*

1. The annexes to this Convention or to any protocol shall form an integral part of the Convention or of such protocol, as the case may be, and, unless expressly provided otherwise, a reference to this Convention or its protocols constitutes at the same time a reference to any annexes thereto. Such annexes shall be restricted to procedural, scientific, technical and administrative matters.

2. Except as may be otherwise provided in any protocol with respect to its annexes, the following procedure shall apply to the proposal, adoption and entry into force of additional annexes to this Convention or of annexes to any protocol:

(a) Annexes to this Convention or to any protocol shall be proposed and adopted according to the procedure laid down in Article 29;

(b) Any Party that is unable to approve an additional annex to this Convention or an annex to any protocol to which it is Party shall so notify the Depositary, in writing, within one year from the date of the communication of the adoption by the Depositary. The Depositary shall without delay notify all Parties of any such notification received. A Party may at any time withdraw a previous declaration of objection and the annexes shall thereupon enter into force for that Party subject to subparagraph (c) below;

(c) On the expiry of one year from the date of the communication of the adoption by the Depositary, the annex shall enter into force for all Parties to this Convention or to any protocol concerned which have not submitted a notification in accordance with the provisions of subparagraph (b) above.

3. The proposal, adoption and entry into force of amendments to annexes to this Convention or to any protocol shall be subject to the same procedure as for the proposal, adoption and entry into force of annexes to the Convention or annexes to any protocol.

4. If an additional annex or an amendment to an annex is related to an amendment to this Convention or to any protocol, the additional annex or amendment shall not enter into force until such time as the amendment to the Convention or to the protocol concerned enters into force.

*Article 31. Right to Vote*

1. Except as provided for in paragraph 2 below, each Contracting Party to this Convention or to any protocol shall have one vote.

2. Regional economic integration organizations, in matters within their competence, shall exercise their right to vote with a number of votes equal to the number of their member States which are Contracting Parties to this Convention or the relevant protocol. Such organizations shall not exercise their right to vote if their member States exercise theirs, and vice versa.

*Article 32. Relationship between this Convention and Its Protocols*

1. A State or a regional economic integration organization may not become a Party to a protocol unless it is, or becomes at the same time, a Contracting Party to this Convention.

2. Decisions under any protocol shall be taken only by the Parties to the protocol concerned. Any Contracting Party that has not ratified, accepted or approved a protocol may participate as an observer in any meeting of the parties to that protocol.

*Article 33. Signature*

This Convention shall be open for signature at Rio de Janeiro by all States and any regional economic integration organization from 5 June 1992 until 14 June 1992, and at the United Nations Headquarters in New York from 15 June 1992 to 4 June 1993.

*Article 34. Ratification, Acceptance or Approval*

1. This Convention and any protocol shall be subject to ratification, acceptance or approval by States and by regional economic integration organizations. Instruments of ratification, acceptance or approval shall be deposited with the Depositary.

2. Any organization referred to in paragraph 1 above which becomes a Contracting Party to this Convention or any protocol without any of its member States being a Contracting Party shall be bound by all the obligations under the Convention or the protocol, as the case may be. In the case of such organizations, one or more of whose member States is a Contracting Party to this Convention or relevant protocol, the organization and its member States shall decide on their respective responsibilities for the performance of their obligations under the Convention or protocol, as the case may be. In such cases, the organization and the member States shall not be entitled to exercise rights under the Convention or relevant protocol concurrently.

3. In their instruments of ratification, acceptance or approval, the organizations referred to in paragraph 1 above shall declare the extent of their competence with respect to the matters governed by the Convention or the relevant protocol. These organizations shall also inform the Depositary of any relevant modification in the extent of their competence.

*Article 35. Accession*

1. This Convention and any protocol shall be open for accession by States and by regional economic integration organizations from the date on which the Convention or the protocol concerned is closed for signature. The instruments of accession shall be deposited with the Depositary.

2. In their instruments of accession, the organizations referred to in paragraph 1 above shall declare the extent of their competence with respect to the matters governed by the Convention or the relevant protocol. These organizations shall also inform the Depositary of any relevant modification in the extent of their competence.

3. The provisions of Article 34, paragraph 2, shall apply to regional economic integration organizations which accede to this Convention or any protocol.

*Article 36. Entry Into Force*

1. This Convention shall enter into force on the ninetieth day after the date of deposit of the thirtieth instrument of ratification, acceptance, approval or accession.

2. Any protocol shall enter into force on the ninetieth day after the date of deposit of the number of instruments of ratification, acceptance, approval or accession, specified in that protocol, has been deposited.

3. For each Contracting Party which ratifies, accepts or approves this Convention or accedes thereto after the deposit of the thirtieth instrument of ratification, acceptance, approval or accession, it shall enter into force on the ninetieth day after the date of deposit by such Contracting Party of its instrument of ratification, acceptance, approval or accession.

4. Any protocol, except as otherwise provided in such protocol, shall enter into force for a Contracting Party that ratifies, accepts or approves that protocol or accedes thereto after its entry into force pursuant to paragraph 2 above, on the ninetieth day after the date on which that Contracting Party deposits its instrument of ratification, acceptance, approval or accession, or on the date on which this Convention enters into force for that Contracting Party, whichever shall be the later.

5. For the purposes of paragraphs 1 and 2 above, any instrument deposited by a regional economic integration organization shall not be counted as additional to those deposited by member States of such organization.

*Article 37. Reservations*

No reservations may be made to this Convention.

*Article 38. Withdrawals*

1. At any time after two years from the date on which this Convention has entered into force for a Contracting Party, that Contracting Party may withdraw from the Convention by giving written notification to the Depositary.

2. Any such withdrawal shall take place upon expiry of one year after the date of its receipt by the Depositary, or on such later date as may be specified in the notification of the withdrawal.

3. Any Contracting Party which withdraws from this Convention shall be considered as also having withdrawn from any protocol to which it is party.

*Article 39. Financial Interim Arrangements*

Provided that it has been fully restructured in accordance with the requirements of Article 21, the Global Environment Facility of the United Nations Development Programme, the United Nations Environment Programme and the International Bank for Reconstruction and Development shall be the institutional structure referred to in Article 21 on an interim basis, for the period between the entry into force of this Convention and the first meeting of the Conference of the Parties or until the Conference of the Parties decides which institutional structure will be designated in accordance with Article 21.

### Article 40. Secretariat Interim Arrangements

The Secretariat to be provided by the Executive Director of the United Nations Environment Programme shall be the Secretariat referred to in Article 24, paragraph 2, on an interim basis for the period between the entry into force of this Convention and the first meeting of the Conference of the Parties.

### Article 41. Depositary

The Secretary-General of the United Nations shall assume the functions of Depositary of this Convention and any protocols.

### Article 42. Authentic Texts

The original of this Convention, of which the Arabic, Chinese, English, French, Russian and Spanish texts are equally authentic, shall be deposited with the Secretary-General of the United Nations.

IN WITNESS WHEREOF the undersigned, being duly authorized to that effect, have signed this Convention.

Done at Rio de Janeiro on this fifth day of June, one thousand nine hundred and ninety-two.

## Annex I
### Identification and Monitoring

1. Ecosystems and habitats: containing high diversity, large numbers of endemic or threatened species, or wilderness; required by migratory species; of social, economic, cultural or scientific importance; or, which are representative, unique or associated with key evolutionary or other biological processes;

2. Species and communities which are: threatened; wild relatives of domesticated or cultivated species; of medicinal, agricultural or other economic value; or social, scientific or cultural importance; or importance for research into the conservation and sustainable use of biological diversity, such as indicator species; and

3. Described genomes and genes of social, scientific or economic importance.

## Annex II
### Part 1  Arbitration

*Article 1*

The claimant party shall notify the Secretariat that the parties are referring a dispute to arbitration pursuant to Article 27. The notification shall state the subject-matter of arbitration and include, in particular, the articles of the Convention or the protocol, the interpretation or application of which are at issue. If the parties do not agree on the subject matter of the dispute before the President of the tribunal is designated, the arbitral tribunal shall determine the subject matter. The Secretariat shall forward the information thus received to all Contracting Parties to this Convention or to the protocol concerned.

*Article 2*

1. In disputes between two parties, the arbitral tribunal shall consist of three members. Each of the parties to the dispute shall appoint an arbitrator and the two arbitrators so appointed shall designate by common agreement the third arbitrator who shall be the President of the tribunal. The latter shall not be a national of one of the parties to the dispute, nor have his or her usual place of residence in the territory of one of these parties, nor be employed by any of them, nor have dealt with the case in any other capacity.

2. In disputes between more than two parties, parties in the same interest shall appoint one arbitrator jointly by agreement.

3. Any vacancy shall be filled in the manner prescribed for the initial appointment.

*Article 3*

1. If the President of the arbitral tribunal has not been designated within two months of the appointment of the second arbitrator, the Secretary-General of the United Nations shall, at the request of a party, designate the President within a further two-month period.

2. If one of the parties to the dispute does not appoint an arbitrator within two months of receipt of the request, the other party may inform the Secretary-General who shall make the designation within a further two-month period.

*Article 4*

The arbitral tribunal shall render its decisions in accordance with the provisions of this Convention, any protocols concerned, and international law.

*Article 5*

Unless the parties to the dispute otherwise agree, the arbitral tribunal shall determine its own rules of procedure.

*Article 6*

The arbitral tribunal may, at the request of one of the parties, recommend essential interim measures of protection.

*Article 7*

The parties to the dispute shall facilitate the work of the arbitral tribunal and, in particular, using all means at their disposal, shall:
(a) Provide it with all relevant documents, information and facilities; and
(b) Enable it, when necessary, to call witnesses or experts and receive their evidence.

*Article 8*

The parties and the arbitrators are under an obligation to protect the confidentiality of any information they receive in confidence during the proceedings of the arbitral tribunal.

*Article 9*

Unless the arbitral tribunal determines otherwise because of the particular circumstances of the case, the costs of the tribunal shall be borne by the parties to the dispute in equal shares. The tribunal shall keep a record of all its costs, and shall furnish a final statement thereof to the parties.

*Article 10*

Any Contracting Party that has an interest of a legal nature in the subject-matter of the dispute which may be affected by the decision in the case, may intervene in the proceedings with the consent of the tribunal.

*Article 11*

The tribunal may hear and determine counterclaims arising directly out of the subject-matter of the dispute.

*Article 12*

Decisions both on procedure and substance of the arbitral tribunal shall be taken by a majority vote of its members.

*Article 13*

If one of the parties to the dispute does not appear before the arbitral tribunal or fails to defend its case, the other party may request the tribunal to continue the proceedings and to make its award. Absence of a party or a failure of a party to defend its case shall not constitute a bar to the proceedings. Before rendering its final decision, the arbitral tribunal must satisfy itself that the claim is well founded in fact and law.

*Article 14*

The tribunal shall render its final decision within five months of the date on which it is fully constituted unless it finds it necessary to extend the time-limit for a period which should not exceed five more months.

*Article 15*

The final decision of the arbitral tribunal shall be confined to the subject-matter of the dispute and shall state the reasons on which it is based. It shall contain the names of the members who have participated and the date of the final decision. Any member of the tribunal may attach a separate or dissenting opinion to the final decision.

*Article 16*

The award shall be binding on the parties to the dispute. It shall be without appeal unless the parties to the dispute have agreed in advance to an appellate procedure.

*Article 17*

Any controversy which may arise between the parties to the dispute as regards the interpretation or manner of implementation of the final decision may be submitted by either party for decision to the arbitral tribunal which rendered it.

## Part 2  Conciliation

*Article 1*

A conciliation commission shall be created upon the request of one of the parties to the dispute. The commission shall, unless the parties otherwise agree, be composed of five members, two appointed by each Party concerned and a President chosen jointly by those members.

*Article 2*

In disputes between more than two parties, parties in the same interest shall appoint their members of the commission jointly by agreement. Where two or more parties have separate interests or there is a disagreement as to whether they are of the same interest, they shall appoint their members separately.

*Article 3*

If any appointments by the parties are not made within two months of the date of the request to create a conciliation commission, the Secretary-General of the United Nations shall, if asked to do so by the party that made the request, make those appointments within a further two-month period.

*Article 4*

If a President of the conciliation commission has not been chosen within two months of the last of the members of the commission being appointed, the Secretary-General of the United Nations shall, if asked to do so by a party, designate a President within a further two-month period.

*Article 5*

The conciliation commission shall take its decisions by majority vote of its members. It shall, unless the parties to the dispute otherwise agree, determine its own procedure. It shall render a proposal for resolution of the dispute, which the parties shall consider in good faith.

*Article 6*

A disagreement as to whether the conciliation commission has competence shall be decided by the commission.

# Contributors

S. JAMES ANAYA, Professor of Law at the University of Iowa College of Law and Special Counsel to the Indian law Resource Center in Helena, Montana, specializes in international law and organization. His recent publications include *Indigenous Peoples in International Law*. Previously, Anaya served as Director of the Law Project of the National Indian Youth Council, an organization with special consultative status at the United Nations. In 1994, Anaya was a Fellow at the Rockefeller Foundation's Study and Conference Center in Bellagio, Italy. He has lectured throughout the United States, Latin America, Europe, and the former Soviet Union. In addition, he has coauthored a major study for the Royal Commission on Aboriginal Peoples of Canada.

GREGORY BENFORD, a physicist, received his B.S. from the University of Oklahoma, and Ph.D. from the University of California at San Diego. He was with the Radiation Laboratory at Lawrence, California, for four years as a postdoctoral fellow and research physicist. Benford is currently a Professor of Physics at the University of California at Davis, and conducts research in plasma turbulence theory and experiment, as well as in astrophysics. He is a Woodrow Wilson Fellow and a Visiting Fellow at Cambridge University, and has served as an adviser to the Department of Energy, NASA, and the White House Council on Space Policy.

GRACIELA CHICHILNISKY is Director of the Program on Information and Resources at Columbia University, and UNESCO Professor of Economics and Mathematics, also at Columbia. She was the 1994–95 Salimbeni Professor at the University of Siena; taught previously at Harvard, Essex, and Stanford; advises or advised the OECD, IPCC, World Bank, United Nations, OPEC, and Global Environmental Facility on international environmental policy; and is a member of the Board of Directors of the National Resources Defence Council. She has published more than 120 scientific articles on economics and mathematics, as well as four books, and authored the widely used concept of *basic needs*.

YVONNE CRIPPS is a member of the Faculty of Law at the University of Cambridge, and the Director of Studies in Law at Emmanuel College, Cambridge. She is also a Visiting Professor of Law at Cornell University, and has been elected a member of the American Law Institute. In addition, she is a member of the Biotechnology Advisory Commission of the Stockholm Environment Institute and has acted as a legal adviser to the Royal Commission on Environmental Pollution. As a member of the Home Office's Animal Procedures Committee, she helps to oversee the regulation of experimentation on animals in Britain.

ROBERT T. FRALEY is President of Monsanto's New Agricultural Products business unit. His responsibilities include the discovery, development, and commercialization of new crop, chemical, and plants biotechnology products. He has gained worldwide recognition for his pioneering contributions to the development of gene transfer technology in plants and its commercial application to crop improvement. Fraley is the author of more than 100 publications and patent applications on technical advances in this area. He is deeply interested in the successful development of plant biotechnology, and the transfer of the technology to developing countries to help provide abundant food and a healthy environment for the future.

ANIL K. GUPTA possesses a degree in Agriculture and Genetics, as well as a Ph.D. in Management. For the last 22 years, he has worked in the field of rural development. He has studied the survival of people in high-risk environments—such as drought and flood-prone regions, forests, and so forth—and the role of indigenous ingenuity in strengthening their coping strategies. Gupta edits a global newsletter, the *Honey Bee*, which serves as an indigenous network of farmers' innovations, and he received the Pew Conservation Scholar Award in 1993.

BRENT HENDRICKS is a graduate of the Harvard Law School, and a Research Fellow at the National Energy-Environment Law and Policy Institute at the University of Tulsa College of Law. He is coauthor, with Lakshman D. Guruswamy, of *International Environmental Law in a Nutshell*. Hendricks has been engaged in numerous interdisciplinary projects involving climate change and biodiversity, and his current research focuses on the institutional and theoretical aspects of international environmental law.

GEOFFREY M. HEAL received both his bachelor's and doctor's degrees from Cambridge University, where he studied physics and economics. He has taught at the universities of Paris, Stockholm, Stanford, Yale, Princeton, and Siena. His publications include 7 books and over 100 articles in scholarly journals. An adviser to corporations and international organizations on environmental issues—recently including the Global Environment Facility of the World Bank, the OECD, and the Clinton administration—he is also Professor of Economics and Finance at the Graduate School of Business, Columbia University, and a member of Columbia's Program on Information and Resources.

ROBERT B. HORSCH is Director of Research and Development of Monsanto's Produce Business Unit. As a pioneer in the development of gene transfer technology, he has lectured around the world on progress in the development of gene transfer methods and on applications of those methods to crop improvement. Horsch has served on the editorial boards of many leading plant science journals, and is author of more than 60 papers and patent applications. He has also served as the principal investigator for several projects that have transferred technology to developing countries in order to change the way food is produced, so as to increase abundance and protect the environment.

LAURA L. JACKSON received her B.A. in biology from Grinnell College in 1983, and her Ph.D. in Ecology and Evolutionary Biology in 1990 from Cornell University. Her research focus is on the effects of agricultural change on biodiversity, and on ecological restoration of agricultural landscapes. She is an Assistant Professor of Biology at the University of Northern Iowa in Cedar Falls, where she teaches courses in conservation biology, applied ecology, and the ecology of agricultural systems.

ANNIE LOVEJOY graduated from Yale University with a degree in history and has worked in the Biodiversity Information Outreach Division of Costa Rica's National Biodiversity Institute since 1993. During her time at INBio, Lovejoy has labored to increase awareness of biodiversity and its intellectual and economic importance, as well as knowledge of INBio and its mission, both inside and outside Costa Rica. She has worked to organize and implement training workshops for biodiversity experts from other tropical developing countries and has published a number of articles about INBio's bioprospecting endeavors.

ARIEL E. LUGO is the Director of the USDA Forest Service International Institute of Tropical Forestry in Rio Piedras, Puerto Rico. He has a Ph.D. in plant ecology from the University of North Carolina at Chapel Hill. Lugo has been active in tropical forest research for some 32 years, with a focus on primary productivity and nutrient cycling in mangroves, forested and nonforested freshwater wetlands, tree plantations, dry to rain forests, and estuaries. Author of over 260 publications, Lugo is currently one of the principal investigators in the Luquillo Experimental Forest Long-Term Ecological Research Program.

BRYAN G. NORTON received his Ph.D. in philosophy from the University of Michigan, and is Professor of Environmental Public Policy at the Georgia Institute of Technology. Much of Norton's past work has focused on questions of biological diversity, culminating in two books, including *Why Preserve Natural Variety?* and *Toward Unity Among Environmentalists*. His current work defines ecosystem health and determines parameters for sustainability. He is a member of the Environmental Economics Advisory Committee of the EPA's Science Advisory Board and a Fellow of the Hastings Center.

ELINOR OSTROM is Codirector of the Workshop in Political Theory and Policy Analysis, and the Arthur F. Bentley Professor of Political Science at Indiana University, Bloomington. She is the author of *Governing the Commons* and *Crafting Institutions for Self-governing Irrigation Systems*; coauthor with Robert Keohane of *Local Commons and Global Interdependence*; coauthor with Larry Schroeder and Susan Wynne of *Institutional Incentives and Sustainable Development*; and coauthor with Roy Gardner and James Walker of *Rules, Games, and Common-pool Resources*.

PETER H. RAVEN is Director of the Missouri Botanical Gardens and Engelmann Professor of Botany at Washington University. He is also Home Secretary of the National Academy of Sciences, Chair of the Report Review Committee of the National Research Council, and a member of the President's Committee of Advisers on Science and Technology. He is the author or editor of 18 books, including textbooks on biology and botany, and over 450 scientific papers. Raven has received numerous prizes and awards, and in 1994, shared with Arturo Gomez Pompa the Tyler Prize for Environmental Achievement.

JOHN W. REID is an economic policy analyst at Conservation International. Before joining Conservation International, he was a research assistant at Resources for the Future. Reid has done extensive fieldwork in the nations of the developing tropics. His current research focuses on providing economic incentives for biodiversity conservation. Reid holds a master's in public policy from the John F. Kennedy School of Government at Harvard University.

WALTER V. REID is Vice President for Program at the World Resources Institute (WRI). He has written or coauthored numerous reports and articles while at WRI, including *Keeping*

*Options Alive: The Scientific Basis for Conserving Biodiversity* (1989) and *Biodiversity Prospecting: Using Genetic Resources for Sustainable Development* (1993), and he is one of the principal authors of the *Global Biodiversity Strategy*. Before joining WRI, Reid was a Gilbert White Fellow with Resources for the Future in Washington, D.C. Reid earned his B.A. in Zoology from the University of California at Berkeley in 1978, and a Ph.D. in Zoology from the University of Washington in 1987.

MARK SAGOFF directs the Institute for Philosophy and Public Policy in the School of Public Affairs at the University of Maryland. He is a Pew Scholar in Conservation and the Environment, President of the International Society of Environmental Ethics, and author of *The Economy of the Earth*. In preparing essays for this volume, Sagoff received financial support from the Pew Charitable Trusts and a National Endowment for the Humanities grant.

ROGER A. SEDJO is a Senior Fellow and Director of the Forest Economics and Policy Program at Resources for the Future. Sedjo has written extensively on forestry economics, renewable resources, and economic development. His current research interests include the economics of ecosystem management, the valuation of biodiversity, the evolution of property rights in ecological resources, and the international spillovers of domestic environmental policies. Sedjo holds a Ph.D. in economics from the University of Washington.

R. DAVID SIMPSON is a Fellow at Resources for the Future. Simpson has conducted extensive research on the economic valuation of biological diversity and ecological systems. A focus of this work has been on the efficacy of market-based incentives for biodiversity and ecosystem conservation. His other research interests include innovation and trade in environmentally friendly technology, and the interrelationships between industrial and environmental policies. Simpson holds a Ph.D. in economics from the Massachusetts Institute of Technology.

ANA M. SITTENFELD, a microbiologist, is the former head of the Center for Research in Cellular and Molecular Biology at the University of Costa Rica. In spring 1991, Sittenfeld joined the National Institute of Biodiversity (INBio) as its Director of Biodiversity Prospecting, with direct responsibility for facilitating the sustainable economic use of biodiversity and strengthening Costa Rica's scientific infrastructure in natural products biochemistry and biotechnology. As the director of this division at INBio, Sittenfeld has been responsible for initiating and managing research collaborations with private companies and academic institutions within Costa Rica and abroad.

CHRISTOPHER D. STONE is Roy P. Crocker Professor of Law at the University of Southern California. His books include *Law, Language, and Ethics: Should Trees Have Standing?*; *Where the Law Ends: The Social Control of Corporate Behavior*; *Earth and Other Ethics: The Case for Moral Pluralism*; and, most recently, *The Gnat Is Older Than Man: Global Environment and Human Agenda*. He is currently examining the legal, moral, and cultural dimensions of global environmental degradation.

GARY H. TOENNIESSEN, a microbiologist, is Deputy Director for Agricultural Sciences at the Rockefeller Foundation. Since 1985, his primary responsibility has been the development and implementation of the foundation's International Program on Rice Biotechnology, a $50 million investment to date, designed to bring the benefits of biotechnology to low-income rice producers and consumers in developing countries.

## The Editors

LAKSHMAN D. GURUSWAMY is Professor of Law and Director of the National Energy-Environment Law and Policy Institute at the University of Tulsa. He has published extensively in legal and scientific journals on international, environmental, and international environmental law. A former environmental law editor of *Lloyd's Maritime and Commercial Law Quarterly*, he is author of *Legal Control of Land Based Sea Pollution*, and coauthor of *International Environmental Law and World Order*, and *International Environmental Law*. At the University of Arizona, he organized two major international conferences—Energy and the Environment: Intersecting Global Issues (1992), and Biological Diversity: Exploring the Complexities (1994)—which developed a number of environmental policy interfaces between the sciences and law within an interdisciplinary framework. He teaches international environmental law and environmental law.

JEFFREY A. MCNEELY is the Chief Scientist and Director of the Biodiversity Programme at the World Conservation Union (IUCN), and is internationally recognized as a leader in the field of biodiversity. Trained as an anthropologist and zoologist at the University of California at Los Angeles and the Los Angeles Zoo, he spent 12 years in Asia advising on conservation issues in Thailand, Indonesia, Nepal, Laos, Cambodia, and Vietnam. He is the author of numerous books, including *Soul of the Tiger: The Relationship between People and Nature in Southeast Asia*; *Mammals of Thailand*; *Economics and Biological Diversity*; *Conserving the World's Biological Diversity*; *People and Protected Areas in the Hindu-kush Himalaya*; *Culture and Conservation*; *National Parks, Conservation, and Development*; *Wildlife Management in Southeast Asia*; *Ecobluff Your Way to Greenism*; and *Conserving Nature: A Regional Overview*. At IUCN, he advises governments on biodiversity policy, organizes international workshops and conferences, and develops new approaches to conservation. He was Secretary-General of the fourth World Congress on National Parks and Protected Areas (Caracas 1992), and serves on the editorial advisory board of five journals.

# Index

adaptive management, 247, 252, 382; generalizing globally, 260

agenda 21, 96–97, 203, 204, 274; biotechnology and, 282

agriculture: aquaculture, 267; biotechnology, 49–64; consumer acceptance of, 58–59; crop rotation, 73, 75; dairy farming, 57, 58, 66, 67, 72, 76–79; developing countries, 60, 62, 69, 190, 192, 195–201; ecological diversity, 80, 271; energy use, 55; family farms, 82–83; farmer's privilege, 193, 196; green revolution, 271; herbicide-resistant crops, 55–56, 66–67, 72–73, 76, 321; high-technology, 383; industrialization of, 67–70; insect-resistant crops, 52–53, 72, 80; international research, 192, 193, 195–201; as a lifestyle, 71, 78, 79, 81–83; postharvest food loss, 56; problems with implementing biotechnology in, 66–84; productivity, 51, 57–58; regulatory process, 60; soil loss and conservation, 54, 55, 67, 69, 70, 75; sustainability, 149–59, 270–71, 283; virus-resistant crops, 53–54, 62, 199; water quality, 70, 76

atmosphere, 122–23

autecology, 249–50

Awas Tingni, 214–17

benefit sharing. *See* intellectual property rights

biodiversity: adaptive management, 247; background, 250–52; consumption, 295; defined, 34, 119, 308; economic (*see* economic responses); economic reasons for protection (*see* economic responses); efficiency, 121–25; emotional reactions to, 35–38; fragmented outlook, 1–2; connection with global change, 106, 122–23, 169; global welfare and, 376; hot spots, 261; human behavior and, 376; intraspecies diversity, 39, 89; legal reasons for protecting biodiversity, 287–88; loss, 16–20, 34, 168–69; management strategies, 39–42; moral reasons for protection, 82, 275, 331; philosophical reasons for protection; pragmatic concerns, 29; preservation, 28; prudential reasons for protection, 269; uncertainty, 260; utility, 57, 288; valuation of, 56–57, 120, 124–26, 133–36, 140. *See also* rationales for protecting biodiversity

biodiversity and conservation, 28, 129, 141–44, 170; *ex situ*, 87–94; *in situ*, 28, 36, 50, 82, 87, 119, 126

biological prospecting, 56, 129, 133, 136, 187, 197, 386, 389; business development, 230; commercial application (*see* biotechnology); efficacy, 274; frameworks, 225–26; macropolicy, 226–28; management, 228; multisectoral collaboration, 233; profitability, 225; property rights (*see* intellectual property rights); risks of, 136, 137; technology access, 230; technology transfer, 232

biological prospecting agreements, 223–34; INBIO agreements (*see* INBIO); legal issues, 235–36

bioprospecting. *See* biological prospecting

biotechnology, 7–8, 50, 51, 57, 59, 62, 66, 67, 68, 84, 191, 197, 382–83; bio-industry, 121–23; commercial application, 134, 136–38, 142; contracts, 137; effect on nature, 277; false assumptions of, 67–71; invalidity as reason for species protection, 126, 276; pharmaceutical (*see* pharmaceutical uses); pharmaceutical research, 129–32, 138–39, 141–44; problems with 66–84, 283; public acceptance, 58–63;

biotechnology (*continued*)
technology transfer, 60–61; threat of, 280; tool for conservation, as a, 92–93. *See also* agriculture; developing countries; genetic resources

Columbia River basin, 266
comprehensive approaches. *See* biodiversity, management strategies
conservation. *See* biodiversity, management strategies
consumption, 193; green accounting, 109; natural resources, 104–5; 168–69
Convention on Biological Diversity (CBD), 131–32, 286, 351–58, 360–72, 378–79; common responsibility to protect biodiversity and, 353–55; forests and, 356–57; framework treaty as, 361–65; implementation, 168, 172–77; legal character of, 351–52; nongovernmental organizations and, 369–70 (*see also* nongovernmental organizations); reinventing CBD, 366–68; state responsibility and, 355; strengthening secretariat, 369; sustainable development and, 351, 372–73; umbrella convention as, 360
cryostress, 94–95

deep ecology, 259
developed countries: prospectus on environmental problems, 103
developing countries, 133–34, 138, 143; benefit sharing, 180–88 (*see also* intellectual property rights); biotechnology access, 60, 62, 63; capacity to conserve biodiversity, 174–75; Endangered Species Act, 247–61; international agreements, 261; patent protection, 316–28 (*see also* intellectual property rights); prospectus on environmental problems, 103. *See also* agriculture
*Diamond v. Chakrabarty*, 334
diversity. *See* species
drugs. *See* pharmaceutical uses

Earth Summit, 101–2, 110, 113, 203. *See also* Agenda 21
ecological management: function, 26–27; values, 26
ecology: agricultural, 80–84; disturbance ecology, 307; evolved processes, 80–84; management of change, 38

economic responses: Asian Tigers, 101, 110, 113; asset conversion of biodiversity, 287–311; biodiversity as commodity, 22, 290, 384; commercial exploitation, 25, 124; comparative advantage, 101, 106–7, 109; conservation through, 25; economics of scale, 111, 114–15; efficiencies, 121–25, 288, 308–9; genetic resources and, 22; institutional development, 228; knowledge-intensive growth, 113; legal remedies and, 294; markets, 27, 118–26; overconsumption, 307–10; potential of biological resources, 24; privatization, 289; property rights, 383; public goods, 384; resource-intensive growth, 108; social benefits, 26–27; subsidization, 306–7, 310; threats to biodiversity, 101–2, 105; technology access, 230–31; technology transfer, 231; trade-offs, 101, 104–10, 115, 288, 307–8; trade sanctions, 294; utility of biological resources, 23, 27, 118–26, 274, 331; valuation, 24–28, 118–20, 124–26, 129–30. *See also* intellectual property rights; international trade
economic responses and market-based remedial actions, 389; subsidies, 389; taxes, 390
economic responses and market development, 230; externalities, 289, 291; private goods, 108–9, 121–81, 288; public goods, 108–9, 121–81, 289, 292–93. *See also* responses
ecosystem, 149–51; agricultural, 71; management, 381–82; preservation, 28; protection, 149–65. *See also* biodiversity; ecology; management of change; management strategies
education, 390
Endangered Species Act, 247; adaptive management, 297
Endangered Species Act, history, 247–52, biodiversity, 250; ecosystem health, 252; single species protection, 248
Endangered Species Act, policies of, 255; working hypothesis as, 257; GOD committee, 257
Endangered Species Act, values, 252–55; biocentric approach, 254; economic approach, 253–54; sustainability approach, 253
energy budgets, 38, 69
environmental policies 33–43, 149–65; developing nations, 180–87, 202–21. *See also* biological prospecting
equitable sharing of benefits, 212–13, 221; CBD requirements, 186–87

ethics and biodiversity, 21, 265–75
European Union, 114–15, 316–28
evolution: adaptability of life systems, 35; environmental pressures, 38; inability of life systems to compensate for change, 72–80
extinction, 59, 381; rates of, 16–20; uncertainties, 34. *See also* species

farmers: plant breeders rights, 346–48; rights under the CBD, 185–87
farming. *See* agriculture
financial mechanisms, 101–15, 120–26
Food and Agricultural Commission on Plant Genetic Resources, 173
Food and Agricultural Organization (FAO). *See* United Nations, FAO
Forests, 390; commercial logging, 220–21; effects of destruction, 17–20; rate of destruction, 19; preservation, 161–65; rain forest, 277; tropical forest, 35, 39
*Funk Bros. Seed Co. v. Kalo Inoculant Co.*, 334

genetic code: agricultural production, 51; engineering. *See* genetic resources
genetic engineering, 49–64; *See* agriculture, insect-resistant crops; agriculture, post-harvest food loss; agriculture, virus-resistant crops
genetic prospecting. *See* biological prospecting
genetic resources, 297–307; access to, 300; *ex situ*, 298; genetic code, 88; governance systems (*see* institutional responses); incremental policies, 377–78; industry, 129, 131–35, 140, 175–76; indigenous peoples, 180–85; institutional responses, 164–74; intellectual property rights, 302; valuation, 133–36, 140
global change research, 8–9, 106, 169
global environment facility, 309
*Graham v. John Deere Co.*, 341

Honey Bee Network, 183–85

INBio (National Biodiversity Institute in Costa Rica), 223–244, 305; research agreements, 236–39
incremental policies, 377–78
indigenous peoples, 180–88, 202–4; Awas Tingni, 214–17; intellectual property rights, 185–87; property rights of, 204, 217; rights of, 218–19; sustainable development, 181–83. *See also* traditional knowledge
industrial controls: public goods, 109. *See* developed countries
institutional responses, 3, 384–86; biological prospecting, 223–241; community-industry partnerships, 202–221; developing countries, 180–87; international responses, 168–77; promoting biological complexity, 149–65; property rights, 190–201
intellectual property rights, 387; Agracetus, 335; benefit sharing, 63; biological prospecting, 119, 129, 131; breeders rights, 346–48; compulsory licensing, 324; Convention on Biological Diversity, 197; developing countries, 191, 192, 305–6; discovery rights, 333; ethics, 337; farmer's privilege, 324; genetic resources, 129, 131–35, 140; Harvard mouse, 336; indigenous peoples, 185–87; international dimensions, 191, 198; justification for, 191, 302–3, 337–40; legal framework, 339; limits to, 305–6; ownership, 302–3; patentable subject matter, 342; patents, 191, 196, 198, 199, 304, 387; plant breeders rights, 303, 340–43; production v. invention, 336; products of nature, 333; proprietary technology, 199, 200; public domain, 337; public policy, 327–28; religious critique, 343–46; requirements, 332–33; takings, 338; trade secrets, 304
International Forestry Resources and Institutions (IFRI), 161–65
International Geosphere Biosphere Program (IGBP), 8
International Labour Organization, 204
International Law: Agreement on Trade-Related Aspects of Intellectual Property Rights (TRIPS), 325; agreements, 171; Biodiversity Convention (*See* Convention on Biological Diversity; Appendix); calls for reform, 174–77; General Agreement on Tariffs and Trade (GATT), 294; implementation, 174–77; political considerations, 170–72; scientific considerations, 768–70; scientific cooperation, 172; social considerations, 168
international trade, 101; comparative advantage, 106–7; optimal economic growth, 106; private goods, 108

Tariffs and Trade (GATT), 106, 114; International Center for Genetic Engineering and Biotechnology (ICGEB), 200; UNCED, 295. *See also* Convention on Biological Diversity

United Nations and the World Bank, 106; International Monetary Fund, 106

utilitarian reasons for biodiversity protection, 66–84, 190–201, 269–76. *See also* agriculture; biological prospecting; economic reasons for biodiversity protection; pharmaceutical uses

World Climate Research Program (WCRP), 8

Library of Congress Cataloging-in-Publication Data
Protection of global biodiversity : converging strategies /
Lakshman D. Guruswamy and Jeffrey A. McNeely, editors.
Includes index.
ISBN 0-8223-2150-5 (cloth : alk. paper).
ISBN 0-8223-2188-2 (pbk. : alk. paper)
1. Biological diversity conservation.
I. Guruswamy, Lakshman D.
II. McNeely, Jeffrey A.
QH 75.P8 1998    833.95'16—dc21    97-46953CIP